Graduate Texts in Mathematics 81

Springer
New York
Berlin
Heidelberg
Barcelona
Hong Kong
London
Milan
Paris
Singapore
Tokyo

Graduate Texts in Mathematics

1 TAKEUTI/ZARING. Introduction to Axiomatic Set Theory. 2nd ed.
2 OXTOBY. Measure and Category. 2nd ed.
3 SCHAEFFER. Topological Vector Spaces.
4 HILTON/STAMMBACH. A Course in Homological Algebra.
5 MAC LANE. Categories for the Working Mathematician.
6 HUGHES/PIPER. Projective Planes.
7 SERRE. A Course in Arithmetic.
8 TAKEUTI/ZARING. Axiometic Set Theory.
9 HUMPHREYS. Introduction to Lie Algebras and Representation Theory.
10 COHEN. A Course in Simple Homotopy Theory.
11 CONWAY. Functions of One Complex Variable. 2nd ed.
12 BEALS. Advanced Mathematical Analysis.
13 ANDERSON/FULLER. Rings and Categories of Modules.
14 GOLUBITSKY GUILEMIN. Stable Mappings and Their Singularities.
15 BERBERIAN. Lectures in Functional Analysis and Operator Theory.
16 WINTER. The Structure of Fields.
17 ROSENBLATT. Random Processes. 2nd ed.
18 HALMOS. Measure Theory.
19 HALMOS. A Hilbert Space Problem Book. 2nd ed., revised.
20 HUSEMOLLER. Fibre Bundles. 2nd ed.
21 HUMPHREYS. Linear Algebraic Groups.
22 BARNES/MACK. An Algebraic Introduction to Mathematical Logic.
23 GREUB. Linear Algebra. 4th ed.
24 HOLMES. Geometric Functional Analysis and Its Applications.
25 HEWITT/STROMBERG. Real and Abstract Analysis.
26 MANES. Algebraic Theories.
27 KELLEY. General Topology.
28 ZARISKI/SAMUEL. Commutative Algebra. Vol. I.
29 ZARISKI/SAMUEL. Commutative Algebra. Vol. II.
30 JACOBSON. Lectures in Abstract Algebra I. Basic Concepts.
31 JACOBSON. Lectures in Abstract Algebra II. Linear Algebra.
32 JACOBSON. Lectures in Abstract Algebra III. Theory of Fields and Galois Theory.
33 HIRSCH. Differential Topology.
34 SPITZER. Principles of Random Walk. 2nd ed.
35 WERMER. Banach Algebras and Several Complex Variables. 2nd ed.
36 KELLEY/NAMIOKA et al. Linear Topological Spaces.
37 MONK. Mathematical Logic.
38 GRAUERT/FRITZSCHE. Several Complex Variables.
39 ARVESON. An Invitation to C^* -Algebras.
40 KEMENY/SNELL/KNAPP. Denumerable Markov Chains. 2nd ed.
41 APOSTOL. Modular Functions and Dirichlet Series in Number Theory. 2nd ed.
42 SERRE. Linear Representations of Finite Groups.
43 GILLMAN/JERISON. Rings of Continuous Functions.
44 KENDIG. Elementary Algebraic Geometry.
45 LOÈVE. Probability Theory I. 4th ed.
46 LOÈVE. Probability Theory II. 4th ed.
47 MOISE. Geometric Topology in Dimentions 2 and 3.

continued after index

Otto Forster

Lectures on Riemann Surfaces

Translated by Bruce Gilligan

With 6 Figures

Springer

Otto Forster
Mathematisches Institut
Univerität München
Theresienstrasse 39
W-8000 München 2
Federal Republic of Germany

Bruce Gilligan *(Translator)*
Department of Mathematics
University of Regina
Regina, Saskatschewan
Canada S4S 0A4

Mathematics Subject Classification (1991): 30-01, 30 Fxx

Library of Congress Cataloging in Publication Data

Forster, Otto, 1937–
Lectures on Riemann surfaces.

(Graduate texts in mathematics; 81)
Translation of: Riemannsche Flächen.
Bibliography: p.
Includes indexes.
1. Riemann surfaces. I. Title. II. Series.
QA333.F6713 515'.223 81-9054
ISBN 0-387-90617-7 AACR2

Title of the Original German Edition: Riemannsche Flächen, Heidelberger Taschenbücher 184, Springer-Verlag, Heidelberg, 1977

Printed and bound by R.R. Donnelley & Sons, Harrisonburg, VA.
Printed in the United States of America.

9 8 7 6 5 4 (Corrected fourth printing, 1999)

ISBN 0-387-90617-7 Springer-Verlag New York Heidelberg Berlin SPIN 10721886
ISBN 3-540-90617-7 Springer-Verlag Berlin Heidelberg New York

Contents

Preface

This book grew out of lectures on Riemann surfaces which the author gave at the universities of Munich, Regensburg and Münster. Its aim is to give an introduction to this rich and beautiful subject, while presenting methods from the theory of complex manifolds which, in the special case of one complex variable, turn out to be particularly elementary and transparent.

The book is divided into three chapters. In the first chapter we consider Riemann surfaces as covering spaces and develop a few basics from topology which are needed for this. Then we construct the Riemann surfaces which arise via analytic continuation of function germs. In particular this includes the Riemann surfaces of algebraic functions. As well we look more closely at analytic functions which display a special multi-valued behavior. Examples of this are the primitives of holomorphic 1-forms and the solutions of linear differential equations.

The second chapter is devoted to compact Riemann surfaces. The main classical results, like the Riemann–Roch Theorem, Abel's Theorem and the Jacobi inversion problem, are presented. Sheaf cohomology is an important technical tool. But only the first cohomology groups are used and these are comparatively easy to handle. The main theorems are all derived, following Serre, from the finite dimensionality of the first cohomology group with coefficients in the sheaf of holomorphic functions. And the proof of this is based on the fact that one can locally solve inhomogeneous Cauchy–Riemann equations and on Schwarz' Lemma.

In the third chapter we prove the Riemann Mapping Theorem for simply connected Riemann surfaces (or Uniformization Theorem) as well as the main theorems of Behnke–Stein for non-compact Riemann surfaces, i.e., the Runge Approximation Theorem and the Theorems of Mittag–Leffler and Weierstrass. This is done using Perron's solution of the Dirichlet problem and Malgrange's method of proof, based on Weyl's Lemma, of the Runge Approximation Theorem. In this chapter we also complete the discussion of Stein's Theorem, begun in Chapter 1, concerning the existence of holomorphic functions with prescribed summands of automorphy and present Röhrl's solution of the Riemann–Hilbert problem on non-compact Riemann surfaces.

We have tried to keep the prerequisites to a bare minimum and to develop the necessary tools as we go along. However the reader is assumed to be familiar with what would generally be covered in one semester courses on one complex variable, on general topology and on algebra. Besides these basics, a few facts from differential topology and functional analysis have been used in Chapters 2 and 3 and these are gathered together in the appendix. Lebesgue integration is not needed, as only holomorphic or differentiable functions (resp. differential forms) are integrated. We have also avoided using, without proof, any theorems on the topology of surfaces.

The material presented corresponds roughly to three semesters of lectures. However, Chapters 2 and 3 presuppose only parts of the preceding chapters. Thus, after §§1, 6 and 9 (the definitions of Riemann surfaces, sheaves and differential forms) the reader could go directly to Chapter 2. And from here, only §§12–14 are needed in Chapter 3 to be able to handle the main theorems on non-compact Riemann surfaces.

The English edition includes exercises which have been added at the end of every section and some additional paragraphs in §§8, 17 and 29. As well, the terminology concerning coverings has been changed. Thanks are due to the many attentive readers of the German edition who helped to eliminate several errors; in particular to G. Elencwajg, who also proposed some of the exercises. Last but not least we would like to thank the translator, B. Gilligan, for his dedicated efforts.

Münster
May, 1981

O. FORSTER

Addendum to Fourth Corrected Printing

For the second and fourth printing a number of misprints and errors have been corrected. I wish to thank B. Gilligan, B. Elsner and O. Hien for preparing lists of errata.

April 1999

O. FORSTER

CHAPTER 1
Covering Spaces

Riemann surfaces originated in complex analysis as a means of dealing with the problem of multi-valued functions. Such multi-valued functions occur because the analytic continuation of a given holomorphic function element along different paths leads in general to different branches of that function. It was the idea of Riemann to replace the domain of the function with a many sheeted covering of the complex plane. If the covering is constructed so that it has as many points lying over any given point in the plane as there are function elements at that point, then on this "covering surface" the analytic function becomes single-valued. Now, forgetting the fact that these surfaces are "spread out" over the complex plane (or the Riemann sphere), we get the notion of an abstract Riemann surface and these may be considered as the natural domain of definition of analytic functions in one complex variable.

We begin this chapter by discussing the general notion of a Riemann surface. Next we consider covering spaces, both from the topological and analytic points of view. Finally, the theory of covering spaces is applied to the problem of analytic continuation, to the construction of Riemann surfaces of algebraic functions, to the integration of differential forms and to finding the solutions of linear differential equations.

§1. The Definition of Riemann Surfaces

In this section we define Riemann surfaces, holomorphic and meromorphic functions on them and also holomorphic maps between Riemann surfaces.

Riemann surfaces are two-dimensional manifolds together with an additional structure which we are about to define. As is well known, an

n-dimensional manifold is a Hausdorff topological space X such that every point $a \in X$ has an open neighborhood which is homeomorphic to an open subset of $\mathbb{R}^n$.

1.1. Definition. Let X be a two-dimensional manifold. A *complex chart* on X is a homeomorphism $\varphi: U \to V$ of an open subset $U \subset X$ onto an open subset $V \subset \mathbb{C}$. Two complex charts $\varphi_i: U_i \to V_i$, $i = 1, 2$ are said to be *holomorphically compatible* if the map

$$\varphi_2 \circ \varphi_1^{-1}: \varphi_1(U_1 \cap U_2) \to \varphi_2(U_1 \cap U_2)$$

is biholomorphic (see Fig. 1).

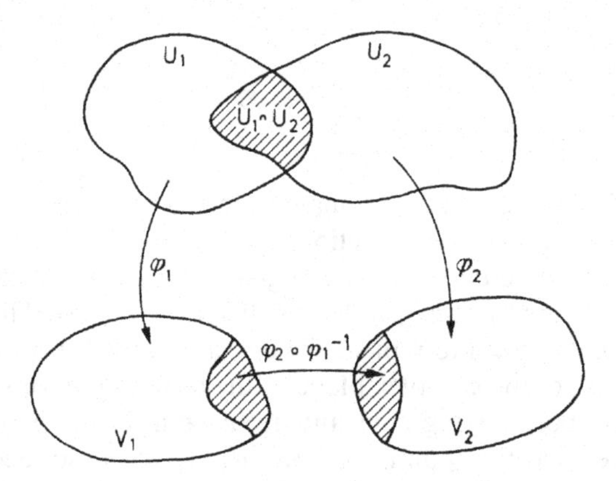

Figure 1

A *complex atlas* on X is a system $\mathfrak{A} = \{\varphi_i: U_i \to V_i, i \in I\}$ of charts which are holomorphically compatible and which cover X, i.e., $\bigcup_{i \in I} U_i = X$.

Two complex atlases $\mathfrak{A}$ and $\mathfrak{A}'$ on X are called *analytically equivalent* if every chart of $\mathfrak{A}$ is holomorphically compatible with every chart of $\mathfrak{A}'$.

1.2. Remarks

(a) If $\varphi: U \to V$ is a complex chart, U_1 is open in U and $V_1 := \varphi(U_1)$, then $\varphi \mid U_1 \to V_1$ is a chart which is holomorphically compatible with $\varphi: U \to V$.

(b) Since the composition of biholomorphic mappings is again biholomorphic, one easily sees that the notion of analytic equivalence of complex atlases is an equivalence relation.

1.3. Definition. By a *complex structure* on a two-dimensional manifold X we mean an equivalence class of analytically equivalent atlases on X.

Thus a complex structure on X can be given by the choice of a complex atlas. Every complex structure Σ on X contains a unique maximal atlas $\mathfrak{A}^*$. If $\mathfrak{A}$ is an arbitrary atlas in Σ, then $\mathfrak{A}^*$ consists of all complex charts on X which are holomorphically compatible with every chart of $\mathfrak{A}$.

1.4. Definition. A *Riemann surface* is a pair (X, Σ), where X is a connected two-dimensional manifold and Σ is a complex structure on X.

One usually writes X instead of (X, Σ) whenever it is clear which complex structure Σ is meant. Sometimes one also writes $(X, \mathfrak{A})$ where $\mathfrak{A}$ is a representative of Σ.

Convention. If X is a Riemann surface, then by a chart on X we always mean a complex chart belonging to the maximal atlas of the complex structure on X.

Remark. Locally a Riemann surface X is nothing but an open set in the complex plane. For, if $\varphi: U \to V \subset \mathbb{C}$ is a chart on X, then φ maps the open set $U \subset X$ bijectively onto V. However, any given point of X is contained in many different charts and no one of these is distinguished from the others. For this reason we may only carry over to Riemann surfaces those notions from complex analysis in the plane which remain invariant under biholomorphic mappings, i.e., those notions which do not depend on the choice of a particular chart.

1.5. Examples of Riemann Surfaces

(a) *The Complex Plane* $\mathbb{C}$. Its complex structure is defined by the atlas whose only chart is the identity map $\mathbb{C} \to \mathbb{C}$.

(b) *Domains.* Suppose X is a Riemann surface and $Y \subset X$ is a domain, i.e., a connected open subset. Then Y has a natural complex structure which makes it a Riemann surface. Namely, one takes as its atlas all those complex charts $\varphi: U \to V$ on X, where $U \subset Y$. In particular, every domain $Y \subset \mathbb{C}$ is a Riemann surface.

(c) *The Riemann sphere* $\mathbb{P}^1$. Let $\mathbb{P}^1 := \mathbb{C} \cup \{\infty\}$, where ∞ is a symbol not contained in $\mathbb{C}$. Introduce the following topology on $\mathbb{P}^1$. The open sets are the usual open sets $U \subset \mathbb{C}$ together with sets of the form $V \cup \{\infty\}$, where $V \subset \mathbb{C}$ is the complement of a compact set $K \subset \mathbb{C}$. With this topology $\mathbb{P}^1$ is a compact Hausdorff topological space, homeomorphic to the 2-sphere S^2. Set

$$U_1 := \mathbb{P}^1 \backslash \{\infty\} = \mathbb{C}$$
$$U_2 := \mathbb{P}^1 \backslash \{0\} = \mathbb{C}^* \cup \{\infty\}.$$

Define maps $\varphi_i: U_i \to \mathbb{C}$, $i = 1, 2$, as follows. φ_1 is the identity map and

$$\varphi_2(z) := \begin{cases} 1/z & \text{for } z \in \mathbb{C}^* \\ 0 & \text{for } z = \infty. \end{cases}$$

Clearly these maps are homeomorphisms and thus $\mathbb{P}^1$ is a two-dimensional manifold. Since U_1 and U_2 are connected and have non-empty intersection, $\mathbb{P}^1$ is also connected.

The complex structure on $\mathbb{P}^1$ is now defined by the atlas consisting of the charts $\varphi_i\colon U_i \to \mathbb{C}$, $i = 1, 2$. We must show that the two charts are holomorphically compatible. But $\varphi_1(U_1 \cap U_2) = \varphi_2(U_1 \cap U_2) = \mathbb{C}^*$ and

$$\varphi_2 \circ \varphi_1^{-1}\colon \mathbb{C}^* \to \mathbb{C}^*, \qquad z \mapsto 1/z,$$

is biholomorphic.

Remark. The notation $\mathbb{P}^1$ comes from the fact that one may consider $\mathbb{P}^1$ as the 1-dimensional projective space over the field of complex numbers.

(d) *Tori.* Suppose $\omega_1, \omega_2 \in \mathbb{C}$ are linearly independent over $\mathbb{R}$. Define

$$\Gamma := \mathbb{Z}\omega_1 + \mathbb{Z}\omega_2 = \{n\omega_1 + m\omega_2\colon n, m \in \mathbb{Z}\}.$$

Γ is called the lattice spanned by ω_1 and ω_2 (Fig. 2). Two complex numbers $z, z' \in \mathbb{C}$ are called equivalent mod Γ if $z - z' \in \Gamma$. The set of all equivalence classes is denoted by $\mathbb{C}/\Gamma$. Let $\pi\colon \mathbb{C} \to \mathbb{C}/\Gamma$ be the canonical projection, i.e., the map which associates to each point $z \in \mathbb{C}$ its equivalence class mod Γ.

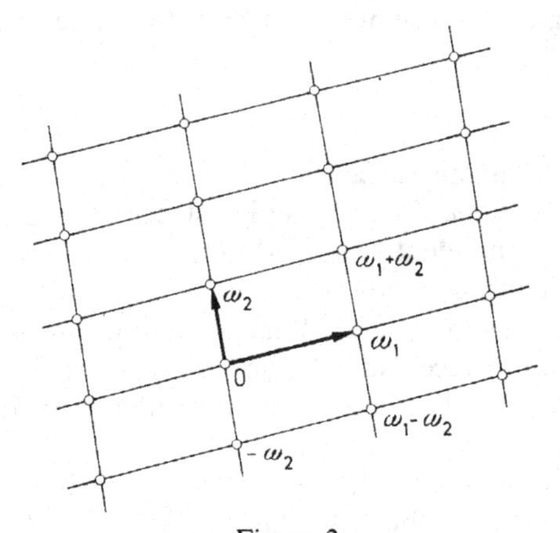

Figure 2

Introduce the following topology (the quotient topology) on $\mathbb{C}/\Gamma$. A subset $U \subset \mathbb{C}/\Gamma$ is open precisely if $\pi^{-1}(U) \subset \mathbb{C}$ is open. With this topology $\mathbb{C}/\Gamma$ is a Hausdorff topological space and the quotient map $\pi\colon \mathbb{C} \to \mathbb{C}/\Gamma$ is continuous. Since $\mathbb{C}$ is connected, $\mathbb{C}/\Gamma$ is also connected. As well $\mathbb{C}/\Gamma$ is compact, for it is covered by the image under π of the compact parallelogram

$$P := \{\lambda\omega_1 + \mu\omega_2\colon \lambda, \mu \in [0, 1]\}.$$

The map π is open, i.e., the image of every open set $V \subset \mathbb{C}$ is open. To see this one has to show that $\hat{V} := \pi^{-1}(\pi(V))$ is open. But

$$\hat{V} = \bigcup_{\omega \in \Gamma} (\omega + V).$$

Since every set $\omega + V$ is open, so is $\hat{V}$.

The complex structure on $\mathbb{C}/\Gamma$ is defined in the following way. Let $V \subset \mathbb{C}$ be an open set such that no two points in V are equivalent under Γ. Then $U := \pi(V)$ is open and $\pi | V \to U$ is a homeomorphism. Its inverse $\varphi: U \to V$ is a complex chart on $\mathbb{C}/\Gamma$. Let $\mathfrak{A}$ be the set of all charts obtained in this fashion. We have to show that any two charts $\varphi_i: U_i \to V_i$, $i = 1, 2$, belonging to $\mathfrak{A}$ are holomorphically compatible. Consider the map

$$\psi := \varphi_2 \circ \varphi_1^{-1}: \varphi_1(U_1 \cap U_2) \to \varphi_2(U_1 \cap U_2).$$

For every $z \in \varphi_1(U_1 \cap U_2)$ one has $\pi(\psi(z)) = \varphi_1^{-1}(z) = \pi(z)$ and thus $\psi(z) - z \in \Gamma$. Since Γ is discrete and ψ is continuous, this implies that $\psi(z) - z$ is constant on every connected component of $\varphi_1(U_1 \cap U_2)$. Thus ψ is holomorphic. Similarly ψ^{-1} is also holomorphic.

Now let $\mathbb{C}/\Gamma$ have the complex structure defined by the complex atlas $\mathfrak{A}$.

Remark. Let $S^1 = \{z \in \mathbb{C}: |z| = 1\}$ be the unit circle. The map which associates to the point of $\mathbb{C}/\Gamma$ represented by $\lambda\omega_1 + \mu\omega_2$, $(\lambda, \mu \in \mathbb{R})$, the point

$$(e^{2\pi i\lambda}, e^{2\pi i\mu}) \in S^1 \times S^1,$$

is a homeomorphism of $\mathbb{C}/\Gamma$ onto the torus $S^1 \times S^1$.

1.6. Definition. Let X be a Riemann surface and $Y \subset X$ an open subset. A function $f: Y \to \mathbb{C}$ is called *holomorphic*, if for every chart $\psi: U \to V$ on X the function

$$f \circ \psi^{-1}: \psi(U \cap Y) \to \mathbb{C}$$

is holomorphic in the usual sense on the open set $\psi(U \cap Y) \subset \mathbb{C}$. The set of all functions holomorphic on Y will be denoted by $\mathcal{O}(Y)$.

1.7. Remarks

(a) The sum and product of holomorphic functions are again holomorphic. Also constant functions are holomorphic. Thus $\mathcal{O}(Y)$ is a $\mathbb{C}$-algebra.

(b) Of course the condition in the definition does not have to be verified for all charts in a maximal atlas on X, just for any family of charts covering Y. Then it is automatically fulfilled for all other charts.

(c) Every chart $\psi: U \to V$ on X is, in particular, a complex-valued function on U. Trivially it is holomorphic. One also calls ψ a local coordinate or a uniformizing parameter and (U, ψ) a *coordinate neighborhood* of any point $a \in U$. In this context one generally uses the letter z instead of ψ.

1.8. Theorem (Riemann's Removable Singularities Theorem). *Let U be an open subset of a Riemann surface and let $a \in U$. Suppose the function $f \in \mathcal{O}(U \backslash \{a\})$ is bounded in some neighborhood of a. Then f can be extended uniquely to a function $\tilde{f} \in \mathcal{O}(U)$.*

This follows directly from Riemann's Removable Singularities Theorem in the complex plane.

We now define holomorphic mappings between Riemann surfaces.

1.9. Definition. Suppose X and Y are Riemann surfaces. A continuous mapping $f: X \to Y$ is called *holomorphic*, if for every pair of charts $\psi_1: U_1 \to V_1$ on X and $\psi_2: U_2 \to V_2$ on Y with $f(U_1) \subset U_2$, the mapping

$$\psi_2 \circ f \circ \psi_1^{-1}: V_1 \to V_2$$

is holomorphic in the usual sense.

A mapping $f: X \to Y$ is called *biholomorphic* if it is bijective and both $f: X \to Y$ and $f^{-1}: Y \to X$ are holomorphic. Two Riemann surfaces X and Y are called *isomorphic* if there exists a biholomorphic mapping $f: X \to Y$.

1.10. Remarks

(a) In the special case $Y = \mathbb{C}$, holomorphic mappings $f: X \to \mathbb{C}$ are clearly the same as holomorphic functions.

(b) If X, Y and Z are Riemann surfaces and $f: X \to Y$ and $g: Y \to Z$ are holomorphic mappings, then the composition $g \circ f: X \to Z$ is also holomorphic.

(c) A continuous mapping $f: X \to Y$ between two Riemann surfaces is holomorphic precisely if for every open set $V \subset Y$ and every holomorphic function $\psi \in \mathcal{O}(V)$, the "pull-back" function $\psi \circ f: f^{-1}(V) \to \mathbb{C}$ is contained in $\mathcal{O}(f^{-1}(V))$. This follows directly from the definitions and the remarks (1.7.c) and (1.10.b).

In this way a holomorphic mapping $f: X \to Y$ induces a mapping

$$f^*: \mathcal{O}(V) \to \mathcal{O}(f^{-1}(V)), \qquad f^*(\psi) = \psi \circ f.$$

One can easily check that f^* is a ring homomorphism. If $g: Y \to Z$ is another holomorphic mapping, W is open in Z, $V := g^{-1}(W)$ and $U := f^{-1}(V)$, then $(g \circ f)^*: \mathcal{O}(W) \to \mathcal{O}(U)$ is the composition of the mappings $g^*: \mathcal{O}(W) \to \mathcal{O}(V)$ and $f^*: \mathcal{O}(V) \to \mathcal{O}(U)$, i.e., $(g \circ f)^* = f^* \circ g^*$.

1.11. Theorem (Identity Theorem). *Suppose X and Y are Riemann surfaces and $f_1, f_2: X \to Y$ are two holomorphic mappings which coincide on a set $A \subset X$ having a limit point $a \in X$. Then f_1 and f_2 are identically equal.*

PROOF. Let G be the set of all points $x \in X$ having an open neighborhood W such that $f_1 \mid W = f_2 \mid W$. By definition G is open. We claim that G is also closed. For, suppose b is a boundary point of G. Then $f_1(b) = f_2(b)$ since f_1 and f_2 are continuous. Choose charts $\varphi: U \to V$ on X and $\psi: U' \to V'$ on Y with $b \in U$ and $f_i(U) \subset U'$. We may also assume that U is connected. The mappings

$$g_i := \psi \circ f_i \circ \varphi^{-1}: V \to V' \subset \mathbb{C}$$

are holomorphic. Since $U \cap G \neq \varnothing$, the Identity Theorem for holomorphic functions on domains in $\mathbb{C}$ implies g_1 and g_2 are identically equal. Thus $f_1 | U = f_2 | U$. Hence $b \in G$ and thus G is closed. Now since X is connected either $G = \varnothing$ or $G = X$. But the first case is excluded since $a \in G$ (using the Identity Theorem in the plane again). Hence f_1 and f_2 coincide on all of X.

□

1.12. Definition. Let X be a Riemann surface and Y be an open subset of X. By a *meromorphic function* on Y we mean a holomorphic function $f: Y' \to \mathbb{C}$, where $Y' \subset Y$ is an open subset, such that the following hold:

(i) $Y \backslash Y'$ contains only isolated points.
(ii) For every point $p \in Y \backslash Y'$ one has

$$\lim_{x \to p} |f(x)| = \infty.$$

The points of $Y \backslash Y'$ are called the *poles* of f. The set of all meromorphic functions on Y is denoted by $\mathcal{M}(Y)$.

1.13. Remarks

(a) Let (U, z) be a coordinate neighborhood of a pole p of f with $z(p) = 0$. Then f may be expanded in a Laurent series

$$f = \sum_{\nu = -k}^{\infty} c_\nu z^\nu$$

in a neighborhood of p.

(b) $\mathcal{M}(Y)$ has the natural structure of a $\mathbb{C}$-algebra. First of all the sum and the product of two meromorphic functions $f, g \in \mathcal{M}(Y)$ are holomorphic functions at those points where both f and g are holomorphic. Then one holomorphically extends, using Riemann's Removable Singularities Theorem, $f + g$ (resp. fg) across any singularities which are removable.

1.14. Example. Suppose $n \geq 1$ and let

$$F(z) = z^n + c_1 z^{n-1} + \cdots + c_n, \qquad c_k \in \mathbb{C},$$

be a polynomial. Then F defines a holomorphic mapping $F: \mathbb{C} \to \mathbb{C}$. If one thinks of $\mathbb{C}$ as a subset of $\mathbb{P}^1$, then $\lim_{z \to \infty} |F(z)| = \infty$. Thus $F \in \mathcal{M}(\mathbb{P}^1)$.

We now interpret meromorphic functions as holomorphic mappings into the Riemann sphere.

1.15. Theorem. *Suppose X is a Riemann surface and $f \in \mathcal{M}(X)$. For each pole p of f, define $f(p) := \infty$. Then $f: X \to \mathbb{P}^1$ is a holomorphic mapping. Conversely, if $f: X \to \mathbb{P}^1$ is a holomorphic mapping, then f is either identically equal to ∞ or else $f^{-1}(\infty)$ consists of isolated points and $f: X \backslash f^{-1}(\infty) \to \mathbb{C}$ is a meromorphic function on X.*

From now on we will identify a meromorphic function $f \in \mathcal{M}(X)$ with the corresponding holomorphic mapping $f: X \to \mathbb{P}^1$.

PROOF

(a) Let $f \in \mathcal{M}(X)$ and let P be the set of poles of f. Then f induces a mapping $f: X \to \mathbb{P}^1$ which is clearly continuous. Suppose $\varphi: U \to V$ and $\psi: U' \to V'$ are charts on X and $\mathbb{P}^1$ resp. with $f(U) \subset U'$. We have to show that

$$g := \psi \circ f \circ \varphi^{-1}: V \to V'$$

is holomorphic. Since f is holomorphic on $X \backslash P$, it follows that g is holomorphic on $V \backslash \varphi(P)$. Hence by Riemann's Removable Singularities Theorem, g is holomorphic on all of V.

(b) The converse follows from the Identity Theorem (1.11). □

1.16. Remark. From (1.11) and (1.15) it follows that the Identity Theorem also holds for meromorphic functions on a Riemann surface. Thus any function $f \in \mathcal{M}(X)$ which is not identically zero has only isolated zeros. This implies that $\mathcal{M}(X)$ is a field.

EXERCISES (§1)

1.1. (a) *One point compactification of* $\mathbb{R}^n$. For $n \geq 1$ let ∞ be a symbol not belonging to $\mathbb{R}^n$. Introduce the following topology on the set $X := \mathbb{R}^n \cup \{\infty\}$. A set $U \subset X$ is open, by definition, if one of the following two conditions is satisfied:

(i) $\infty \notin U$ and U is open in $\mathbb{R}^n$ with respect to the usual topology on $\mathbb{R}^n$.

(ii) $\infty \in U$ and $K := X \backslash U$ is compact in $\mathbb{R}^n$ with respect to the usual topology on $\mathbb{R}^n$.

Show that X is a compact Hausdorff topological space.

(b) *Stereographic projection.* Consider the unit n-sphere

$$S^n = \{(x_1, \ldots, x_{n+1}) \in \mathbb{R}^{n+1}: x_1^2 + \cdots + x_{n+1}^2 = 1\}$$

and the stereographic projection

$$\sigma: S^n \to \mathbb{R}^n \cup \{\infty\}$$

given by

$$\sigma(x_1, \ldots, x_{n+1}) := \begin{cases} \dfrac{1}{1 - x_{n+1}}(x_1, \ldots, x_n) & \text{if } x_{n+1} \neq 1 \\ \infty, & \text{if } x_{n+1} = 1. \end{cases}$$

Show that σ is a homeomorphism of S^n onto X.

1.2. Suppose

$$\begin{pmatrix} a & b \\ c & d \end{pmatrix} \in \mathrm{GL}(2, \mathbb{C}).$$

Show that the linear fractional transformation

$$f(z) = \frac{az + b}{cz + d},$$

which is holomorphic on $\{z \in \mathbb{C} : cz + d \neq 0\}$, can be extended to a meromorphic function on $\mathbb{P}^1$ (also denoted by f). Show that $f\colon \mathbb{P}^1 \to \mathbb{P}^1$ is biholomorphic, i.e., f is an automorphism of $\mathbb{P}^1$.

1.3. Identify $\mathbb{P}^1$ with the unit sphere in $\mathbb{R}^3$ using the stereographic projection

$$\sigma\colon S^2 \to \mathbb{C} \cup \{\infty\} \cong \mathbb{P}^1$$

defined in Ex. 1.1. Let SO(3) be the group of orthogonal 3×3-matrices having determinant 1, i.e.,

$$\mathrm{SO}(3) = \{A \in \mathrm{GL}(3, \mathbb{R}) : A^T A = I, \det A = 1\}.$$

For every $A \in \mathrm{SO}(3)$, show that the map

$$\sigma \circ A \circ \sigma^{-1}\colon \mathbb{P}^1 \to \mathbb{P}^1$$

is biholomorphic.
[*Hint*: Use the fact that every matrix $A \in \mathrm{SO}(3)$ may be written as a product $A = A_1 \cdots A_k$, where

$$A_j = \begin{pmatrix} 0 & 1 & 0 \\ 0 & 0 & 1 \\ 1 & 0 & 0 \end{pmatrix}$$

or else is a matrix of the form

$$\begin{pmatrix} B & 0 \\ 0 & 1 \end{pmatrix}$$

with $B \in \mathrm{SO}(2)$.]

1.4. Let $\Gamma = \mathbb{Z}\omega_1 + \mathbb{Z}\omega_2$ and $\Gamma' = \mathbb{Z}\omega_1' + \mathbb{Z}\omega_2'$ be two lattices in $\mathbb{C}$. Show that $\Gamma = \Gamma'$ if and only if there exists a matrix $A \in \mathrm{SL}(2, \mathbb{Z}) := \{A \in \mathrm{GL}(2, \mathbb{Z}) : \det A = 1\}$ such that

$$\begin{pmatrix} \omega_1' \\ \omega_2' \end{pmatrix} = A \begin{pmatrix} \omega_1 \\ \omega_2 \end{pmatrix}.$$

1.5. (a) Let $\Gamma, \Gamma' \subset \mathbb{C}$ be two lattices. Suppose $\alpha \in \mathbb{C}^*$ such that $\alpha\Gamma \subset \Gamma'$. Show that the map $\mathbb{C} \to \mathbb{C}$, $z \mapsto \alpha z$ induces a holomorphic map

$$\mathbb{C}/\Gamma \to \mathbb{C}/\Gamma',$$

which is biholomorphic if and only if $\alpha\Gamma = \Gamma'$.

(b) Show that every torus $X = \mathbb{C}/\Gamma$ is isomorphic to a torus of the form

$$X(\tau) := \mathbb{C}/(\mathbb{Z} + \mathbb{Z}\tau),$$

where $\tau \in \mathbb{C}$ satisfies $\mathrm{Im}(\tau) > 0$.

(c) Suppose $\begin{pmatrix} a & b \\ c & d \end{pmatrix} \in \mathrm{SL}(2, \mathbb{Z})$ and $\mathrm{Im}(\tau) > 0$. Let

$$\tau' := \frac{a\tau + b}{c\tau + d}.$$

Show that the tori $X(\tau)$ and $X(\tau')$ are isomorphic.

§2. Elementary Properties of Holomorphic Mappings

In this section we note some of the elementary topological properties of holomorphic mappings between Riemann surfaces. Using these we show that one can easily derive some of the famous theorems of complex analysis, e.g., Liouville's Theorem and the Fundamental Theorem of Algebra.

2.1. Theorem (Local Behavior of Holomorphic Mappings). *Suppose X and Y are Riemann surfaces and $f\colon X \to Y$ is a non-constant holomorphic mapping. Suppose $a \in X$ and $b := f(a)$. Then there exists an integer $k \geq 1$ and charts $\varphi\colon U \to V$ on X and $\psi\colon U' \to V'$ on Y with the following properties:*

(i) $a \in U$, $\varphi(a) = 0$; $\quad b \in U'$, $\psi(b) = 0$.
(ii) $f(U) \subset U'$.
(iii) *The map $F := \psi \circ f \circ \varphi^{-1}\colon V \to V'$ is given by*

$$F(z) = z^k \quad \textit{for all } z \in V.$$

PROOF. First we note that there exist charts $\varphi_1\colon U_1 \to V_1$ on X and $\psi\colon U' \to V'$ on Y such that properties (i) and (ii) are satisfied if one replaces (U, φ) by (U_1, φ_1). Now it follows from the Identity Theorem that the function

$$f_1 := \psi \circ f \circ \varphi_1^{-1}\colon V_1 \to V' \subset \mathbb{C}$$

is non-constant. Since $f_1(0) = 0$, there is a $k \geq 1$ such that $f_1(z) = z^k g(z)$, where g is a holomorphic function on V_1 with $g(0) \neq 0$. Hence there exists a neighborhood of 0 and a holomorphic function h on this neighborhood such that $h^k = g$. The correspondence $z \mapsto zh(z)$ defines a biholomorphic mapping $\alpha\colon V_2 \to V$ of an open neighborhood $V_2 \subset V_1$ of zero onto an open neighborhood V of zero. Let $U := \varphi_1^{-1}(V_2)$. Now replace the chart $\varphi_1\colon U_1 \to V_1$ by the chart $\varphi\colon U \to V$ where $\varphi = \alpha \circ \varphi_1$. Then by construction the mapping $F = \psi \circ f \circ \varphi^{-1}$ satisfies $F(z) = z^k$. □

2.2. Remark. The number k in Theorem (2.1) can be characterized in the following way. For every neighborhood U_0 of a there exist neighborhoods $U \subset U_0$ of a and W of $b = f(a)$ such that the set $f^{-1}(y) \cap U$ contains exactly k elements for every point $y \in W$, $y \neq b$. One calls k the *multiplicity* with which the mapping f takes the value b at the point a or one just says that f has *multiplicity* k at the point a.

2.3. Example. Let $f(z) = z^k + c_1 z^{k-1} + \cdots + c_k$ be a polynomial of degree k. Then f can be considered as a holomorphic mapping $f\colon \mathbb{P}^1 \to \mathbb{P}^1$ where $f(\infty) = \infty$ (cf. §1). Using a chart about ∞, one can easily check that ∞ is taken with multiplicity k.

2.4. Corollary. *Let X and Y be Riemann surfaces and let $f: X \to Y$ be a non-constant holomorphic mapping. Then f is open, i.e., the image of every open set under f is open.*

PROOF. It follows directly from Theorem (2.1) that if U is a neighborhood of a point $a \in X$ then $f(U)$ is a neighborhood of the point $f(a)$. This implies f is open. □

2.5. Corollary. *Let X and Y be Riemann surfaces and let $f: X \to Y$ be an injective holomorphic mapping. Then f is a biholomorphic mapping of X onto $f(X)$.*

PROOF. Since f is injective, in the local description of f given by Theorem (2.1), one always has $k = 1$. Hence the inverse mapping $f^{-1}: f(X) \to X$ is holomorphic. □

2.6. Corollary (Maximum Principle). *Suppose X is a Riemann surface and $f: X \to \mathbb{C}$ is a non-constant holomorphic function. Then the absolute value of f does not attain its maximum.*

PROOF. Suppose that there were a point $a \in X$ such that

$$R := |f(a)| = \sup\{|f(x)|: x \in X\}.$$

Then

$$f(X) \subset K := \{z \in \mathbb{C}: |z| \leq R\}.$$

Since $f(X)$ is open, it lies in the interior of K. This contradicts the assumption that $f(a) \in \partial K$. □

2.7. Theorem. *Suppose X and Y are Riemann surfaces. Suppose X is compact and $f: X \to Y$ is a non-constant holomorphic mapping. Then Y is compact and f is surjective.*

PROOF. By (2.4) $f(X)$ is open. Since X is compact, $f(X)$ is compact and thus closed. Since the only subsets of a connected topological space which are both open and closed are the empty set and the whole space, it follows that $f(X) = Y$. Thus f is surjective and Y compact. □

2.8. Corollary. *Every holomorphic function on a compact Riemann surface is constant.*

This follows from Theorem (2.7) since $\mathbb{C}$ is not compact.

2.9. Corollary. *Every meromorphic function f on $\mathbb{P}^1$ is rational, i.e., can be written as the quotient of two polynomials.*

PROOF. The function f has only finitely many poles. For if it did have infinitely many poles then they would have a limit point and by the Identity Theorem f would be identically equal to ∞. We may assume the point ∞ is not a pole of f. Otherwise consider $1/f$ instead of f. Now suppose $a_1, \ldots, a_n \in \mathbb{C}$ are the poles of f and

$$h_\nu(z) = \sum_{j=-k_\nu}^{-1} c_{\nu j}(z - a_\nu)^j,$$

is the principal part of f at the pole a_ν, for $\nu = 1, \ldots, n$. Then the function $g := f - (h_1 + \cdots + h_n)$ is holomorphic on $\mathbb{P}^1$ and thus a constant by Corollary (2.8). From this it follows that f is rational. □

2.10. Liouville's Theorem. *Every bounded holomorphic function $f\colon \mathbb{C} \to \mathbb{C}$ is constant.*

PROOF. By Riemann's Removable Singularities Theorem (2.8) f can be analytically continued to a holomorphic mapping $f\colon \mathbb{P}^1 \to \mathbb{C}$. By Corollary (2.8) f is constant. □

2.11. The Fundamental Theorem of Algebra. *Let $n \geq 1$ and let*

$$f(z) = z^n + c_1 z^{n-1} + \cdots + c_n$$

be a polynomial with coefficients $c_\nu \in \mathbb{C}$. Then there exists at least one point $a \in \mathbb{C}$ such that $f(a) = 0$.

PROOF. The polynomial f may be considered as a holomorphic mapping $f\colon \mathbb{P}^1 \to \mathbb{P}^1$, where $f(\infty) = \infty$. By Theorem (2.7) this mapping is surjective and thus $0 \in f(\mathbb{C})$. □

2.12. Doubly Periodic Functions. Suppose $\omega_1, \omega_2 \in \mathbb{C}$ are linearly independent over $\mathbb{R}$ and $\Gamma := \mathbb{Z}\omega_1 + \mathbb{Z}\omega_2$ is the lattice spanned by them. A meromorphic function $f\colon \mathbb{C} \to \mathbb{P}^1$ is called doubly periodic with respect to Γ, if

$$f(z) = f(z + \omega) \quad \text{for every } z \in \mathbb{C} \text{ and } \omega \in \Gamma.$$

Clearly, for this to hold it suffices that $f(z) = f(z + \omega_1) = f(z + \omega_2)$ for every $z \in \mathbb{C}$. Let $\pi\colon \mathbb{C} \to \mathbb{C}/\Gamma$ be the canonical map. Then the doubly periodic function f induces a function $F\colon \mathbb{C}/\Gamma \to \mathbb{P}^1$ such that $f = F \circ \pi$. It follows directly from the definition of the complex structure on $\mathbb{C}/\Gamma$ that F is a meromorphic function on $\mathbb{C}/\Gamma$. Conversely, for any meromorphic function $F\colon \mathbb{C}/\Gamma \to \mathbb{P}^1$, the composition $f = F \circ \pi\colon \mathbb{C} \to \mathbb{P}^1$ is a meromorphic function which is doubly periodic with respect to Γ. Thus the meromorphic functions on the torus $\mathbb{C}/\Gamma$ are in one-to-one correspondence with the meromorphic functions on $\mathbb{C}$, doubly periodic with respect to Γ. Hence from Theorem (2.7) we have:

2.13. Theorem. *Every doubly periodic holomorphic function $f: \mathbb{C} \to \mathbb{C}$ is constant. Every non-constant doubly periodic meromorphic function $f: \mathbb{C} \to \mathbb{P}^1$ attains every value $c \in \mathbb{P}^1$.*

EXERCISES (§2)

2.1. Let $\Gamma \subset \mathbb{C}$ be a lattice. The Weierstrass $\wp$-function with respect to Γ is defined by

$$\wp_\Gamma(z) = \frac{1}{z^2} + \sum_{\omega \in \Gamma \setminus 0} \left(\frac{1}{(z-\omega)^2} - \frac{1}{\omega^2} \right).$$

(a) Prove that $\wp_\Gamma$ is a doubly periodic meromorphic function with respect to Γ which has poles at the points of Γ. [*Hint*: First consider the derivative

$$\wp'_\Gamma(z) = -2 \sum_{\omega \in \Gamma} \frac{1}{(z-\omega)^3}.\Bigg]$$

(b) Let $f \in \mathcal{M}(\mathbb{C})$ be a doubly periodic function with respect to Γ which has its poles at the points of Γ and which has the following Laurent expansion about the origin

$$f(z) = \sum_{k=-2}^{\infty} c_k z^k, \quad \text{where } c_{-2} = 1,\ c_{-1} = c_0 = 0.$$

Prove that $f = \wp_\Gamma$.

2.2. Suppose X is a Riemann surface and $f: X \to \mathbb{C}$ is a non-constant holomorphic function. Show that $\mathrm{Re}(f)$ does not attain its maximum.

2.3. Suppose $f: \mathbb{C} \to \mathbb{C}$ is a holomorphic function, whose real part is bounded from above. Then f is constant.

2.4. Suppose $f: X \to Y$ is a non-constant holomorphic map and

$$f^*: \mathcal{O}(Y) \to \mathcal{O}(X), \qquad f^*(\varphi) := \varphi \circ f.$$

Show that f^* is a ring monomorphism.

2.5. Suppose $p_1, \ldots, p_n$ are points on the compact Riemann surface X and $X' := X \setminus \{p_1, \ldots, p_n\}$. Suppose

$$f: X' \to \mathbb{C}$$

is a non-constant holomorphic function. Show that the image of f comes arbitrarily close to every $c \in \mathbb{C}$.

§3. Homotopy of Curves. The Fundamental Group

In this section we present some of the topological results connected with the notion of homotopy of curves.

By a *curve* in a topological space X we mean a continuous mapping $u: I \to X$, where $I := [0, 1] \subset \mathbb{R}$ is the unit interval. The point $a := u(0)$ is called the initial point and $b := u(1)$ the end point of u. One also says that u is a curve from a to b or that the curve u joins a to b.

Let us recall that a topological space X is called *arcwise connected* or *pathwise connected* if any two points $a, b \in X$ can be joined by a curve. An arcwise connected space is also connected, i.e., there does not exist a decomposition $X = U_1 \cup U_2$ where U_1 and U_2 are non-empty disjoint open sets. A topological space is called *locally arcwise connected* if every point has a neighborhood basis of arcwise connected sets. In particular this is always the case for manifolds. A connected, locally arcwise connected space X is (globally) arcwise connected. For one can easily show that the set of all points $x \in X$ which can be joined with a given point $a \in X$ is both open and closed.

3.1. Definition. Suppose X is a topological space and $a, b \in X$. Two curves u, $v: I \to X$ from a to b are called *homotopic*, denoted $u \sim v$, if there exists a continuous mapping $A: I \times I \to X$ with the following properties:

(i) $A(t, 0) = u(t)$ for every $t \in I$,
(ii) $A(t, 1) = v(t)$ for every $t \in I$,
(iii) $A(0, s) = a$ and $A(1, s) = b$ for every $s \in I$.

Remark. If one sets $u_s(t) := A(t, s)$, then every u_s is a curve from a to b and $u_0 = u$, $u_1 = v$. The family of curves $(u_s)_{0 \le s \le 1}$ is said to be a deformation of the curve u into the curve v or a homotopy from u to v, cf. Fig. 3.

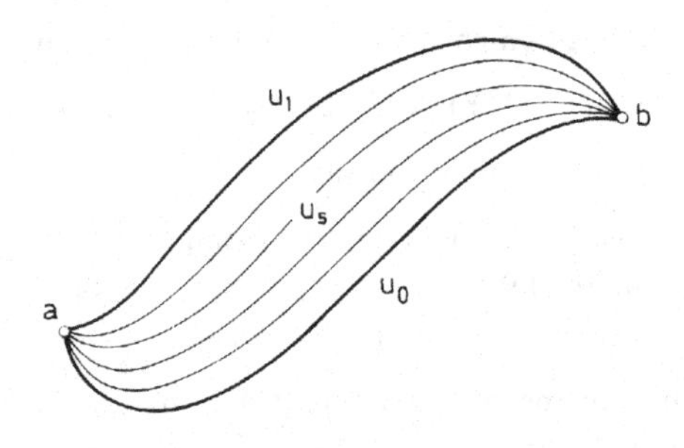

Figure 3

3.2. Theorem. *Suppose X is a topological space and $a, b \in X$. Then the notion of homotopy is an equivalence relation on the set of all curves from a to b.*

PROOF. Reflexitivity and symmetry are clear. As to the transitivity, suppose $u, v, w: I \to X$ are three curves from a to b with $u \sim v$ and $v \sim w$. We must show that $u \sim w$. By assumption there exist continuous mappings A, $B: I \times I \to X$ such that for every $t, s \in I$ the following hold:

$$A(t, 0) = u(t),$$
$$A(t, 1) = B(t, 0) = v(t),$$
$$B(t, 1) = w(t),$$
$$A(0, s) = B(0, s) = a,$$
$$A(1, s) = B(1, s) = b.$$

Define $C: I \times I \to X$ by

$$C(t, s) := \begin{cases} A(t, 2s) & \text{for } 0 \leq s \leq \frac{1}{2}, \\ B(t, 2s - 1) & \text{for } \frac{1}{2} \leq s \leq 1. \end{cases}$$

Then C is continuous and is a homotopy from u to w. □

3.3. Lemma. *Suppose $u: I \to X$ is a curve in the topological space X and $\varphi: I \to I$ is a continuous mapping such that $\varphi(0) = 0$ and $\varphi(1) = 1$. Then the curves u and $u \circ \varphi$ are homotopic.*

PROOF. Define $A: I \times I \to X$ by

$$A(t, s) := u((1 - s)t + s\varphi(t)).$$

Then A is continuous and

$$A(t, 0) = u(t), \qquad A(t, 1) = (u \circ \varphi)(t)$$
$$A(0, s) = u(0) \qquad A(1, s) = u(1)$$

for every $t, s \in I$. Thus u and $u \circ \varphi$ are homotopic. □

3.4. Definition. Suppose a, b and c are three points in a topological space X, $u: I \to X$ is a curve from a to b and $v: I \to X$ is a curve from b to c.

(i) The *product curve* $u \cdot v: I \to X$ from a to c is defined by

$$(u \cdot v)(t) := \begin{cases} u(2t) & \text{for } 0 \leq t \leq \frac{1}{2}, \\ v(2t - 1) & \text{for } \frac{1}{2} \leq t \leq 1. \end{cases}$$

(ii) The *inverse curve* $u^-: I \to X$ from a to b is defined by

$$u^-(t) := u(1 - t) \quad \text{for every } t \in I.$$

The product curve $u \cdot v$ first traces the points of the curve u and then those of the curve v but at twice the speed. The inverse curve u^- passes along the same points as u but in the opposite direction.

One can easily check that if $u_1, u_2: I \to X$ are homotopic curves from a to b and $v_1, v_2: I \to X$ are homotopic curves from b to c, then $u_1 \cdot v_1 \sim u_2 \cdot v_2$ and $u_1^- \sim u_2^-$.

3.5. Definition. Suppose X is a topological space and $a \in X$. By the *constant curve at* a is meant the constant mapping $u_0 \colon I \to X$, i.e., $u_0(t) = a$ for every $t \in I$.

3.6. Theorem. *Suppose X is a topological space and $a, b, c \in X$. Suppose u, v, $w \colon I \to X$ are curves in X such that*

$$u(0) = a, \qquad u(1) = b = v(0), \qquad v(1) = c = w(0), \qquad w(1) = d.$$

Further let u_0 be the constant curve at a, v_0 the constant curve at b. Then the following homotopies exist:

(i) $u_0 \cdot u \sim u \sim u \cdot v_0$,
(ii) $u \cdot u^- \sim u_0$,
(iii) $(u \cdot v) \cdot w \sim u \cdot (v \cdot w)$.

PROOF
(i) By the definition of the product of curves

$$(u_0 \cdot u)(t) = \begin{cases} u_0(2t) = u(0) & \text{for } 0 \le t \le \frac{1}{2}, \\ u(2t - 1) & \text{for } \frac{1}{2} \le t \le 1. \end{cases}$$

Thus $u_0 \cdot u = u \circ \psi$, where $\psi \colon I \to I$ is the parameter transformation defined by $\psi(t) = 0$ for $0 \le t \le \frac{1}{2}$, $\psi(t) = 2t - 1$ for $\frac{1}{2} \le t \le 1$. Thus it follows from Lemma (3.3) that $u_0 \cdot u \sim u$. Similarly $u \cdot v_0 \sim u$.
(ii) By definition

$$(u \cdot u^-)(t) = \begin{cases} u(2t) & \text{for } 0 \le t \le \frac{1}{2}, \\ u(2 - 2t) & \text{for } \frac{1}{2} \le t \le 1. \end{cases}$$

Now define $A \colon I \times I \to X$ by

$$A(t, s) := \begin{cases} u(2t(1 - s)) & \text{for } 0 \le t \le \frac{1}{2}, \\ u(2(1 - t)(1 - s)) & \text{for } \frac{1}{2} \le t \le 1. \end{cases}$$

Then A is a homotopy from $u \cdot u^-$ to the point curve u_0.
(iii) One can easily check that

$$(u \cdot v) \cdot w = (u \cdot (v \cdot w)) \circ \psi,$$

where $\psi \colon I \to I$ is the parameter transformation given by:

(a) $\psi(0) = 0$, $\psi(\frac{1}{4}) = \frac{1}{2}$, $\psi(\frac{1}{2}) = \frac{3}{4}$, $\psi(1) = 1$
(b) ψ is affine linear on each of the intervals $[0, \frac{1}{4}]$, $[\frac{1}{4}, \frac{1}{2}]$, $[\frac{1}{2}, 1]$.

Hence the result follows from Lemma (3.3). □

Remark. Analogous to (iii) is the following fact. If $u_1, \ldots, u_n$ are curves in X such that the initial point of each u_{k+1} equals the end point of u_k, then bracketing the product $u_1 \cdot u_2 \cdot \cdots \cdot u_n$ in various ways corresponds to taking various parameter transformations $\psi \colon I \to I$ such that $\psi(0) = 0$ and $\psi(1) = 1$. In particular all such bracketings are homotopic.

3.7. Definition. A curve $u: I \to X$ in a topological space X is called *closed* if $u(0) = u(1)$. A closed curve $u: I \to X$ with initial and end point a is said to be *null-homotopic* if it is homotopic to the constant curve at a.

3.8. Theorem and Definition. *Suppose X is a topological space and $a \in X$ is a point. The set $\pi_1(X, a)$ of homotopy classes of closed curves in X with initial and end point a forms a group under the operation induced by the product of curves. This group is called the fundamental group of X with base point a.*

Notation. For any closed curve u denote by $\text{cl}(u)$ its homotopy class. Thus the group operation in $\pi_1(X, a)$ is by definition $\text{cl}(u)\,\text{cl}(v) = \text{cl}(u \cdot v)$.

PROOF. The fact that the group operation is well-defined follows from the remark at the end of Definition (3.4). Theorem (3.6) implies that the operation is associative and the class of null homotopic curves is the identity element. Inverses satisfy

$$\text{cl}(u)^{-1} = \text{cl}(u^{-}). \qquad \square$$

3.9. Dependence on the Base Point. Suppose X is a topological space and a, $b \in X$ are points which are joined by a curve w. Then a mapping

$$f: \pi_1(X, a) \to \pi_1(X, b)$$

can be defined as follows:

$$f(\text{cl}(u)) := \text{cl}(w^{-} \cdot u \cdot w).$$

One easily sees that this mapping is an isomorphism. Thus for an arcwise connected space X the fundamental group is essentially independent of the base point and we often just write $\pi_1(X)$ instead of $\pi_1(X, a)$. Note however that the isomorphism $\pi_1(X, a) \to \pi_1(X, b)$ depends in general on the curve w joining a to b used in its construction. If w_1 is another curve from a to b and $f_1: \pi_1(X, a) \to \pi_1(X, b)$ is defined by

$$f_1(\text{cl}(u)) := \text{cl}(w_1^{-} \cdot u \cdot w_1),$$

then the automorphism

$$F := f_1^{-1} \circ f: \pi_1(X, a) \to \pi_1(X, a)$$

satisfies $F(\text{cl}(u)) = \text{cl}(w_1 \cdot w^{-} \cdot u \cdot w \cdot w_1^{-})$, i.e.,

$$F(\alpha) = \gamma \cdot \alpha \cdot \gamma^{-1} \quad \text{for every } \alpha \in \pi_1(X, a),$$

where γ denotes the homotopy class of the closed curve $w_1 \cdot w^{-}$. Thus if $\pi_1(X, a)$ is abelian, then this shows that $\pi_1(X, a)$ and $\pi_1(X, b)$ are canonically isomorphic.

3.10. Definition. An arcwise connected topological space X is called *simply connected* if $\pi_1(X) = 0$.

Remark. Although the group operation in $\pi_1(X)$ is written multiplicatively, one writes $\pi_1(X) = 0$ if $\pi_1(X)$ only contains the identity.

3.11. Theorem. *Suppose X is an arcwise connected, simply connected topological space and $a, b \in X$. Then any two curves $u, v: I \to X$ from a to b are homotopic.*

PROOF. Let u_0 (resp. v_0) be the constant curve at a (resp. b). Now $\pi_1(X, b) = 0$ implies $v^- \cdot u \sim v_0$ and thus $v \cdot (v^- \cdot u) \sim v \cdot v_0$. But $v \cdot (v^- \cdot u) \sim (v \cdot v^-) \cdot u \sim u_0 \cdot u \sim u$ and $v \cdot v_0 \sim v$ by Theorem (3.6), i.e., $u \sim v$. □

3.12. Examples

(a) A subset $X \subset \mathbb{R}^n$ is called *star-shaped* with respect to a point $a \in X$ if for every point $x \in X$ the straight line segment $\lambda a + (1 - \lambda)x$, $0 \le \lambda \le 1$, is contained in X. Every star-shaped subset $X \subset \mathbb{R}^n$ is simply connected. For suppose $u: I \to X$ is a closed curve with initial and end point a (with a as above). Then

$$A: I \times I \to X, \qquad A(t, s) := sa + (1 - s)u(t)$$

is a homotopy from u to the point curve at a. Thus $\pi_1(X, a) = 0$. In particular, the complex plane $\mathbb{C}$ and every disk in $\mathbb{C}$ are simply connected. As well $\mathbb{C}\backslash\mathbb{R}_+$ and $\mathbb{C}\backslash\mathbb{R}_-$ are simply connected, where $\mathbb{R}_+$ (resp. $\mathbb{R}_-$) denotes the positive (resp. negative) real axis.

(b) The *Riemann sphere* $\mathbb{P}^1$ is also simply connected. One can see this as follows. Let $U_1 := \mathbb{P}^1\backslash\{\infty\}$ and $U_2 := \mathbb{P}^1\backslash\{0\}$. Since U_1 and U_2 are homeomorphic to $\mathbb{C}$, they are simply connected. Now suppose $u: I \to \mathbb{P}^1$ is any closed curve starting and ending at 0. Since I is compact and u is continuous, one can find finitely many, not necessarily closed, curves $u_1, \ldots, u_{2n+1}: I \to \mathbb{P}^1$ with the following properties:

(i) The product

$$v := u_1 \cdot u_2 \cdot \cdots \cdot u_{2n+1}$$

is, up to a parameter transformation, equal to the curve u and thus is homotopic to u.

(ii) The curves u_{2k+1}, $k = 0, \ldots, n$, lie entirely in U_1, and the curves u_{2k}, $k = 1, \ldots, n$, lie entirely in U_2. The initial and end points of the u_{2k} are different from ∞. Now by Theorem (3.11) one can find curves u'_{2k}, homotopic to u_{2k}, lying entirely in $U_2\backslash\{\infty\}$. Then

$$v' := u_1 \cdot u'_2 \cdot u_3 \cdot \cdots \cdot u'_{2n} \cdot u_{2n+1}$$

is homotopic to v and thus to u as well and lies in U_1. Since $\pi_1(U_1) = 0$, v' is null homotopic. Thus u is null homotopic too.

3.13. Definition. Suppose X is a topological space and $u, v: I \to X$ are two closed curves in X, which do not necessarily have the same initial point.

Then the curves u and v are called *free homotopic* as closed curves, if there exists a continuous mapping $A: I \times I \to X$ with the following properties:

(i) $A(t, 0) = u(t)$ for every $t \in I$,
(ii) $A(t, 1) = v(t)$ for every $t \in I$,
(iii) $A(0, s) = A(1, s)$ for every $s \in I$.

Remark. If one sets $u_s(t) := A(t, s)$, then each u_s is a closed curve in X and $u_0 = u$, $u_1 = v$. The family of curves u_s, $0 \le s \le 1$, gives a deformation of the curve u into the curve v. Let $w(t) := A(0, t)$, $0 \le t \le 1$. Then w is a curve which joins $a := u(0) = u(1)$ to $b := v(0) = v(1)$. Note that for each s the point $w(s)$ is the initial and end point of the curve u_s. It is easy to see (cf. Fig. 4) that u is homotopic, while keeping the initial and end point a fixed, to the curve $w \cdot v \cdot w^-$.

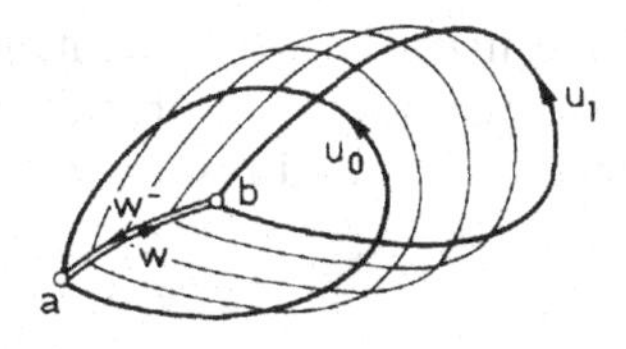

Figure 4

3.14. Theorem. *A pathwise connected topological space X is simply connected if and only if any two closed curves in X are free homotopic as closed curves.*

The proof is simple and is left to the reader.

3.15. Functorial Behavior. Suppose $f: X \to Y$ is a continuous mapping between the topological spaces X and Y. If $u: I \to X$ is a curve in X, then $f \circ u: I \to Y$ is a curve in Y. If $u, u': I \to X$ are homotopic, then $f \circ u, f \circ u'$ are also homotopic. Hence f induces a mapping

$$f_*: \pi_1(X, a) \to \pi_1(Y, f(a))$$

of the fundamental groups. This mapping is a group homomorphism, since $f \circ (u \cdot v) = (f \circ u) \cdot (f \circ v)$. If $g: Y \to Z$ is another continuous mapping, then $(g \circ f)_* = g_* \circ f_*$.

Exercises (§3)

3.1. (a) Suppose X is a manifold and $U_1, U_2 \subset X$ are two open, connected and simply connected subsets such that $U_1 \cap U_2$ is connected. Show that $U_1 \cup U_2$ is simply connected.
(b) Using (a) show that S^n for $n \ge 2$ is simply connected.

3.2. Suppose X and Y are arcwise connected topological spaces. Prove $\pi_1(X \times Y) \cong \pi_1(X) \times \pi_1(Y)$.

3.3. Let (X, a) and (Y, b) be topological spaces with base points $a \in X$ and $b \in Y$. Let $f, g \colon X \to Y$ be two continuous maps with $f(a) = g(a) = b$. Then f and g are called homotopic if there exists a continuous map

$$F \colon X \times [0, 1] \to Y$$

such that $F(x, 0) = f(x)$ and $F(x, 1) = g(x)$ for every $x \in X$ and $F(a, t) = b$ for every $t \in [0, 1]$. Consider the induced maps

$$f_*, g_* \colon \pi_1(X, a) \to \pi_1(Y, b).$$

Show that $f_* = g_*$ if f and g are homotopic.

§4. Branched and Unbranched Coverings

Non-constant holomorphic maps between Riemann surfaces are covering maps, possibly having branch points. For this reason we now gather together the most important ideas and results from the theory of covering spaces.

4.1. Definition. Suppose X and Y are topological spaces and $p \colon Y \to X$ is a continuous map. For $x \in X$, the set $p^{-1}(x)$ is called the *fiber* of p over x. If $y \in p^{-1}(x)$, then one says that the point y *lies over* x. If $p \colon Y \to X$ and $q \colon Z \to X$ are continuous maps, then a map $f \colon Y \to Z$ is called *fiber-preserving* if $p = q \circ f$. This means that any point $y \in Y$, lying over the point $x \in X$, is mapped to a point which also lies over x.

A subset A of a topological space is called discrete if every point $a \in A$ has a neighborhood V such that $V \cap A = \{a\}$. A mapping $p \colon Y \to X$, between topological spaces X and Y, is said to be discrete if the fiber $p^{-1}(x)$ of every point $x \in X$ is a discrete subset of Y.

4.2. Theorem. *Suppose X and Y are Riemann surfaces and $p \colon Y \to X$ is a non-constant holomorphic map. Then p is open and discrete.*

PROOF. By (2.4) the map p is open. If the fiber of some point $a \in X$ were not discrete, then, by the Identity Theorem (1.11), p would be identically equal to a. □

If $p \colon Y \to X$ is a non-constant holomorphic map, then we will say that Y is a *domain over* X.

A holomorphic (resp. meromorphic) function $f \colon Y \to \mathbb{C}$ (resp. $f \colon Y \to \mathbb{P}^1$) may also be considered as a multi-valued holomorphic (meromorphic) function on X. If $x \in X$ and $p^{-1}(x) = \{y_j \colon j \in J\}$, then the $f(y_j)$, $j \in J$, are the different values of this multi-valued function at the point x. Of course it might turn out that $p^{-1}(x)$ is a single point or is empty.

As an example, suppose $Y = \mathbb{C}$, $X = \mathbb{C}^*$ and $p = \exp: \mathbb{C} \to \mathbb{C}^*$. Then the identity mapping $\mathrm{id}: \mathbb{C} \to \mathbb{C}$ corresponds to the multi-valued logarithm on $\mathbb{C}^*$. For, the set $\exp^{-1}(b)$, where $b \in \mathbb{C}^*$, consists of exactly the various values of the logarithm of b. The following diagram illustrates this.

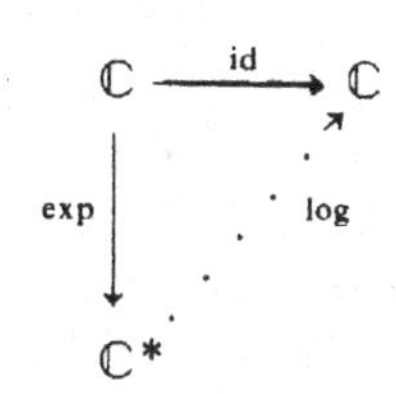

4.3 Definition. Suppose X and Y are Riemann surfaces and $p: Y \to X$ is a non-constant holomorphic map. A point $y \in Y$ is called a *branch point* or *ramification point* of p, if there is no neighborhood V of y such that $p \mid V$ is injective. The map p is called an *unbranched holomorphic map* if it has no branch points.

4.4. Theorem. *Suppose X and Y are Riemann surfaces. A non-constant holomorphic map $p: Y \to X$ has no branch points if and only if p is a local homeomorphism, i.e., every point $y \in Y$ has an open neighborhood V which is mapped homeomorphically by p onto an open set U in X.*

PROOF. Suppose $p: Y \to X$ has no branch points and $y \in Y$ is arbitrary. Since y is not a branch point, there exists an open neighborhood V of y such that $p \mid V$ is injective. Since p is continuous and open, p maps the set V homeomorphically onto the open set $U := p(V)$.

Conversely, assume $p: Y \to X$ is a local homeomorphism. Then for any $y \in Y$ there exists an open neighborhood V of y which is mapped homeomorphically by p onto an open set in X. In particular, $p \mid V$ is injective and y is not a branch point of p. □

4.5. Examples

(a) Suppose k is a natural number ≥ 2 and let $p_k: \mathbb{C} \to \mathbb{C}$ be the mapping defined by $p_k(z) := z^k$. Then $0 \in \mathbb{C}$ is a branch point of p_k and the mapping $p_k \mid \mathbb{C}^* \to \mathbb{C}$ is unbranched.

(b) Suppose $p: Y \to X$ is a non-constant holomorphic map, $y \in Y$ and $x := p(y)$. Then y is a branch point of p precisely if the mapping p takes the value x at the point y with multiplicity ≥ 2, cf. (2.2). By Theorem (2.1) the local behavior of p near y is just the same as the local behavior of the mapping p_k in example (a) near the origin.

(c) The mapping $\exp: \mathbb{C} \to \mathbb{C}^*$ is an unbranched holomorphic map. For exp is injective on every subset $V \subset \mathbb{C}$ which does not contain two points differing by an integral multiple of $2\pi i$.

(d) Suppose $\Gamma \subset \mathbb{C}$ is a lattice and $\pi: \mathbb{C} \to \mathbb{C}/\Gamma$ is the canonical quotient mapping, cf. (1.5.d). Then π is unbranched.

4.6. Theorem. *Suppose X is a Riemann surface, Y is a Hausdorff topological space and $p: Y \to X$ is a local homeomorphism. Then there is a unique complex structure on Y such that p is holomorphic.*

Remark. By (2.5) it follows that p is even locally biholomorphic.

PROOF. Suppose $\varphi_1: U_1 \to V \subset \mathbb{C}$ is a chart of the complex structure of X such that there exists an open subset $U \subset Y$ with $p | U \to U_1$ a homeomorphism. Then $\varphi := \varphi_1 \circ p: U \to V$ is a complex chart on Y. Let $\mathfrak{A}$ be the set of all complex charts on Y obtained in this way. It is easy to see that the charts of $\mathfrak{A}$ cover Y and are holomorphically compatible with one another. Now let Y have the complex structure defined by $\mathfrak{A}$. Then the projection p is locally biholomorphic and so, in particular, is a holomorphic mapping.

Uniqueness may be proved as follows. Suppose $\mathfrak{A}'$ is another complex atlas on Y such that the mapping $p: (Y, \mathfrak{A}') \to X$ is holomorphic and thus locally biholomorphic. Then the identity mapping $(Y, \mathfrak{A}) \to (Y, \mathfrak{A}')$ is locally biholomorphic and thus is a biholomorphic mapping. Hence $\mathfrak{A}$ and $\mathfrak{A}'$ define the same complex structure. □

4.7. The Lifting of Mappings. Suppose X, Y and Z are topological spaces and $p: Y \to X$ and $f: Z \to X$ are continuous maps. Then by a *lifting* of f with respect to p is meant a continuous mapping $g: Z \to Y$ such that $f = p \circ g$, i.e., the following diagram commutes.

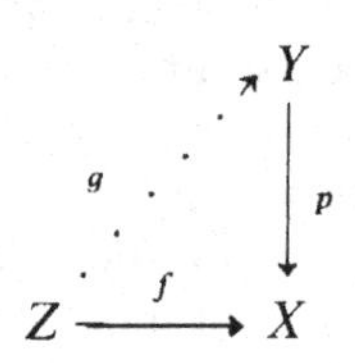

4.8. Theorem (Uniqueness of Lifting). *Suppose X and Y are Hausdorff spaces and $p: Y \to X$ is a local homeomorphism. Suppose Z is a connected topological space and $f: Z \to X$ is a continuous mapping. If $g_1, g_2: Z \to Y$ are two liftings of f and $g_1(z_0) = g_2(z_0)$ for some point $z_0 \in Z$ then $g_1 = g_2$.*

PROOF. Let $T := \{z \in Z: g_1(z) = g_2(z)\}$. The set T is closed, since it is the preimage of the diagonal $\Delta \subset Y \times Y$ under the mapping $(g_1, g_2): Z \to Y \times Y$. We claim that T is also open. Let $z \in T$ and let $g_1(z) = g_2(z) =: y$. Since p is a local homeomorphism, there exists a neighborhood V of y which is mapped by p homeomorphically onto a neighborhood U of $p(y) = f(z)$. Since g_1 and g_2 are both continuous, there is a neighborhood W of z with $g_i(W) \subset V$. Now let $\varphi: U \to V$ be the inverse of $p | V \to U$ and note that φ is continuous. Because $p \circ g_i = f$, one has $g_i | W = \varphi \circ (f | W)$ for $i = 1, 2$. Thus $g_1 | W = g_2 | W$ and $W \subset T$. Hence T is open. Since Z is connected and T is non-empty, $T = Z$ and thus $g_1 = g_2$. □

4.9. Theorem. *Suppose X, Y and Z are Riemann surfaces, $p: Y \to X$ is an unbranched holomorphic map and $f: Z \to X$ is any holomorphic map. Then every lifting $g: Z \to Y$ of f is holomorphic.*

PROOF. Suppose $c \in Z$ is an arbitrary point and let $b := g(c)$ and $a := p(b) = f(c)$. There exist open neighborhoods V of b and U of a such that $p | V \to U$ is biholomorphic. Suppose $\varphi: U \to V$ is the inverse map. Since g is continuous, there is an open neighborhood W of c such that $g(W) \subset V$. But $f = p \circ g$ implies $g | W = \varphi \circ (f | W)$ and thus g is holomorphic at the point c. □

Consequence. Suppose X, Y and Z are Riemann surfaces and $p: Y \to X$ and $q: Z \to X$ are unbranched holomorphic maps. Then every continuous fiber-preserving map $f: Y \to Z$ is holomorphic. For f is a lifting of p with respect to q.

Lifting of Curves. Suppose X and Y are Hausdorff spaces and $p: Y \to X$ is a local homeomorphism. We are particularly interested in the lifting of curves $u: [0, 1] \to X$. By Theorem (4.8) a lifting $\hat{u}: [0, 1] \to Y$ of u, if it exists at all, is uniquely determined once the lifting of the initial point is specified.

In the following we again let $I := [0, 1]$.

4.10. Theorem (Lifting of Homotopic Curves). *Suppose X and Y are Hausdorff spaces and $p: Y \to X$ is a local homeomorphism. Suppose $a, b \in X$ and $\hat{a} \in Y$ is a point such that $p(\hat{a}) = a$. Further suppose a continuous mapping $A: I \times I \to X$ is given such that $A(0, s) = a$ and $A(1, s) = b$ for every $s \in I$. Set*

$$u_s(t) := A(t, s).$$

If every curve u_s can be lifted to a curve $\hat{u}_s$ with initial point $\hat{a}$, then $\hat{u}_0$ and $\hat{u}_1$ have the same end point and are homotopic.

PROOF. Define a mapping $\hat{A}: I \times I \to Y$ by $\hat{A}(t, s) := \hat{u}_s(t)$.

Claim (a) There exists $\varepsilon_0 > 0$ such that $\hat{A}$ is continuous on $[0, \varepsilon_0[\times I$.

Proof. There are neighborhoods V of $\hat{a}$ and U of a such that $p | V \to U$ is a homeomorphism. Let $\varphi: U \to V$ be the inverse map. Since $A(0 \times I) = \{a\}$ and A is continuous, there exists $\varepsilon_0 > 0$ such that $A([0, \varepsilon_0] \times I) \subset U$. Because of the uniqueness of the lifting of curves, one has

$$\hat{u}_s | [0, \varepsilon_0] = \varphi \circ u_s | [0, \varepsilon_0] \quad \text{for every } s \in I.$$

Thus $\hat{A} = \varphi \circ A$ on $[0, \varepsilon_0] \times I$ and this implies $\hat{A}$ is continuous on $[0, \varepsilon_0[\times I$.

Claim (b) The mapping $\hat{A}$ is continuous on all of $I \times I$.

Proof. Suppose to the contrary that there is a point $(t_0, \sigma) \in I \times I$ at which $\hat{A}$ is not continuous. Let τ be the infinum of all those t such that $\hat{A}$ is not continuous at (t, σ). By (a) $\tau \geq \varepsilon_0$. Let $x := A(\tau, \sigma)$ and $y := \hat{A}(\tau, \sigma) = \hat{u}_\sigma(\tau)$.

There are neighborhoods V of y and U of x such that $p|V \to U$ is a homeomorphism. Let $\varphi: U \to V$ be the inverse. Since A is continuous, there exists $\varepsilon > 0$ such that $A(I_\varepsilon(\tau) \times I_\varepsilon(\sigma)) \subset U$, where

$$I_\varepsilon(\xi) = \{t \in I: |t - \xi| < \varepsilon\}.$$

In particular $u_\sigma(I_\varepsilon(\tau)) \subset U$ and thus

$$\hat{u}_\sigma|I_\varepsilon(\tau) = \varphi \circ u_\sigma|I_\varepsilon(\tau).$$

Choose $t_1 \in I_\varepsilon(\tau)$ with $t_1 < \tau$. Then

$$\hat{A}(t_1, \sigma) = \hat{u}_\sigma(t_1) \in V.$$

Since $\hat{A}$ is continuous at (t_1, σ), there exists $\delta > 0$, $\delta \leq \varepsilon$, such that

$$\hat{A}(t_1, s) = \hat{u}_s(t_1) \in V \quad \text{for every } s \in I_\delta(\sigma).$$

Because of the uniqueness of liftings it now follows that for every $s \in I_\delta(\sigma)$

$$\hat{u}_s|I_\varepsilon(\tau) = \varphi \circ u_s|I_\varepsilon(\tau).$$

Thus $\hat{A} = \varphi \circ A$ on $I_\varepsilon(\tau) \times I_\delta(\sigma)$. But this contradicts the definition of (τ, σ). Thus $\hat{A}$ is continuous on $I \times I$.

Since $A = p \circ \hat{A}$ and $A(\{1\} \times I) = \{b\}$, it follows that $\hat{A}(\{1\} \times I) \subset p^{-1}(b)$. Since $p^{-1}(b)$ is discrete and $\{1\} \times I$ is connected, $\hat{A}(\{1\} \times I)$ consists of a single point. This implies that the curves $\hat{u}_0$ and $\hat{u}_1$ have the same end point and, by means of $\hat{A}$, they are homotopic. □

Covering Maps. We would now like to give a condition which will ensure that the lifting of curves is always possible.

4.11. Definition. Suppose X and Y are topological spaces. A mapping $p: Y \to X$ is called a *covering map* if the following holds.

Every point $x \in X$ has an open neighborhood U such that its preimage $p^{-1}(U)$ can be represented as

$$p^{-1}(U) = \bigcup_{j \in J} V_j,$$

where the V_j, $j \in J$, are disjoint open subsets of Y, and all the mappings $p|V_j \to U$ are homeomorphisms. In particular, p is a local homeomorphism.

4.12. Examples

(a) Let $D = \{z \in \mathbb{C}: |z| < 1\}$ be the unit disk in the complex plane and let $p: D \to \mathbb{C}$ be the canonical injection. Then p is a local homeomorphism, but not a covering map. For, no point $a \in \mathbb{C}$ with $|a| = 1$ has a neighborhood U with the property required in the definition.

(b) Let k be a natural number ≥ 2 and let

$$p_k: \mathbb{C}^* \to \mathbb{C}^*, \qquad z \mapsto z^k.$$

Then p_k is a covering map. For, suppose $a \in \mathbb{C}^*$ is arbitrary and choose $b \in \mathbb{C}^*$ with $p_k(b) = a$. Since p_k is a local homeomorphism, there are open neighborhoods V_0 of b and U of a such that $p_k | V_0 \to U$ is a homeomorphism. Then

$$p_k^{-1}(U) = V_0 \cup \omega V_0 \cup \cdots \cup \omega^{k-1} V_0,$$

where ω is a kth primitive root of unity, say $\omega = \exp(2\pi i/k)$. It is clear that the sets $V_j := \omega^j V_0$, $j = 0, \ldots, k-1$, are pairwise disjoint and each $p_k | V_j \to U$ is a homeomorphism.

(c) The mapping $\exp: \mathbb{C} \to \mathbb{C}^*$ is a covering map.

PROOF. Suppose $a \in \mathbb{C}^*$ and $b \in \mathbb{C}$ with $\exp(b) = a$. Since exp is a local homeomorphism, there exist open neighborhoods V_0 of b and U of a such that $\exp | V_0 \to U$ is a homeomorphism. Then

$$\exp^{-1}(U) = \bigcup_{n \in \mathbb{Z}} V_n, \quad \text{where } V_n := 2\pi i n + V_0.$$

Clearly the V_n are pairwise disjoint and each map $\exp | V_n \to U$ is a homeomorphism.

(d) Suppose $\Gamma \subset \mathbb{C}$ is a lattice and $\pi: \mathbb{C} \to \mathbb{C}/\Gamma$ is the canonical quotient mapping. In the same way as in example (c) one can show that π is a covering map.

4.13. Definition. A continuous map $p: Y \to X$ is said to have the *curve lifting property* if the following condition holds. For every curve $u: [0, 1] \to X$ and every point $y_0 \in Y$ with $p(y_0) = u(0)$ there exists a lifting $\hat{u}: [0, 1] \to Y$ of u such that $\hat{u}(0) = y_0$.

4.14. Theorem. *Every covering map $p: Y \to X$ of topological spaces X and Y has the curve lifting property.*

PROOF. Suppose $u: [0, 1] \to X$ is a curve and $y_0 \in Y$ with $p(y_0) = u(0)$. Because of the compactness of $[0, 1]$ there exists a partition

$$0 = t_0 < t_1 < \cdots < t_n = 1$$

and open sets $U_k \subset X$, $k = 1, \ldots, n$, with the following properties:

(i) $u([t_{k-1}, t_k]) \subset U_k$,

(ii) $p^{-1}(U_k) = \bigcup_{j \in J_k} V_{kj}$,

where the $V_{kj} \subset Y$ are open sets such that $p | V_{kj} \to U_k$ are homeomorphisms. Now we shall prove by induction on $k = 0, 1, \ldots, n$ the existence of a lifting $\hat{u} | [0, t_k] \to X$ with $\hat{u}(0) = y_0$. For $k = 0$ this is trivial. So suppose $k \geq 1$ and $\hat{u} | [0, t_{k-1}] \to X$ is already constructed and let $\hat{u}(t_{k-1}) =: y_{k-1}$. Since $p(y_{k-1}) = u(t_{k-1}) \in U_k$, there exists $j \in J_k$ such that $y_{k-1} \in V_{kj}$. Let

$\varphi: U_k \to V_{kj}$ be the inverse of the homeomorphism $p \mid V_{kj} \to U_k$. Then if we set

$$\hat{u} \mid [t_{k-1}, t_k] := \varphi \circ (u \mid [t_{k-1}, t_k]),$$

we obtain a continuous extension of the lifting $\hat{u}$ to the interval $[0, t_k]$. □

4.15. Remark. Suppose X and Y are Hausdorff spaces, $p: Y \to X$ is a covering map and $x_0 \in X$, $y_0 \in Y$ are points with $p(y_0) = x_0$. Then by (4.14) and (4.8) for every curve $u: [0, 1] \to X$ with $u(0) = x_0$ there exists exactly one lifting $\hat{u}: [0, 1] \to Y$ such that $\hat{u}(0) = y_0$. When the curve u is closed, the lifting $\hat{u}$ need not be closed. An example of this is the following. Let $X = Y = \mathbb{C}^*$,

$$p: \mathbb{C}^* \to \mathbb{C}^*, \qquad z \mapsto z^2,$$

and $x_0 = y_0 = 1$. Define the curve $u: [0, 1] \to \mathbb{C}^*$ by $u(t) = e^{2\pi i t}$. Then u has initial and end point 1 and is thus closed. But $\hat{u}(t) := e^{\pi i t}$ defines a lifting $\hat{u}: [0, 1] \to \mathbb{C}^*$ of u with respect to p which has initial point 1 and end point -1.

However from Theorem (4.10) it follows that every lifting of a closed null-homotopic curve is again closed and null-homotopic.

4.16. Theorem. *Suppose X and Y are Hausdorff spaces with X pathwise connected and $p: Y \to X$ is a covering map. Then for any two points $x_0, x_1 \in X$ the sets $p^{-1}(x_0)$ and $p^{-1}(x_1)$ have the same cardinality. In particular, if Y is non-empty, then p is surjective.*

The cardinality of $p^{-1}(x)$ for $x \in X$ is called the *number of sheets* of the covering and may be either finite or infinite.

PROOF. Construct a mapping $\varphi: p^{-1}(x_0) \to p^{-1}(x_1)$ in the following way. Choose a curve $u: [0, 1] \to X$ joining x_0 to x_1. If $y \in p^{-1}(x_0)$ is an arbitrary point, then there exists precisely one lifting $\hat{u}: [0, 1] \to Y$ of u such that $\hat{u}(0) = y$. Set $\varphi(y) := \hat{u}(1) \in p^{-1}(x_1)$. The uniqueness of liftings then implies that the mapping just constructed is bijective. □

Remark. In general the bijective mapping constructed in the proof depends on the choice of the curve u. Thus in general there is no well-defined way to enumerate globally the "sheets" of a covering.

4.17. Theorem. *Suppose X and Y are Hausdorff spaces and $p: Y \to X$ is a covering map. Further, suppose Z is a simply connected, pathwise connected and locally pathwise connected topological space and $f: Z \to X$ is a continuous mapping. Then for every choice of points $z_0 \in Z$ and $y_0 \in Y$ with $f(z_0) = p(y_0)$ there exists precisely one lifting $\hat{f}: Z \to Y$ such that $\hat{f}(z_0) = y_0$.*

Remark. In the following proof the only properties of the mapping p that are used are that it is a local homeomorphism and has the curve lifting property.

PROOF. Define the mapping $\hat{f}: Z \to Y$ in the following way. Suppose $z \in Z$ is an arbitrary point and $u: I \to Z$ is a curve from z_0 to z. Then $v := f \circ u$ is a curve in X with initial point $f(z_0)$ and end point $f(z)$. Let $\hat{v}: I \to Y$ be the unique lifting of v which has initial point y_0. Then set $\hat{f}(z) := \hat{v}(1)$. This definition is independent of the choice of curve u from z_0 to z. For, suppose u_1 is another curve from z_0 to z. Then u_1 is homotopic to u. Thus $v_1 := f \circ u_1$ and $v = f \circ u$ are also homotopic. By Theorem (4.10) the liftings $\hat{v}_1$ of v_1 and $\hat{v}$ of v with $\hat{v}_1(0) = \hat{v}(0) = y_0$ have the same end point. Hence $\hat{f}(z)$ is well-defined. Also by construction $f = p \circ \hat{f}$.

All that remains to be proved is that the mapping $\hat{f}: Z \to Y$ is continuous. Let $z \in Z$, $y = \hat{f}(z)$ and suppose V is a neighborhood of y. We must show that there exists a neighborhood W of z such that $\hat{f}(W) \subset V$. Since p is a local homeomorphism, we may assume, possibly by shrinking V, that there is a neighborhood U of $p(y) = f(z)$ such that $p \mid V \to U$ is a homeomorphism. Let $\varphi: U \to V$ be its inverse. Since f is continuous and Z is locally pathwise connected, there exists a pathwise connected neighborhood W of z such that $f(W) \subset U$.

Now we claim that $\hat{f}(W) \subset V$. To see this suppose that the curves u, v and $\hat{v}$ are defined as above. Let $z' \in W$ be an arbitrary point and let u' be a curve from z to z' which lies entirely in W. Then the curve $v' := f \circ u'$ lies entirely in U and $\hat{v}' := \varphi \circ v'$ is a lifting of v' with initial point y. Hence the product $\hat{v} \cdot \hat{v}'$ is a lifting of $v \cdot v' = f \circ (u \cdot u')$ with initial point y_0. Thus

$$\hat{f}(z') = (\hat{v} \cdot \hat{v}')(1) = \hat{v}'(1) \in V. \qquad \square$$

4.18. Example (The Logarithm of a Function). Suppose X is a simply connected Riemann surface and $f: X \to \mathbb{C}^*$ is a nowhere vanishing holomorphic function on X. We would like to find the logarithm of f, i.e., find a holomorphic function $F: X \to \mathbb{C}$ such that $\exp(F) = f$. But this just means that F is a lifting of f with respect to the covering $\exp: \mathbb{C} \to \mathbb{C}^*$, i.e.,

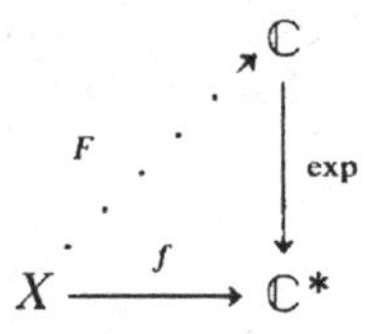

If $x_0 \in X$ and $c \in \mathbb{C}$ is any solution of the equation $e^c = f(x_0)$, then by Theorem (4.17) there exists a lifting $F: X \to \mathbb{C}$ of the required kind with $F(x_0) = c$. By Theorem (4.9), F is holomorphic. Also any other solution of the problem differs from F by an additive constant $2\pi i n$, $n \in \mathbb{Z}$.

As a special case suppose X is a simply connected domain in $\mathbb{C}^*$ and $j: X \to \mathbb{C}^*$ is the canonical injection, i.e., $j(z) = z$. Then every lifting of j with respect to exp is nothing more than a branch of the function log on X. Analogously one can construct various roots of a nowhere vanishing holomorphic function $f: X \to \mathbb{C}^*$ on any simply connected Riemann surface X. To do this one uses the covering in Example (4.12.b).

4.19. Theorem. *Suppose X is a manifold, Y is a Hausdorff space and $p: Y \to X$ is a local homeomorphism with the curve lifting property. Then p is a covering map.*

PROOF. Suppose $x_0 \in X$ is an arbitrary point and $y_j, j \in J$, are the preimages of x_0 with respect to p. Take U to be an open neighborhood of x_0 which is homeomorphic to a ball and let $f: U \to X$ be the canonical injection. From the remark in Theorem (4.17) it follows that for every $j \in J$ there is a lifting $\hat{f}_j: U \to Y$ of f such that $\hat{f}_j(x_0) = y_j$. Let $V_j := \hat{f}_j(U)$. Now one can easily convince oneself that

$$p^{-1}(U) = \bigcup_{j \in J} V_j,$$

that the V_j are pairwise disjoint open sets and that every mapping $p \mid V_j \to U$ is a homeomorphism. □

4.20. Proper Mappings. Recall that a locally compact topological space is a Hausdorff space such that every point has a compact neighborhood. A continuous mapping $f: X \to Y$ between two locally compact spaces is called *proper* if the preimage of every compact set is compact. For example this is always so if X is compact. A proper mapping is closed, i.e., the image of every closed set is closed. This follows from the fact that in a locally compact space a subset is closed precisely if its intersection with every compact set is compact.

4.21. Lemma. *Suppose X and Y are locally compact spaces and $p: Y \to X$ is a proper, discrete map. Then the following hold:*

(a) *For every point $x \in X$ the set $p^{-1}(x)$ is finite.*

(b) *If $x \in X$ and V is a neighborhood of $p^{-1}(x)$, then there exists a neighborhood U of x with $p^{-1}(U) \subset V$.*

PROOF

(a) This follows from the fact that $p^{-1}(x)$ is a compact discrete subset of Y.

(b) We may assume that V is open and thus $Y \backslash V$ is closed. Then $p(Y \backslash V) =: A$ is also closed and $x \notin A$. Thus $U := X \backslash A$ is an open neighborhood of x such that $p^{-1}(U) \subset V$. □

4.22. Theorem. *Suppose X and Y are locally compact spaces and $p: Y \to X$ is a proper local homeomorphism. Then p is a covering map.*

PROOF. Suppose $x \in X$ is arbitrary and let $p^{-1}(x) = \{y_1, \ldots, y_n\}$, where $y_i \neq y_j$ for $i \neq j$. Since p is a local homeomorphism, for every $j = 1, \ldots, n$ there exists an open neighborhood W_j of y_j and an open neighborhood U_j of x, such that $p \mid W_j \to U_j$ is a homeomorphism. We may assume that the W_j are pairwise disjoint. Now $W_1 \cup \cdots \cup W_n$ is a neighborhood of $p^{-1}(x)$. Thus by (4.21.b) there exists an open neighborhood $U \subset U_1 \cap \cdots \cap U_n$ of x with $p^{-1}(U) \subset W_1 \cup \cdots \cup W_n$. If we let $V_j := W_j \cap p^{-1}(U)$, then the V_j are disjoint open sets with

$$p^{-1}(U) = V_1 \cup \cdots \cup V_n$$

and all the mappings $p \mid V_j \to U$, $j = 1, \ldots, n$ are homeomorphisms. □

4.23. Proper Holomorphic Mappings. Suppose X and Y are Riemann surfaces and $f: X \to Y$ is a proper, non-constant, holomorphic mapping. It follows from Theorem (2.1) that the set A of branch points of f is closed and discrete. Since f is proper, $B := f(A)$ is also closed and discrete. One calls B the set of *critical values of f*.

Let $Y' := Y \backslash B$ and $X' := X \backslash f^{-1}(B) \subset X \backslash A$. Then $f \mid X' \to Y'$ is a proper unbranched holomorphic covering and by (4.22), (4.16) and (4.21.a) it has a well-defined finite number of sheets n. This means that every value $c \in Y'$ is taken exactly n times. In order to be able to extend this statement to the critical values $b \in B$ as well, we have to consider the multiplicities.

For $x \in X$ denote by $v(f, x)$ the multiplicity, in the sense of (2.2), with which f takes the values $f(x)$ at the point x. Then we will say that f takes the value $c \in Y$, counting multiplicities, m times on X, if

$$m = \sum_{x \in p^{-1}(c)} v(f, x).$$

4.24. Theorem. *Suppose X and Y are Riemann surfaces and $f: X \to Y$ is a proper non-constant holomorphic map. Then there exists a natural number n such that f takes every value $c \in Y$, counting multiplicities, n times.*

PROOF. Using the same notation as in (4.23) let n be the number of sheets of the unbranched covering $f \mid X' \to Y'$. Suppose $b \in B$ is a critical value, $f^{-1}(b) = \{x_1, \ldots, x_r\}$ and $k_j := v(f, x_j)$. By (2.1) and (2.2) there exist disjoint neighborhoods U_j of x_j and V_j of b such that for every $c \in V_j \backslash \{b\}$ the set $f^{-1}(c) \cap U_j$ consists of exactly k_j points ($j = 1, \ldots, r$). By Lemma (4.21.b) we can find a neighborhood $V \subset V_1 \cap \cdots \cap V_r$ of b such that $f^{-1}(V) \subset U_1 \cup \cdots \cup U_r$. Then for every point $c \in V \cap Y'$ we have that $f^{-1}(c)$ consists of $k_1 + \cdots + k_r$ points. On the other hand, for $c \in Y'$ the cardinality of $p^{-1}(c)$ is equal to n. Thus $n = k_1 + \cdots + k_r$. □

Remark. A proper non-constant holomorphic map will be called an *n-sheeted holomorphic covering map*, where n is the integer found in the previous Theorem. Note that holomorphic covering maps are allowed to have branch points. If we wish to emphasize that there are none, then we will specifically say that the map is *unbranched*. If we speak of a topological covering map or if there is no complex structure, then we mean a covering map in the sense of (4.11).

4.25. Corollary. *On any compact Riemann surface X every non-constant meromorphic function $f: X \to \mathbb{P}^1$ has as many zeros as poles, where each is counted according to multiplicities.*

Proof. The mapping $f: X \to \mathbb{P}^1$ is proper. □

4.26. Corollary. *Any polynomial of nth degree*

$$f(z) = z^n + a_1 z^{n-1} + \cdots + a_n \in \mathbb{C}[z]$$

has, counting multiplicities, exactly n zero.

Proof. By (2.3) we may consider f as a holomorphic mapping $f: \mathbb{P}^1 \to \mathbb{P}^1$ which, counting multiplicities, takes the value ∞ exactly n times. □

Exercises (§4)

4.1. Let $X := \mathbb{C} \setminus \{\pm 1\}$, $Y := \mathbb{C} \setminus \{(\pi/2) + k\pi,\ k \in \mathbb{Z}\}$. Show that

$$\sin\colon Y \to X$$

is a topological covering map. Consider the following curves in X.

$$u\colon [0, 1] \to X,\ u(t) := 1 - e^{2\pi i t}$$

$$v\colon [0, 1] \to X,\ v(t) := -1 + e^{2\pi i t}.$$

Let $w_1\colon [0, 1] \to Y$ be the lifting of $u \cdot v$ with $w_1(0) = 0$ and $w_2\colon [0, 1] \to Y$ be the lifting of $v \cdot u$ with $w_2(0) = 0$. Show that

$$w_1(1) = 2\pi$$

$$w_2(1) = -2\pi.$$

Conclude that $\pi_1(X)$ is not abelian.

4.2. Let X and Y be arcwise connected Hausdorff topological spaces and $f\colon Y \to X$ be a covering map. Show that the induced map

$$f_*\colon \pi_1(Y) \to \pi_1(X)$$

is injective.

4.3. Let X and Y be Hausdorff spaces and $p\colon Y \to X$ be a covering map. Let Z be a connected, locally arcwise connected topological space and $f\colon Z \to X$ a continuous map. Let $c \in Z$, $a := f(c)$ and $b \in Y$ such that $p(b) = a$. Prove that there exists a lifting $\tilde{f}: Z \to Y$ of f with $\tilde{f}(c) = b$ if and only if $f_*\pi_1(Z,c) \subset p_*\pi_1(Y,b)$.

4.4. (a) Show that

$$\tan\colon \mathbb{C} \to \mathbb{P}^1$$

is a local homeomorphism.

(b) Show that $\tan(\mathbb{C}) = \mathbb{P}^1 \setminus \{\pm i\}$ and

$$\tan\colon \mathbb{C} \to \mathbb{P}^1 \setminus \{\pm i\}$$

is a covering map.

(c) Let $X = \mathbb{C} \setminus \{it\colon t \in \mathbb{R},\ |t| \geq 1\}$. Show that for every $k \in \mathbb{Z}$ there exists a unique holomorphic function $\arctan_k\colon X \to \mathbb{C}$ with

$$\tan \circ \arctan_k = \mathrm{id}_X$$

and

$$\arctan_k(0) = k$$

(the kth branch of arctan).

4.5. Determine the ramification points of the map

$$f\colon \mathbb{C} \to \mathbb{P}^1, \qquad f(z) := \frac{1}{2}\left(z + \frac{1}{z}\right).$$

§5. The Universal Covering and Covering Transformations

Amongst all the covering spaces of a manifold X, there is one which deserves to be called the "largest," namely, the universal covering. All other covering spaces can be obtained from this one as quotients, and what happens to the universal covering when it is acted on by the group of "covering transformations" is closely related to the fundamental group of X. An investigation of these ideas is the focus of attention in this section.

5.1. Definition. Suppose X and Y are connected topological spaces and $p\colon Y \to X$ is a covering map. $p\colon Y \to X$ is called the *universal covering* of X if it satisfies the following universal property. For every covering map $q\colon Z \to X$, with Z connected, and every choice of points $y_0 \in Y$, $z_0 \in Z$ with $p(y_0) = q(z_0)$ there exists exactly one continuous fiber-preserving mapping $f\colon Y \to Z$ such that $f(y_0) = z_0$.

A connected topological space X has up to isomorphism at most one universal covering. For, with the above notation, suppose $q\colon Z \to X$ is also a universal covering. Then there exists a fiber-preserving continuous mapping

$g: Z \to Y$ such that $g(z_0) = y_0$. The compositions $g \circ f: Y \to Y$ and $f \circ g: Z \to Z$ are continuous fiber-preserving mappings such that $g \circ f(y_0) = y_0$ and $f \circ g(z_0) = z_0$. Because of the universality condition there can exist only one continuous fiber-preserving mapping in each case which satisfies these conditions. Thus $g \circ f = \mathrm{id}_Y$ and $f \circ g = \mathrm{id}_Z$. Hence $f: Y \to Z$ is a fiber-preserving homeomorphism.

5.2. Theorem. *Suppose X and Y are connected manifolds, Y is simply connected and $p: Y \to X$ is a covering map. Then p is the universal covering of X.*

PROOF. This follows directly from the definition and Theorem (4.17). □

5.3. Theorem. *Suppose X is a connected manifold. Then there exists a connected, simply connected manifold $\tilde{X}$ and a covering map $p: \tilde{X} \to X$.*

By Theorem (5.2) $\tilde{X} \to X$ is the universal covering of X.

PROOF. Pick a point $x_0 \in X$. For $x \in X$ let $\pi(x_0, x)$ denote the set of homotopy classes of curves having initial point x_0 and end point x. Let

$$\tilde{X} := \{(x, \alpha): x \in X, \quad \alpha \in \pi(x_0, x)\}.$$

Define the mapping $p: \tilde{X} \to X$ by $p(x, \alpha) := x$. We will now define a topology on $\tilde{X}$ so that $\tilde{X}$ becomes a connected, simply connected Hausdorff manifold and $p: \tilde{X} \to X$ is a covering map.

Suppose $(x, \alpha) \in \tilde{X}$ and $U \subset X$ is an open, connected, simply connected neighborhood of x. Define a subset $[U, \alpha] \subset \tilde{X}$ as follows: $[U, \alpha]$ consists of all points $(y, \beta) \in \tilde{X}$ such that $y \in U$ and $\beta = \mathrm{cl}(u \cdot v)$, where u is a curve from x_0 to x such that $\alpha = \mathrm{cl}(u)$ and v is a curve from x to y which lies completely in U. (Since U is simply connected, β is independent of the choice of the curve v.) Now let $\mathfrak{B}$ be the system of all such sets $[U, \alpha]$.

Claim (*a*) $\mathfrak{B}$ is the basis for a topology on $\tilde{X}$.

Proof

(i) Clearly every point of $\tilde{X}$ lies in at least one $[U, \alpha]$.

(ii) Suppose $(z, \gamma) \in [U, \alpha] \cap [V, \beta]$. Then $z \in U \cap V$ and there exists an open, connected and simply connected neighborhood $W \subset U \cap V$ of z. Then, as one can easily check,

$$(z, \gamma) \in [W, \gamma] \subset [U, \alpha] \cap [V, \beta].$$

From (i) and (ii) the claim follows.

Claim (*b*) The mapping $p: \tilde{X} \to X$ is a local homeomorphism and in particular is continuous. This follows from the fact that for every $[U, \alpha] \in \mathfrak{B}$ the mapping $p \mid [U, \alpha] \to U$ is a homeomorphism.

Claim (*c*) $\tilde{X}$ is Hausdorff.

It suffices to show that any two points $(x, \alpha), (x, \beta) \in \tilde{X}$, where $\alpha \neq \beta$, have disjoint neighborhoods. Suppose $U \subset X$ is an open, connected, simply connected neighborhood of x. Then $[U, \alpha] \cap [U, \beta] = \emptyset$. Otherwise there would be an element (y, γ) in the intersection. Suppose w is a curve in U from x to y and $\alpha = \text{cl}(u)$, $\beta = \text{cl}(v)$. Then by definition $\gamma = \text{cl}(u \cdot w) = \text{cl}(v \cdot w)$. Thus $\text{cl}(u) = \text{cl}(v)$. But this contradicts the assumption that $\alpha \neq \beta$.

Claim (*d*) $\tilde{X}$ is connected and $p: \tilde{X} \to X$ has the curve lifting property and thus by (4.19) is a covering map. Suppose $u: [0, 1] \to X$ is a curve with initial point x_0. For $s \in [0, 1]$ let $u_s: [0, 1] \to X$ be the curve defined by $u_s(t) := u(st)$. (The curve u_s runs along the points of the curve u corresponding to parameter values $t \in [0, s]$.) Further suppose v is a closed curve with initial and end point x_0. Then the mapping

$$\hat{u}: [0, 1] \to \tilde{X}, \qquad t \mapsto (u(t), \text{cl}(v \cdot u_t))$$

is continuous and is a lifting of u with $\hat{u}(0) = (x_0, \text{cl}(v))$. This follows directly from the definition of the topology on $\tilde{X}$. Finally, suppose $w: [0, 1] \to X$ is a curve with *arbitrary* initial point $x_1 := w(0)$, $\alpha \in \pi(x_0, x_1)$ and v is a curve from x_0 to x_1 with $\text{cl}(v) = \alpha$. Then it is easy to see that the lifting of $u := v \cdot w$ with $\hat{u}(0) = (x_0, \varepsilon)$, where ε is the homotopy class of the constant curve at x_0, gives rise to a lifting of w with $\hat{w}(0) = (x_1, \alpha)$.

Claim (*e*) $\tilde{X}$ is simply connected.

Let $w: [0, 1] \to \tilde{X}$ be a closed curve with initial and end point (x_0, ε). Then $u := p \circ w$ is a closed curve in X with $u(0) = x_0$. Now let $\tilde{u}: [0, 1] \to \tilde{X}$ be the lifting of u, which exists by claim (*d*), where v is chosen to be the constant curve at x_0. Because of the uniqueness of liftings, $\tilde{u} = w$. Thus $\tilde{u}(1) = (x_0, \text{cl}(u)) = (x_0, \varepsilon)$ and hence u is null-homotopic. By Theorem (4.10) w is also null-homotopic and thus $\tilde{X}$ is simply connected.

This completes the proof of Theorem (5.3). □

Remark. In particular, one can construct the universal covering of any Riemann surface and by (4.6) this universal covering is, in a natural way, a Riemann surface as well.

5.4. Definition. Suppose X and Y are topological spaces and $p: Y \to X$ is a covering map. By a *covering transformation* or *deck transformation* of this covering we mean a fiber-preserving homeomorphism $f: Y \to Y$. With operation the composition of mappings, the set of all covering transformation of $p: Y \to X$ forms a group which we denote by $\text{Deck}(Y/X)$. If there is any chance of confusion, then we will write $\text{Deck}(Y \xrightarrow{p} X)$ instead of $\text{Deck}(Y/X)$.

5.5. Definition. Suppose X and Y are connected Hausdorff spaces and $p: Y \to X$ is a covering map. The covering is called *Galois* (the terms *normal* and *regular* are also in common usage) if for every pair of points $y_0, y_1 \in Y$

with $p(y_0) = p(y_1)$ there exists a covering transformation $f: Y \to Y$ such that $f(y_0) = y_1$.

Remark. By Theorem (4.8) there exists at most one covering transformation $f: Y \to Y$ with $f(y_0) = y_1$, for f is a lifting of $p: Y \to X$.

Example. The mapping $p: \mathbb{C}^* \to \mathbb{C}^*$, $z \mapsto z^k$, is a covering map. It is *Galois* since for any $z_1, z_2 \in \mathbb{C}^*$ with $p(z_1) = p(z_2)$, one has $z_2 = \omega z_1$ where ω is a kth root of unity and the mapping $z \mapsto \omega z$ is a covering transformation.

There is a connection between *Galois* coverings and *Galois* field extensions, cf. (8.12).

5.6. Theorem. *Suppose X is a connected manifold and $p: \tilde{X} \to X$ is its universal covering. Then p is Galois and* $\operatorname{Deck}(\tilde{X}/X)$ *is isomorphic to the fundamental group $\pi_1(X)$.*

PROOF

(a) Suppose $y_0, y_1 \in \tilde{X}$ with $p(y_0) = p(y_1)$. By the definition of the universal covering there exists a continuous fiber-preserving mapping $f: \tilde{X} \to \tilde{X}$ with $f(y_0) = y_1$. We have to show that f is a homeomorphism. This can be seen as follows. As above there exists a continuous fiber-preserving mapping $g: \tilde{X} \to \tilde{X}$ with $g(y_1) = y_0$. But then $f \circ g$ and $g \circ f$ are continuous fiber-preserving mappings of $\tilde{X}$ into itself such that $f \circ g(y_1) = y_1$ and $g \circ f(y_0) = y_0$. Again from the definition of the universal covering it follows that $f \circ g$ and $g \circ f$ are both the identity map of $\tilde{X}$. Thus f is a homeomorphism and hence a covering transformation. This shows the covering $\tilde{X} \to X$ is *Galois.*

(b) Suppose $x_0 \in X$ and $y_0 \in \tilde{X}$ is a point with $p(y_0) = x_0$. Define a mapping

$$\Phi: \operatorname{Deck}(\tilde{X}/X) \to \pi_1(X, x_0)$$

as follows: Suppose $\sigma \in \operatorname{Deck}(\tilde{X}/X)$ and v is a curve in $\tilde{X}$ with initial point y_0 and end point $\sigma(y_0)$. (The homotopy class of v is uniquely determined since $\tilde{X}$ is simply connected.) The curve $p \circ v$ in X has initial and end point x_0. Let $\Phi(\sigma)$ be the homotopy class of $p \circ v$.

(i) Φ is a group homomorphism. Suppose $\sigma, \tau \in \operatorname{Deck}(\tilde{X}/X)$ and v (resp. w) is a curve in $\tilde{X}$ with initial point y_0 and end point $\sigma(y_0)$ (resp. $\tau(y_0)$). Then $\sigma \circ w$ is a curve with initial point $\sigma(y_0)$ and end point $\sigma\tau(y_0)$. Also $p \circ (\sigma \circ w) = p \circ w$. The product curve $v \cdot (\sigma \circ w)$ has initial point y_0 and end point $\sigma\tau(y_0)$. Thus

$$\begin{aligned}\Phi(\sigma\tau) &= \operatorname{cl}(p \circ (v \cdot (\sigma \circ w)) = \operatorname{cl}(p \circ v)\operatorname{cl}(p \circ (\sigma \circ w)) \\ &= \operatorname{cl}(p \circ v)\operatorname{cl}(p \circ w) = \Phi(\sigma)\Phi(\tau).\end{aligned}$$

(ii) Φ is injective.

Suppose $\sigma \in \text{Deck}(\tilde{X}/X)$ and v is a curve in $\tilde{X}$ from y_0 to $\sigma(y_0)$. Assume $\Phi(\sigma) = \varepsilon$, i.e., $p \circ v$ is null-homotopic. Since v is a lifting of $p \circ v$, it follows from (4.10) that the end point $\sigma(y_0)$ of v is the same as the initial point y_0. This implies $\sigma = \text{id}_{\tilde{X}}$.

(iii) Φ is surjective.

Suppose $\alpha \in \pi_1(X, x_0)$ and u is a curve representing α. Let v be a lifting of u to $\tilde{X}$ with initial point y_0 and suppose the end point of v is y_1. Then there exists $\sigma \in \text{Deck}(\tilde{X}/X)$ such that $\sigma(y_0) = y_1$. From the definition of Φ one has $\Phi(\sigma) = \alpha$. This completes the proof. □

5.7. Examples

(a) $\exp: \mathbb{C} \to \mathbb{C}^*$ is the universal covering of $\mathbb{C}^*$, since $\mathbb{C}$ is simply connected. For $n \in \mathbb{Z}$ let $\tau_n: \mathbb{C} \to \mathbb{C}$ be translation by $2\pi in$. Then $\exp(\tau_n(z)) = \exp(z + 2\pi in) = \exp(z)$ for every $z \in \mathbb{C}$ and thus τ_n is a covering transformation. If σ is any covering transformation, then $\exp(\sigma(0)) = \exp(0) = 1$ and thus there exists $n \in \mathbb{Z}$ such that $\sigma(0) = 2\pi in$. Since $\tau_n(0) = 2\pi in$ as well, $\sigma = \tau_n$. Thus

$$\text{Deck}(\mathbb{C} \xrightarrow{\exp} \mathbb{C}^*) = \{\tau_n: n \in \mathbb{Z}\}.$$

Since the last group is isomorphic to $\mathbb{Z}$,

$$\pi_1(\mathbb{C}^*) \cong \mathbb{Z}.$$

(b) Let

$$H = \{z \in \mathbb{C}: \text{Re}(z) < 0\}$$

be the left half plane and

$$D^* = \{z \in \mathbb{C}: 0 < |z| < 1\}.$$

Then $\exp: H \to D^*$ is the universal covering of the punctured unit disk. As in Example (a) one can show that the group of covering transformations consists of all translations by integral multiples of $2\pi i$ and that $\pi_1(D^*) \cong \mathbb{Z}$.

(c) Suppose $\Gamma = \mathbb{Z}\gamma_1 + \mathbb{Z}\gamma_2$ is a lattice in $\mathbb{C}$. Then the canonical quotient mapping $\mathbb{C} \to \mathbb{C}/\Gamma$ is the universal covering of the torus $\mathbb{C}/\Gamma$. For $\gamma \in \Gamma$ denote by $\tau_\gamma: \mathbb{C} \to \mathbb{C}$ translation by γ. Analogous to Example (a) one can show that $\text{Deck}(\mathbb{C} \to \mathbb{C}/\Gamma) \cong \{\tau_\gamma: \gamma \in \Gamma\}$. Thus

$$\pi_1(\mathbb{C}/\Gamma) \cong \Gamma \cong \mathbb{Z} \times \mathbb{Z}.$$

Consequence. There does not exist any meromorphic function on $\mathbb{C}$ doubly-periodic with respect to Γ which mod Γ has a single pole of first order.

PROOF. Such a function would define a holomorphic mapping $f: \mathbb{C}/\Gamma \to \mathbb{P}^1$ which takes the value ∞ only once. By (4.24) and (2.5) f would be biholomorphic and in particular $\pi_1(\mathbb{C}/\Gamma) \cong \pi_1(\mathbb{P}^1) = 0$, a contradiction! □

Remark. Later (18.3) we will give necessary and sufficient conditions for the existence of a doubly periodic meromorphic function with prescribed principal parts. However it is worth noting that one can make the above observation using only topological reasons.

5.8. Definition. Suppose X and Y are topological spaces, $p: Y \to X$ is a covering map and G is a subgroup of $\operatorname{Deck}(Y/X)$. Two points $y, y' \in Y$ are called *equivalent modulo* G, if there exists $\sigma \in G$ such that $\sigma(y) = y'$. Clearly this really is an equivalence relation on Y.

5.9. Theorem. *Suppose X and Y are connected manifolds, $q: Y \to X$ is a covering map and $p: \tilde{X} \to X$ is the universal covering. Let $f: \tilde{X} \to Y$ be a continuous fiber-preserving mapping, which by the definition of the universal covering exists. Then f is a covering map and there exists a subgroup $G \subset \operatorname{Deck}(\tilde{X}/X)$ such that two points $x, x' \in \tilde{X}$ are mapped onto the same point by f precisely if they are equivalent modulo G. Moreover $G \cong \pi_1(Y)$.*

PROOF. First we will show that f is a local homeomorphism. Suppose $x \in \tilde{X}$, $p(x) =: s$ and $f(x) =: y$. Since p is a local homeomorphism, there exist open neighborhoods W_1 of x and U_1 of s, such that $p \mid W_1 \to U_1$ is a homeomorphism. Since q is a covering map, there exists an open connected neighborhood U of s contained in U_1 and pairwise disjoint open sets V_i, $i \in I$, such that $q^{-1}(U) = \bigcup V_i$ and $q \mid V_i \to U$ is a homeomorphism for every $i \in I$. Let V be the particular V_i containing the point y and let $W := p^{-1}(U) \cap W_1$. Then $y \in f(W) \subset q^{-1}(U)$ and since $f(W)$ is connected, it follows that $f(W) = V$. Since $p \mid W \to U$ and $q \mid V \to U$ are homeomorphisms, $f \mid W \to V$ is also a homeomorphism. Thus f is a local homeomorphism.

In order to prove that f is a covering map, consider a curve v in Y with initial point y_0 and a point $x_0 \in \tilde{X}$ with $f(x_0) = y_0$. We have to show that the curve v can be lifted to $\tilde{X}$ with initial point x_0. Since $p: \tilde{X} \to X$ is a covering map the curve $q \circ v$ in X may be lifted to a curve u in $\tilde{X}$ with initial point x_0. Then the curves $f \circ u$ and v in Y are both liftings of the curve $q \circ v$ and have the same initial point y_0. Thus they coincide. But this means that u is the desired lifting of v. Thus f is a covering map by Theorem (4.19).

Let $G := \operatorname{Deck}(\tilde{X}/Y)$. This is a subgroup of $\operatorname{Deck}(\tilde{X}/X)$. Since $\tilde{X}$ is simply connected, $f: \tilde{X} \to Y$ is the universal covering of Y and so is Galois. Hence $G \cong \pi_1(Y)$ and $f(x) = f(x')$ precisely if there exists $\sigma \in G$ such that $\sigma(x) = x'$. This completes the proof of Theorem (5.9). □

We will now use Theorem (5.9) to determine all the covering spaces of the punctured unit disk $D^* = \{z \in \mathbb{C} : 0 < |z| < 1\}$.

5.10. Theorem. *Suppose X is a Riemann surface and $f: X \to D^*$ is an unbranched holomorphic covering map. Then one of the following holds:*

(i) *If the covering has an infinite number of sheets, then there exists a biholomorphic mapping $\varphi\colon X \to H$ of X onto the left half plane such that diagram* (1) *is commutative.*

$$\begin{array}{ccc} X & \xrightarrow{\varphi} & H \\ & {\scriptstyle f}\searrow \quad \swarrow{\scriptstyle \exp} & \\ & D^* & \end{array} \tag{1}$$

(ii) *If the covering is k-sheeted* ($k < \infty$), *then there exists a biholomorphic mapping $\varphi\colon X \to D^*$ such that diagram* (2) *is commutative, where $p_k\colon D^* \to D^*$ is the mapping $z \mapsto z^k$.*

$$\begin{array}{ccc} X & \xrightarrow{\varphi} & D^* \\ & {\scriptstyle f}\searrow \quad \swarrow{\scriptstyle p_k} & \\ & D^* & \end{array} \tag{2}$$

Thus every covering map of D^* is either isomorphic to the covering given by the logarithm or else by the kth root.

PROOF. Since $\exp\colon H \to D^*$ is the universal covering, there exists a holomorphic mapping $\psi\colon H \to X$ such that $\exp = f \circ \psi$. Let $G \subset \operatorname{Deck}(H/D^*)$ be the corresponding subgroup.

(i) If G consists only of the identity, then $\psi\colon H \to X$ is a biholomorphic map. Then the mapping $\varphi\colon X \to H$, which we are looking for, is the inverse mapping of ψ.

(ii) Now

$$\operatorname{Deck}(H/D^*) = \{\tau_n\colon n \in \mathbb{Z}\},$$

where $\tau_n\colon H \to H$ denotes the translation $z \mapsto z + 2\pi i n$. Thus for every subgroup $G \subset \operatorname{Deck}(H/D^*)$ which is not the identity, there exists a natural number $k \geq 1$ so that

$$G = \{\tau_{nk}\colon n \in \mathbb{Z}\}.$$

Let $g\colon H \to D^*$ be the covering map defined by $g(z) = \exp(z/k)$. Then $g(z) = g(z')$ precisely if z and z' are equivalent modulo G. Hence there exists a bijective mapping $\varphi\colon X \to D^*$ such that the diagram

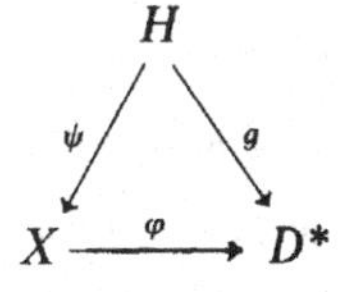

is commutative. Since ψ and g are locally biholomorphic, φ is biholomorphic. It is now easy to check that diagram (2) is commutative and the Theorem is proved. □

5.11. Theorem. *Suppose X is a Riemann surface, D is the unit disk and $f\colon X \to D$ is a proper non-constant holomorphic map which is unbranched over $D^* = D\backslash\{0\}$. Then there exists a natural number $k \geq 1$ and a biholomorphic mapping $\varphi\colon X \to D$ such that the diagram*

$$\begin{array}{ccc} X & \xrightarrow{\varphi} & D \\ & {\scriptstyle f}\searrow \quad \swarrow{\scriptstyle p_k} & \\ & D & \end{array} \qquad (*)$$

is commutative, where $p_k(z) := z^k$.

PROOF. Let $X^* := f^{-1}(D^*)$. Then $f \,|\, X^* \to D^*$ is an unbranched proper holomorphic covering map. By the previous Theorem there is a commutative diagram

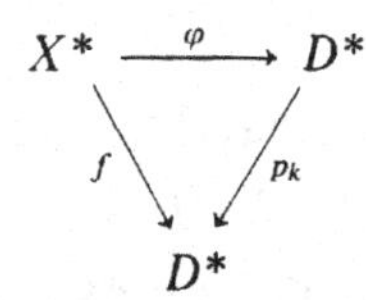

for some biholomorphic mapping $\varphi\colon X^* \to D^*$. We claim that $f^{-1}(0)$ consists of only one point. To the contrary suppose $f^{-1}(0)$ consists of n points $b_1, \ldots, b_n$ where $n \geq 2$. Then there exist disjoint open neighborhoods V_i of b_i and a disk $D(r) = \{z \in \mathbb{C}\colon |z| < r\}$, $0 < r \leq 1$, such that

$$f^{-1}(D(r)) \subset V_1 \cup \cdots \cup V_n. \qquad (**)$$

Let $D^*(r) = D(r)\backslash\{0\}$. Since $f^{-1}(D^*(r))$ is homeomorphic to $p_k^{-1}(D^*(r)) = D^*(\sqrt[k]{r})$, it is connected. Since every point b_i is an accumulation point of $f^{-1}(D^*(r))$, $f^{-1}(D(r))$ is also connected. But this contradicts (**). Thus $f^{-1}(0)$ consists of a single point $b \in X$. Hence by defining $\varphi(b) := 0$ one can continue the mapping $\varphi\colon X^* \to D^*$ to a biholomorphic mapping $\varphi\colon X \to D$ which makes the diagram (*) commutative. □

EXERCISES (§5)

5.1. Let $X = \mathbb{C}\backslash\{\pm 1\}$, and $Y = \mathbb{C}\backslash\{(\pi/2) + k\pi\colon k \in \mathbb{Z}\}$ (cf. Ex. 4.1.). Prove that $\operatorname{Deck}(Y \xrightarrow{\sin} X)$ consists of the following transformations

(i) $f_k(z) = z + 2k\pi, \quad k \in \mathbb{Z}$
(ii) $g_k(z) = -z + (2k+1)\pi, \quad k \in \mathbb{Z}$.

Calculate the products $f_k \circ f_l$, $f_k \circ g_l$, $g_l \circ f_k$, $g_k \circ g_l$.

5.2. Let X be a connected manifold and $p: \tilde{X} \to X$ be its universal covering. Let $G \subset \mathrm{Deck}(\tilde{X}/X)$ be a subgroup, $Y := \tilde{X}/G$ be the quotient of $\tilde{X}$ by the equivalence relation defined in 5.8 and $q: Y \to X$ be the map induced by p. Show that q is a covering map which is Galois if and only if G is a normal subgroup of $\mathrm{Deck}(\tilde{X}/X)$. In the latter case

$$\mathrm{Deck}(Y/X) \cong \mathrm{Deck}(\tilde{X}/X)/G.$$

5.3. Determine the covering transformations of

$$\tan: \mathbb{C} \to \mathbb{P}^1 \setminus \{i, -i\}$$

(cf. Ex. 4.4).

5.4. Let $\Gamma, \Gamma' \subset \mathbb{C}$ be lattices and

$$f: \mathbb{C}/\Gamma \to \mathbb{C}/\Gamma'$$

a non-constant holomorphic map with $f(0) = 0$. Show that there exists a unique $\alpha \in \mathbb{C}^*$ such that $\alpha\Gamma \subset \Gamma'$ and the following diagram is commutative

$$\begin{array}{ccc} \mathbb{C} & \xrightarrow{F} & \mathbb{C} \\ \downarrow \pi & & \downarrow \pi' \\ \mathbb{C}/\Gamma & \xrightarrow{f} & \mathbb{C}/\Gamma' \end{array}$$

where $F(z) = \alpha z$ and π and π' are the canonical projections. Prove that f is an unbranched covering map and

$$\mathrm{Deck}(\mathbb{C}/\Gamma \xrightarrow{f} \mathbb{C}/\Gamma') \cong \Gamma'/\alpha\Gamma.$$

5.5. Let $X := \mathbb{C} \setminus \{2, -2\}$, $Y := \mathbb{C} \setminus \{\pm 1, \pm 2\}$, and let $p: Y \to X$ be the map

$$p(z) := z^3 - 3z.$$

Prove that p is an unbranched 3-sheeted holomorphic covering map. Calculate $\mathrm{Deck}(Y/X)$ and show that the covering $Y \to X$ is not Galois.
[*Hint*: Use the fact that every biholomorphic map $f: Y \to Y$ extends to an automorphism of $\mathbb{P}^1$.]

5.6. Let $X := \mathbb{C} \setminus \{0, 1\}$, $Y := \mathbb{C} \setminus \{0, \pm i, \pm i\sqrt{2}\}$ and let $p: Y \to X$ be the map

$$p(z) := (z^2 + 1)^2.$$

Prove that p is an unbranched 4-sheeted covering map, which is not Galois and that

$$\mathrm{Deck}(Y/X) = \{\mathrm{id}, \varphi\},$$

where $\varphi(z) := -z$.

5.7. Suppose X and Y are connected Hausdorff spaces. Show that every 2-sheeted covering map $p: Y \to X$ is Galois.

§6. Sheaves

In complex analysis one frequently has to deal with functions which have various domains of definition. The notion of a sheaf gives a suitable formal setting to handle this situation.

6.1. Definition. Suppose X is a topological space and $\mathfrak{T}$ is the system of open sets in X. A *presheaf* of abelian groups on X is a pair $(\mathcal{F}, \rho)$ consisting of

(i) a family $\mathcal{F} = (\mathcal{F}(U))_{U \in \mathfrak{T}}$ of abelian groups,
(ii) a family $\rho = (\rho_V^U)_{U, V \in \mathfrak{T}, V \subset U}$ of group homomorphisms

$$\rho_V^U: \mathcal{F}(U) \to \mathcal{F}(V), \quad \text{where } V \text{ is open in } U,$$

with the following properties:

$$\rho_U^U = \mathrm{id}_{\mathcal{F}(U)} \quad \text{for every } U \in \mathfrak{T},$$
$$\rho_W^V \circ \rho_V^U = \rho_W^U \quad \text{for } W \subset V \subset U.$$

Remark. Generally one just writes $\mathcal{F}$ instead of $(\mathcal{F}, \rho)$. The homomorphisms ρ_V^U are called *restriction homomorphisms.* Instead of $\rho_V^U(f)$ for $f \in \mathcal{F}(U)$ one writes just $f \mid V$. Analogous to presheaves of abelian groups one can also define presheaves of vector spaces, rings, sets, etc.

6.2. Example. Suppose X is an arbitrary topological space. For any open subset $U \subset X$ let $\mathscr{C}(U)$ be the vector space of all continuous functions $f: U \to \mathbb{C}$. For $V \subset U$ let $\rho_V^U: \mathscr{C}(U) \to \mathscr{C}(V)$ be the usual restriction mapping. Then $(\mathscr{C}, \rho)$ is a presheaf of vector spaces on X.

6.3. Definition. A presheaf $\mathcal{F}$ on a topological space X is called a *sheaf* if for every open set $U \subset X$ and every family of open subsets $U_i \subset U$, $i \in I$, such that $U = \bigcup_{i \in I} U_i$ the following conditions, which we will call the Sheaf Axioms, are satisfied:

(I) If $f, g \in \mathcal{F}(U)$ are elements such that $f \mid U_i = g \mid U_i$ for every $i \in I$, then $f = g$.

(II) Given elements $f_i \in \mathcal{F}(U_i)$, $i \in I$, such that

$$f_i \mid U_i \cap U_j = f_j \mid U_i \cap U_j \quad \text{for all } i, j \in I,$$

then there exists an $f \in \mathcal{F}(U)$ such that $f \mid U_i = f_i$ for every $i \in I$.

Remark. The element f, whose existence is assured by (II), is by (I) uniquely determined.

Applying (I) and (II) to the case $U = \varnothing = \bigcup_{i \in \varnothing} U_i$ implies $\mathcal{F}(\varnothing)$ consists of exactly one element.

6.4. Examples

(a) For every topological space X the presheaf $\mathscr{C}$ defined in (6.2) is a sheaf. Both Sheaf Axioms (I) and (II) are trivially fulfilled.

(b) Suppose X is a Riemann surface and $\mathcal{O}(U)$ is the ring of holomorphic functions defined on the open set $U \subset X$. Taking the usual restriction mapping $\mathcal{O}(U) \to \mathcal{O}(V)$ for $V \subset U$ one gets the sheaf $\mathcal{O}$ of holomorphic functions on X. The sheaf $\mathcal{M}$ of meromorphic functions on X is defined analogously.

(c) For an open subset U of a Riemann surface X let $\mathcal{O}^*(U)$ be the multiplicative group of all holomorphic maps $f: U \to \mathbb{C}^*$. With the usual restriction maps $\mathcal{O}^*$ is a sheaf of (multiplicative) abelian groups. The sheaf $\mathcal{M}^*$ is defined analogously: For any open set $U \subset X$, $\mathcal{M}^*(U)$ consists of all meromorphic functions $f \in \mathcal{M}(U)$ which do not vanish identically on any connected component of U.

(d) Suppose X is an arbitrary topological space and G is an abelian group. Define a presheaf $\mathscr{G}$ on X as follows: For any non-empty open subset $U \subset X$ let $\mathscr{G}(U) := G$ and let $\mathscr{G}(\varnothing) := 0$. As for the restriction mappings, let $\rho_V^U = \mathrm{id}_G$ if $V \neq \varnothing$ and let $\rho_\varnothing^U$ be the zero homomorphism. If G contains at least two distinct elements g_1, g_2 and if X has two disjoint non-empty open subsets U_1, U_2, then $\mathscr{G}$ is not a sheaf. This is because Sheaf Axion (II) does not hold. For, since $U_1 \cap U_2 = \varnothing$, one has $g_1 | U_1 \cap U_2 = 0 = g_2 | U_1 \cap U_2$ but there is no $f \in \mathscr{G}(U_1 \cup U_2) = G$ such that $f | U_1 = g_1$ and $f | U_2 = g_2$.

(e) One can easily modify the previous example to obtain a sheaf. For any open set U, let $\tilde{\mathscr{G}}(U)$ be the abelian group of all locally constant mappings $g: U \to G$. Then if U is a non-empty connected open set, one has $\tilde{\mathscr{G}}(U) = G$. For $V \subset U$ let $\tilde{\mathscr{G}}(U) \to \tilde{\mathscr{G}}(V)$ be the usual restriction. Then $\tilde{\mathscr{G}}$ is a sheaf on X which is called the sheaf of locally constant functions with values in G. Often it is just denoted by G.

6.5. The Stalk of a Presheaf. Suppose $\mathscr{F}$ is a presheaf of sets on a topological space X and $a \in X$ is a point. On the disjoint union

$$\dot{\bigcup_{U \ni a}} \mathscr{F}(U),$$

where the union is taken over all the open neighborhoods U of a, introduce an equivalence relation $\underset{a}{\sim}$ as follows: Two elements $f \in \mathscr{F}(U)$ and $g \in \mathscr{F}(V)$ are related $f \underset{a}{\sim} g$ precisely if there exists an open set W with $a \in W \subset U \cap V$ such that $f | W = g | W$. One can easily check that this really is an equivalence relation. The set $\mathscr{F}_a$ of all equivalence classes, the so-called inductive limit of $\mathscr{F}(U)$, is given by

$$\mathscr{F}_a := \varinjlim_{U \ni a} \mathscr{F}(U) := \left(\dot{\bigcup_{U \ni a}} \mathscr{F}(U) \right) \Big/ \underset{a}{\sim},$$

and is called the *stalk* of $\mathscr{F}$ at the point a. If $\mathscr{F}$ is a presheaf of abelian groups (resp. vector spaces, rings), then the stalk $\mathscr{F}_a$ with the operation defined on the equivalence classes by means of the operation defined on representatives, is also an abelian group (resp. vector space, ring).

For any open neighborhood U of a, let

$$\rho_a: \mathscr{F}(U) \to \mathscr{F}_a$$

be the mapping which assigns to each element $f \in \mathscr{F}(U)$ its equivalence class modulo $\underset{a}{\sim}$. One calls $\rho_a(f)$ the *germ* of f at a. As an example consider the sheaf $\mathcal{O}$ of holomorphic functions on a domain $X \subset \mathbb{C}$. Let $a \in X$. A *germ of a holomorphic function* $\varphi \in \mathcal{O}_a$ is represented by a holomorphic function in an open neighborhood of a and thus has a Taylor series expansion $\sum_{\nu=0}^{\infty} c_\nu(z-a)^\nu$ with a positive radius of convergence. Two holomorphic functions on neighborhoods of a determine the same germ at a precisely if they have the same Taylor series expansion about a. Thus there is an isomorphism between the stalk $\mathcal{O}_a$ and the ring $\mathbb{C}\{z-a\}$ of all convergent power series in $z-a$ with complex coefficients. In an analogous way, the ring $\mathcal{M}_a$ of *germs of meromorphic functions* at a is isomorphic to the ring of all convergent Laurent series

$$\sum_{\nu=k}^{\infty} c_\nu(z-a)^\nu, \qquad k \in \mathbb{Z}, \qquad c_\nu \in \mathbb{C},$$

which have finite principal parts.

For any germ of a function $\varphi \in \mathcal{O}_a$ the value of the function, $\varphi(a) \in \mathbb{C}$, is well-defined, i.e., is independent of the choice of representative.

6.6. Lemma. *Suppose $\mathscr{F}$ is a sheaf of abelian groups on the topological space X and $U \subset X$ is an open subset. Then an element $f \in \mathscr{F}(U)$ is zero precisely if all germs $\rho_x(f) \in \mathscr{F}_x$, $x \in U$, vanish.*

This follows directly from Sheaf Axiom (I).

6.7. The Topological Space Associated to a Presheaf. Suppose X is a topological space and $\mathscr{F}$ is a presheaf on X. Let

$$|\mathscr{F}| := \dot{\bigcup_{x \in X}} \mathscr{F}_x$$

be the disjoint union of all the stalks. Denote by

$$p: |\mathscr{F}| \to X$$

the mapping which assigns to each element $\varphi \in \mathscr{F}_x$ the point x. Now introduce a topology on $|\mathscr{F}|$ as follows: For any open subset $U \subset X$ and an element $f \in \mathscr{F}(U)$, let

$$[U, f] := \{\rho_x(f): x \in U\} \subset |\mathscr{F}|.$$

6.8. Theorem. *The system $\mathfrak{B}$ of all sets $[U, f]$, where U is open in X and $f \in \mathscr{F}(U)$, is a basis for a topology on $|\mathscr{F}|$. The projection $p: |\mathscr{F}| \to X$ is a local homeomorphism.*

PROOF

(a) To see that $\mathfrak{B}$ forms a basis for a topology on $|\mathscr{F}|$, one has to verify the following two conditions:

(i) Every element $\varphi \in |\mathscr{F}|$ is contained in at least one $[U, f]$. This is trivial.

(ii) If $\varphi \in [U, f] \cap [V, g]$, then there exists a $[W, h] \in \mathfrak{B}$ such that $\varphi \in [W, h] \subset [U, f] \cap [V, g]$. For suppose $p(\varphi) = x$. Then $x \in U \cap V$ and $\varphi = \rho_x(f) = \rho_x(g)$. Hence there exists an open neighborhood $W \subset U \cap V$ of x such that $f \mid W = g \mid W =: h$. This implies $\varphi \in [W, h] \subset [U, f] \cap [V, g]$.

(b) Now we will show that $p: |\mathscr{F}| \to X$ is a local homeomorphism. Suppose $\varphi \in |\mathscr{F}|$ and $p(\varphi) = x$. There exists a $[U, f] \in \mathfrak{B}$ with $\varphi \in [U, f]$. Then $[U, f]$ is an open neighborhood of φ and U is an open neighborhood of x. The mapping $p \mid [U, f] \to U$ is bijective and also continuous and open as one sees immediately from the definition. Thus $p: |\mathscr{F}| \to X$ is a local homeomorphism. □

6.9. Definition. A presheaf $\mathscr{F}$ on a topological space X is said to satisfy the *Identity Theorem* if the following holds. If $Y \subset X$ is a domain and $f, g \in \mathscr{F}(Y)$ are elements whose germs $\rho_a(f)$ and $\rho_a(g)$ coincide at a point $a \in Y$, then $f = g$.

For example, this condition is satisfied by the sheaf $\mathscr{O}$ (resp. $\mathscr{M}$) of holomorphic (resp. meromorphic) functions on a Riemann surface X.

6.10. Theorem. *Suppose X is a locally connected Hausdorff space and $\mathscr{F}$ is a presheaf on X which satisfies the Identity Theorem. Then the topological space $|\mathscr{F}|$ is Hausdorff.*

PROOF. Suppose $\varphi_1, \varphi_2 \in |\mathscr{F}|$ and $\varphi_1 \neq \varphi_2$. We have to find disjoint neighborhoods of φ_1 and φ_2.

Case 1. Suppose $p(\varphi_1) =: x \neq y := p(\varphi_2)$. Since X is Hausdorff, there exist disjoint neighborhoods U and V of x and y respectively. Then $p^{-1}(U)$ and $p^{-1}(V)$ are disjoint neighborhoods of φ_1 and φ_2, respectively.

Case 2. Suppose $p(\varphi_1) = p(\varphi_2) =: x$. Suppose the germs $\varphi_i \in \mathscr{F}_x$ are represented by elements $f_i \in \mathscr{F}(U_i)$, where the U_i are open neighborhoods of x, $i = 1, 2$. Let $U \subset U_1 \cap U_2$ be a connected open neighborhood of x. Then $[U, f_i|U]$ are open neighborhoods of φ_i. Now suppose there exists $\psi \in [U, f_1|U] \cap [U, f_2|U]$. Let $p(\psi) = y$. Then $\psi = \rho_y(f_1) = \rho_y(f_2)$. From the Identity Theorem it follows that $f_1 | U = f_2 | U$, thus $\varphi_1 = \varphi_2$. Contradiction! Hence $[U, f_1 | U]$ and $[U, f_2 | U]$ are disjoint. □

EXERCISES (§6)

6.1. Suppose X is a Riemann surface. For $U \subset X$ open, let $\mathscr{B}(U)$ be the vector space of all bounded holomorphic functions $f: U \to \mathbb{C}$. For $V \subset U$ let $\mathscr{B}(U) \to \mathscr{B}(V)$ be the usual restriction map. Show that $\mathscr{B}$ is a presheaf which satisfies sheaf axiom (I) but not sheaf axiom (II).

6.2. Suppose X is a Riemann surface. For $U \subset X$ open, let

$$\mathscr{F}(U) := \mathcal{O}^*(U)/\exp \mathcal{O}(U).$$

Show that $\mathscr{F}$ with the usual restriction maps is a presheaf which does not satisfy sheaf axiom (I).

6.3. Suppose $\mathscr{F}$ is a presheaf on the topological space X and $p: |\mathscr{F}| \to X$ is the associated covering space. For $U \subset X$ open, let $\tilde{\mathscr{F}}(U)$ be the space of all sections of p over U, i.e., the space of all continuous maps

$$f: U \to |\mathscr{F}|$$

with $p \circ f = \mathrm{id}_U$. Prove the following:

(a) $\tilde{\mathscr{F}}$ together with the natural restriction maps is a sheaf,
(b) There is a natural isomorphism of the stalks

$$\mathscr{F}_x \xrightarrow{\sim} \tilde{\mathscr{F}}_x, \quad \text{for every } x \in X.$$

§7. Analytic Continuation

Next we consider the construction of Riemann surfaces which arise from the analytic continuation of germs of functions.

7.1. Definition. Suppose X is a Riemann surface, $u: [0, 1] \to X$ is a curve and $a := u(0)$, $b := u(1)$. The holomorphic function germ $\psi \in \mathcal{O}_b$ is said to result from the *analytic continuation along the curve u* of the holomorphic function germ $\varphi \in \mathcal{O}_a$ if the following holds. There exists a family $\varphi_t \in \mathcal{O}_{u(t)}$, $t \in [0, 1]$ of function germs with $\varphi_0 = \varphi$ and $\varphi_1 = \psi$ with the property that for every $\tau \in [0, 1]$ there exists a neighborhood $T \subset [0, 1]$ of τ, an open set $U \subset X$ with $u(T) \subset U$ and a function $f \in \mathcal{O}(U)$ such that

$$\rho_{u(t)}(f) = \varphi_t \quad \text{for every } t \in T.$$

Here $\rho_{u(t)}(f)$ is the germ of f at the point $u(t)$. Because of the compactness of $[0, 1]$ this condition is equivalent to the following (see Fig. 5). There exist a partition $0 = t_0 < t_1 < \cdots < t_{n-1} < t_n = 1$ of the interval $[0, 1]$, domains $U_i \subset X$ with $u([t_{i-1}, t_i]) \subset U_i$ and holomorphic functions $f_i \in \mathcal{O}(U_i)$ for $i = 1, \ldots, n$ such that:

(i) φ is the germ of f_1 at the point a and ψ is the germ of f_n at the point b.
(ii) $f_i | V_i = f_{i+1} | V_i$ for $i = 1, \ldots, n-1$, where V_i denotes the connected component of $U_i \cap U_{i+1}$ containing the point $u(t_i)$.

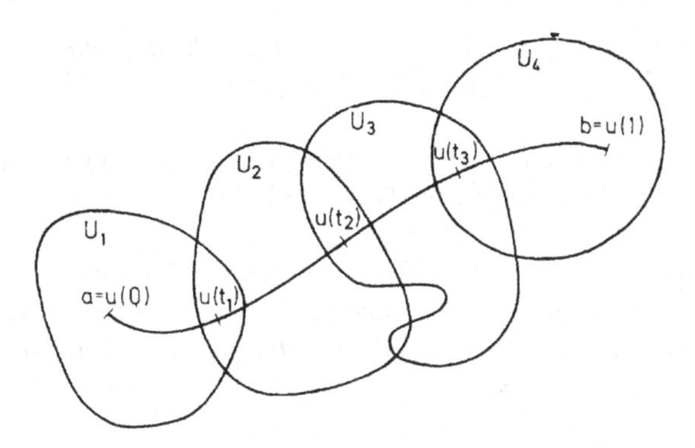

Figure 5

If one carries out the construction given in (6.7) for the sheaf $\mathcal{O}$ of holomorphic functions, then one gets a map $p\colon |\mathcal{O}| \to X$. The next Lemma shows that one can interpret analytic continuation along a curve by means of this map.

7.2. Lemma. *Suppose X is a Riemann surface and $u\colon [0, 1] \to X$ is a curve in X with $u(0) =: a$ and $u(1) =: b$. Then a function germ $\psi \in \mathcal{O}_b$ is the analytic continuation of a function germ $\varphi \in \mathcal{O}_a$ along u precisely if there exists a lifting $\hat{u}\colon [0, 1] \to |\mathcal{O}|$ of the curve u such that $\hat{u}(0) = \varphi$ and $\hat{u}(1) = \psi$.*

PROOF

(a) Suppose $\psi \in \mathcal{O}_b$ is the analytic continuation of $\varphi \in \mathcal{O}_a$ along u. Let $\varphi_t \in \mathcal{O}_{u(t)}$ for $t \in [0, 1]$ be the family of function germs as given in the Definition (7.1). It follows directly from the definition of the topology of $|\mathcal{O}|$ that the correspondence $t \mapsto \varphi_t$ represents a continuous mapping $\hat{u}\colon [0, 1] \to |\mathcal{O}|$. Thus $\hat{u}$ is a lifting of u with $\hat{u}(0) = \varphi_0 = \varphi$ and $\hat{u}(1) = \varphi_1 = \psi$.

(b) Suppose there is a lifting $\hat{u}\colon [0, 1] \to \mathcal{O}$ of u with $\hat{u}(0) = \varphi$ and $\hat{u}(1) = \psi$. For $t \in [0, 1]$, let $\varphi_t := \hat{u}(t)$. Then $\varphi_t \in \mathcal{O}_{u(t)}$ and $\varphi_0 = \varphi$, $\varphi_1 = \psi$. Let $\tau \in [0, 1]$ and suppose $[U, f] \subset |\mathcal{O}|$ is an open neighborhood of $\hat{u}(\tau)$. Then there exists a neighborhood $T \subset [0, 1]$ of τ such that $\hat{u}(T) \subset [U, f]$. This implies $u(T) \subset U$ and $\varphi_t = \hat{u}(t) = \rho_{u(t)}(f)$ for every $t \in T$. But this means that ψ is the analytic continuation of φ along u. □

Because of the uniqueness of liftings (Theorem 4.8) it follows from the lemma that if the analytic continuation of a function germ exists, then it is uniquely determined. Another consequence of the lemma is the Monodromy Theorem.

7.3. Monodromy Theorem. *Suppose X is a Riemann surface and u_0, $u_1\colon [0, 1] \to X$ are homotopic curves from a to b. Suppose u_s, $0 \le s \le 1$, is a deformation of u_0 into u_1 and $\varphi \in \mathcal{O}_a$ is a function germ which admits an* *analytic continuation along every curve u_s. Then the analytic continuations of φ along u_0 and u_1 yield the same function germ $\psi \in \mathcal{O}_b$.*

PROOF. Apply Theorem (4.10) to the local homeomorphism $|\mathcal{O}| \to X$, noting that $|\mathcal{O}|$ is Hausdorff by Theorem (6.10). □

7.4. Corollary. *Suppose X is a simply connected Riemann surface, $a \in X$ and $\varphi \in \mathcal{O}_a$ is a function germ which admits an analytic continuation along every curve starting at a. Then there exists a globally defined holomorphic function $f \in \mathcal{O}(X)$ such that $\rho_a(f) = \varphi$.*

Remark. Because of the Identity Theorem, f is uniquely determined.

PROOF. For any $x \in X$ let $\psi_x \in \mathcal{O}_x$ be the function germ which results from the analytical continuation of φ along any curve from a to x. Since X is simply connected, ψ_x is independent of which curve is chosen. Set $f(x) := \psi_x(x)$. Then f is a holomorphic function on X such that $\rho_a(f) = \varphi$. □

7.5. In general, even if the analytic continuation of some function germ is possible along two curves with the same initial and end points, then the resulting germs at the end point may be different. Thus if we consider all the germs arising by analytic continuation from the given function germ we get a multi-valued function. Our next task is to look at this situation and to make the details precise.

Suppose X and Y are Riemann surfaces and $\mathcal{O}_X$ and $\mathcal{O}_Y$ are the sheaves of holomorphic functions on them. Suppose $p: Y \to X$ is an unbranched holomorphic map. Since p is locally biholomorphic, for each $y \in Y$ it induces an isomorphism $p^*: \mathcal{O}_{X,\,p(y)} \to \mathcal{O}_{Y,\,y}$. Let

$$p_*: \mathcal{O}_{Y,\,y} \to \mathcal{O}_{X,\,p(y)}$$

be the inverse of p^*.

7.6. Definition. Suppose X is a Riemann surface, $a \in X$ is a point and $\varphi \in \mathcal{O}_a$ is a function germ. A quadrupel (Y, p, f, b) is called an *analytic continuation* of φ if:

(i) Y is a Riemann surface and $p: Y \to X$ is an unbranched holomorphic map.

(ii) f is a holomorphic function on Y.

(iii) b is a point of Y such that $p(b) = a$ and

$$p_*(\rho_b(f)) = \varphi.$$

An analytic continuation (Y, p, f, b) of φ is said to be *maximal* if it has the following universal property. If (Z, q, g, c) is any other analytic continuation of φ, then there exists a fiber-preserving holomorphic mapping $F: Z \to Y$ such that $F(c) = b$ and $F^*(f) = g$.

A maximal analytic continuation is unique up to isomorphism. Namely, using the above notation, if (Y, p, f, b) and (Z, q, g, c) are two maximal analytic continuations of φ, then there exists a fiber-preserving holomorphic mapping $G\colon Y \to Z$ such that $G(b) = c$ and $G^*(g) = f$. The composition $F \circ G$ is a fiber-preserving holomorphic mapping of Y onto itself which leaves the point b fixed. Hence by Theorem (4.8) one has $F \circ G = \mathrm{id}_Y$. Similarly $G \circ F = \mathrm{id}_Z$ and thus $G\colon Y \to Z$ is biholomorphic.

7.7. Lemma. *Suppose X is a Riemann surface, $a \in X$, $\varphi \in \mathcal{O}_a$ and (Y, p, f, b) is an analytic continuation of φ. Then if $v\colon [0, 1] \to Y$ is a curve with $v(0) = b$ and $v(1) =: y$, then the function germ $\psi := p_*(\rho_y(f)) \in \mathcal{O}_{p(y)}$ is an analytic continuation of φ along the curve $u := p \circ v$.*

PROOF. For $t \in [0, 1]$ let $\varphi_t := p_*(\rho_{v(t)}(f)) \in \mathcal{O}_{p(v(t))} = \mathcal{O}_{u(t)}$. Then $\varphi_0 = \varphi$ and $\varphi_1 = p_*(f_y) = \psi$. Suppose $t_0 \in [0, 1]$. Since $p\colon Y \to X$ is a local homeomorphism, there exist open neighborhoods $V \subset Y$ and $U \subset X$ of $v(t_0)$ and $p(v(t_0)) = u(t_0)$ resp. such that $p \mid V \to U$ is biholomorphic. Let $q\colon U \to V$ be the inverse mapping and let $g := q^*(f \mid V) \in \mathcal{O}(U)$. Then $p_*(\rho_\eta(f)) = \rho_{p(\eta)}(g)$ for every $\eta \in V$. There exists a neighborhood T of t_0 in $[0, 1]$ such that $v(T) \subset V$, i.e., $u(T) \subset U$. For every $t \in T$

$$\rho_{u(t)}(g) = p_*(\rho_{v(t)}(f)) = \varphi_t.$$

This proves that ψ is an analytic continuation of φ along u. □

7.8. Theorem. *Suppose X is a Riemann surface, $a \in X$ and $\varphi \in \mathcal{O}_a$ is a holomorphic function germ at the point a. Then there exists a maximal analytic continuation (Y, p, f, b) of φ.*

PROOF. Let Y be the connected component of $|\mathcal{O}|$ containing φ. Let p also denote the restriction of the mapping $p\colon |\mathcal{O}| \to X$ to Y. Then $p\colon Y \to X$ is a local homeomorphism. By Theorem (4.6) there is a complex structure on Y so that it becomes a Riemann surface and the mapping $p\colon Y \to X$ is holomorphic. Now define a holomorphic function $f\colon Y \to \mathbb{C}$ as follows. By definition every $\eta \in Y$ is a function germ at the point $p(\eta)$. Set $f(\eta) := \eta(p(\eta))$. One easily sees that f is holomorphic and $p_*(\rho_\eta(f)) = \eta$ for every $\eta \in Y$. Thus if one lets $b := \varphi$, then (Y, p, f, b) is an analytic continuation of φ.

Now we will show that (Y, p, f, b) is a maximal analytic continuation of φ. Suppose (Z, q, g, c) is another analytic continuation of φ. Define the map $F\colon Z \to Y$ as follows. Suppose $\zeta \in Z$ and $q(\zeta) =: x$. By Lemma (7.7) the function germ $q_*(\rho_\zeta(g)) \in \mathcal{O}_x$ arises by analytic continuation along a curve from a to x from the function germ φ. By Lemma (7.2) Y consists of all function germs which are obtained by the analytic continuation of φ along curves. Hence there exists exactly one $\eta \in Y$ such that $q_*(\rho_\zeta(g)) = \eta$. Let $F(\zeta) = \eta$. It is easy to check that $F\colon Z \to Y$ is a fiber-preserving holomorphic map such that $F(c) = b$ and $F^*(f) = g$. □

Remark. The analytic continuation of meromorphic function germs can be handled by using the techniques employed in this section for holomorphic function germs. One just looks at the map $|\mathcal{M}| \to X$. So far we have disregarded branch points but in the next section we will also consider these for the special case of algebraic functions.

EXERCISES (§7)

7.1. Suppose X and Y are Riemann surfaces, $p: Y \to X$ is a holomorphic (unbranched) covering map and $f: Y \to \mathbb{C}$ is a holomorphic function. Let $b \in Y$, $a := p(b)$ and $\varphi := p_*(\rho_b(f)) \in \mathcal{O}_a$. Prove that (Y, p, f, b) is a maximal analytic continuation of φ if and only if the following condition is satisfied: For any two distinct points $b_1, b_2 \in p^{-1}(a)$ the germs $\varphi_1 := p_*(\rho_{b_1}(f))$ and $\varphi_2 := p_*(\rho_{b_2}(f))$ are different.

7.2. Suppose X is a Riemann surface and $a \in X$. Suppose $\varphi \in \mathcal{O}_a$ admits an analytic continuation along every curve in X which starts at a. Let (Y, p, f, b) be the maximal analytic continuation of φ. Prove that $p: Y \to X$ is a covering map.

§8. Algebraic Functions

One of the first examples of a multi-valued function which one encounters in complex analysis is the square root $w = \sqrt{z}$. This is a particular case of an algebraic function, i.e., a function $w = w(z)$ which satisfies an algebraic equation $w^n + a_1(z)w^{n-1} + \cdots + a_n(z) = 0$, where the coefficients a_ν are given meromorphic functions of z. In this section we present the construction of the Riemann surfaces of algebraic functions. It turns out that they are proper coverings such that the number of sheets equals the degree of the algebraic equation.

8.1. The Elementary Symmetric Functions. Suppose X and Y are Riemann surfaces, $\pi: Y \to X$ is an n-sheeted unbranched holomorphic covering map and f is a meromorphic function on Y. Every point $x \in X$ has an open neighborhood U such that $\pi^{-1}(U)$ is the disjoint union of open sets $V_1, \ldots, V_n$ and $\pi: V_\nu \to U$ is biholomorphic for $\nu = 1, \ldots, n$. Let $\tau_\nu: U \to V_\nu$ be the inverse mapping of $\pi | V_\nu \to U$ and let $f_\nu := \tau_\nu^* f = f \circ \tau_\nu$. Suppose T is an indeterminate and consider

$$\prod_{\nu=1}^{n} (T - f_\nu) = T^n + c_1 T^{n-1} + \cdots + c_n.$$

Then the c_ν are meromorphic functions in U and

$$c_\nu = (-1)^\nu s_\nu(f_1, \ldots, f_n),$$

where s_ν denotes the νth elementary symmetric function in n variables. If one carries out this same construction in a neighborhood U' of another point

$x' \in X$, then one gets the same functions $c_1, \ldots, c_n$. Thus these functions piece together to give global meromorphic functions $c_1, \ldots, c_n \in \mathcal{M}(X)$, which we call the elementary symmetric functions of f with respect to the covering $Y \to X$.

8.2. Theorem. *Suppose X and Y are Riemann surfaces and $\pi\colon Y \to X$ is an n-sheeted branched holomorphic covering map. Suppose $A \subset X$ is a closed discrete subset which contains all the critical values of π and let $B = \pi^{-1}(A)$. Suppose f is a holomorphic (resp. meromorphic) function on $Y \backslash B$ and $c_1, \ldots, c_n \in \mathcal{O}(X \backslash A)$ (resp. $\in \mathcal{M}(X \backslash A)$) are the elementary symmetric functions of f. Then f may be continued holomorphically (resp. meromorphically) to Y precisely if all the c_ν may be continued holomorphically (resp. meromorphically) to X.*

The Theorem ensures that the elementary symmetric functions of a function $f \in \mathcal{M}(Y)$ are also defined when the map $Y \to X$ is a branched holomorphic covering.

PROOF. Suppose $a \in A$ and $b_1, \ldots, b_m$ are the preimages of a. Suppose (U, z) is a relatively compact coordinate neighborhood of a with $z(a) = 0$ and $U \cap A = \{a\}$. Then $V := \pi^{-1}(U)$ is a relatively compact neighborhood of each of the b_μ.

1. First consider the case $f \in \mathcal{O}(Y \backslash B)$.

(a) Assume f may be continued holomorphically to all the points b_μ. Then f is bounded on $V \backslash \{b_1, \ldots, b_m\}$. This implies that all the c_ν are bounded on $U \backslash \{a\}$. By Riemann's Removable Singularities Theorem they may all be continued holomorphically to a.

(b) Suppose all the c_ν may be continued holomorphically to a. Then all the c_ν are bounded on $U \backslash \{a\}$. But this implies f is bounded on $V \backslash \{b_1, \ldots, b_m\}$, for, if $y \in V \backslash \{b_1, \ldots, b_m\}$ and $x = \pi(y)$, then

$$f(y)^n + c_1(x) f(y)^{n-1} + \cdots + c_n(x) = 0.$$

Again Riemann's Removable Singularities Theorem implies that f may be continued holomorphically to every point b_μ.

2. Now suppose $f \in \mathcal{M}(Y \backslash B)$.

(a) Assume f may be continued meromorphically to all points b_μ. The function $\varphi := \pi^* z \in \mathcal{O}(V)$ vanishes at all the points b_μ. Thus $\varphi^k f$ may be continued holomorphically to all the points b_μ if k is sufficiently large. The elementary symmetric functions of $\varphi^k f$ are $z^{k\nu} c_\nu$ and by the first part of the proof they may be continued holomorphically to a. Thus all the c_ν may be continued meromorphically to a.

(b) Suppose all the c_ν may be continued meromorphically to a. Using the above notation one has: For k sufficiently large all the $z^{k\nu} c_\nu$ admit holomorphic continuations to a. Thus $\varphi^k f$ admits a holomorphic continuation to all the points b_μ. This implies that f may be continued meromorphically to all of the points b_μ. □

For later use note that the proof does not use the fact that Y is connected. Thus the Theorem also holds in the case that Y is a disjoint union of finitely many Riemann surfaces.

If $\pi\colon Y \to X$ is a non-constant holomorphic map between Riemann surfaces X and Y, then for any meromorphic function f on X the function $\pi^* f := f \circ \pi$ is a meromorphic function on Y. Thus there is a map

$$\pi^*\colon \mathcal{M}(X) \to \mathcal{M}(Y)$$

which is a monomorphism of fields.

8.3. Theorem. *Suppose X and Y are Riemann surfaces and $\pi\colon Y \to X$ is a branched holomorphic n-sheeted covering map. If $f \in \mathcal{M}(Y)$ and $c_1, \ldots, c_n \in \mathcal{M}(X)$ are the elementary symmetric functions of f, then*

$$f^n + (\pi^* c_1) f^{n-1} + \cdots + (\pi^* c_{n-1}) f + \pi^* c_n = 0.$$

The monomorphism $\pi^*\colon \mathcal{M}(X) \to \mathcal{M}(Y)$ is an algebraic field extension of degree $\leq n$. Moreover, if there exist an $f \in \mathcal{M}(Y)$ and an $x \in X$ with preimages $y_1, \ldots, y_n \in Y$ such that the values $f(y_\nu)$ for $\nu = 1, \ldots, n$ are all distinct, then the field extension $\pi^*\colon \mathcal{M}(X) \to \mathcal{M}(Y)$ has degree n.

Remark. We will see later (cf. (14.13) and (26.6)) that the last statement of the Theorem is always fulfilled.

PROOF. The existence of the equation

$$f^n + \sum_{\nu=1}^{n} (\pi^* c_\nu) f^{n-\nu} = 0$$

follows directly from the definition of the elementary symmetric functions of f.

Let $L := \mathcal{M}(Y)$ and $K := \pi^* \mathcal{M}(X) \subset L$. Then every $f \in L$ is algebraic over K and the minimal polynomial of f over K has degree $\leq n$. Suppose $f_0 \in L$ is an element for which the degree n_0 of its minimal polynomial is maximal. We claim $L = K(f_0)$. Choose an arbitrary element $f \in L$ and consider the field $K(f_0, f)$. By the Theorem of the Primitive Element there exists $g \in L$ such that $K(f_0, f) = K(g)$. By the definition of n_0 one has $\dim_K K(g) \leq n_0$. On the other hand,

$$\dim_K K(f_0, f) \geq \dim_K K(f_0) = n_0.$$

Thus $K(f_0) = K(f_0, f)$ and $f \in K(f_0)$.

Finally if the degree of the minimal polynomial of f over K were equal to $m < n$, then f would be able to take at most m different values over every point $x \in X$. □

8.4. Theorem. *Suppose X is a Riemann surface, $A \subset X$ is a closed discrete subset and let $X' = X \setminus A$. Suppose Y' is another Riemann surface and π': $Y' \to X'$ is a proper unbranched holomorphic covering. Then π' extends to a branched covering of X, i.e., there exists a Riemann surface Y, a proper holomorphic mapping π: $Y \to X$ and a fiber-preserving biholomorphic mapping*

$$\varphi\colon Y \setminus \pi^{-1}(A) \to Y'.$$

PROOF. For every $a \in A$ choose a coordinate neighborhood (U_a, z_a) on X with the following properties: $z_a(a) = 0$, $z_a(U_a)$ is the unit disk in $\mathbb{C}$ and $U_a \cap U_{a'} = \varnothing$ if $a \neq a'$. Let $U_a^* = U_a \setminus \{a\}$. Since π': $Y' \to X'$ is proper, $\pi'^{-1}(U_a^*)$ consists of a finite number of connected components $V_{a\nu}^*$, $\nu = 1, \ldots, n(a)$. For every ν the mapping $\pi' \mid V_{a\nu}^* \to U_a^*$ is an unbranched covering. Let its covering number be $k_{a\nu}$. By Theorem (5.10) there exist biholomorphic mappings $\zeta_{a\nu}$: $V_{a\nu}^* \to D^*$ of $V_{a\nu}^*$ onto the punctured unit disk $D^* = D \setminus \{0\}$ such that the diagram

$$\begin{array}{ccc} V_{a\nu}^* & \xrightarrow{\zeta_{a\nu}} & D^* \\ \downarrow{\scriptstyle \pi'} & & \downarrow{\scriptstyle \pi_{a\nu}} \\ U_a^* & \xrightarrow{z_a} & D^* \end{array}$$

is commutative, where $\pi_{a\nu}(\zeta) = \zeta^{k_{a\nu}}$.

Now choose "ideal points" $p_{a\nu}$, $a \in A$, $\nu = 1, \ldots, n(a)$, i.e., pairwise distinct elements of some set disjoint from Y'. Then on

$$Y := Y' \cup \{p_{a\nu}\colon a \in A, \nu = 1, \ldots, n(a)\}$$

there exists precisely one topology with the following property. If W_i, $i \in I$ is a neighborhood basis of a, then

$$\{p_{a\nu}\} \cup (\pi'^{-1}(W_i) \cap V_{a\nu}^*), \qquad i \in I,$$

is a neighborhood basis of $p_{a\nu}$ and on Y' it induces the given topology. This makes Y into a Hausdorff space. Define π: $Y \to X$ by $\pi(y) = \pi'(y)$ for $y \in Y'$ and $\pi(p_{a\nu}) = a$. Then, as one easily checks, π is proper.

In order to make Y into a Riemann surface, add to the charts of the complex structure of Y' the following charts. Let $V_{a\nu} = V_{a\nu}^* \cup \{p_{a\nu}\}$ and let

$$\zeta_{a\nu}\colon V_{a\nu} \to D$$

be the continuation of the mapping $\zeta_{a\nu}$: $V_{a\nu}^* \to D^*$ described above which is obtained by defining $\zeta_{a\nu}(p_{a\nu}) := 0$. Since the last mapping is biholomorphic with respect to the complex structure of Y', the new charts $\zeta_{a\nu}$: $V_{a\nu} \to D$ are holomorphically compatible with the charts of the complex structure of Y'. The mapping π: $Y \to X$ is holomorphic. Since $Y \setminus \pi^{-1}(A) = Y'$ by construction, we may choose φ: $Y \setminus \pi^{-1}(A) \to Y'$ to be the identity mapping. This then shows the existence of a continuation of the covering π': $Y' \to X'$. □

The following Theorem shows that the continuation of the covering, whose existence was just proven, is uniquely determined up to isomorphism.

8.5. Theorem. *Suppose X, Y and Z are Riemann surfaces and $\pi: Y \to X$, $\tau: Z \to X$ are proper holomorphic covering maps. Let $A \subset X$ be a closed discrete subset and let $X' := X \backslash A$, $Y' := \pi^{-1}(X')$ and $Z' := \tau^{-1}(X')$. Then every fiber-preserving biholomorphic mapping $\sigma': Y' \to Z'$ can be extended to a fiber-preserving biholomorphic mapping $\sigma: Y \to Z$. In particular every covering transformation $\sigma' \in \mathrm{Deck}(Y'/X')$ can be extended to a covering transformation $\sigma \in \mathrm{Deck}(Y/X)$.*

PROOF. Suppose $a \in A$ and (U, z) is a coordinate neighborhood of a such that $z(a) = 0$ and $z(U)$ is the unit disk. Let $U^* = U\backslash\{a\}$. Moreover we may assume that U is so small that π and τ are unbranched over U^*. Let $V_1, \ldots, V_n$ (resp. $W_1, \ldots, W_m$) be the connected components of $\pi^{-1}(U)$ (resp. $\tau^{-1}(U)$). Then $V_\nu^* := V_\nu \backslash \pi^{-1}(a)$ (resp. $W_\mu^* := W_\mu \backslash \tau^{-1}(a)$) are the connected components of $\pi^{-1}(U^*)$ (resp. $\tau^{-1}(U^*)$).

Since $\sigma' | \pi^{-1}(U^*) \to \tau^{-1}(U^*)$ is biholomorphic, $n = m$ and one may renumber so that $\sigma'(V_\nu^*) = W_\nu^*$. Since $\pi | V_\nu^* \to U^*$ is a finite sheeted unbranched covering, $V_\nu \cap \pi^{-1}(a)$ (resp. $W_\nu \cap \tau^{-1}(a)$) consists by Theorem (5.11) of exactly one point b_ν (resp. c_ν). Hence $\sigma' | \pi^{-1}(U^*) \to \tau^{-1}(U^*)$ can be continued to a bijective mapping $\pi^{-1}(U) \to \tau^{-1}(U)$ which assigns to b_ν the point c_ν. Since $\pi | V_\nu \to U$ and $\tau | W_\nu \to U$ are proper, the continuation is a homeomorphism and by Riemann's Removable Singularities Theorem it is biholomorphic as well. (The Removable Singularities Theorem applies since V_ν and W_ν are isomorphic to the unit disk by Theorem (5.11).) If one now applies this construction to every exceptional point $a \in A$, then one gets the desired continuation $\sigma: Y \to Z$. □

Theorem (8.5) makes the following definition meaningful (cf. Definition 5.5).

8.6. Definition. Suppose X and Y are Riemann surfaces and $\pi: Y \to X$ is a branched holomorphic covering. Let $A \subset X$ be the set of critical values of π and let $X' := X \backslash A$ and $Y' := \pi^{-1}(X')$. Then the covering $Y \to X$ is called *Galois* if the covering $Y' \to X'$ is Galois.

8.7. Lemma. *Suppose $c_1, \ldots, c_n$ are holomorphic functions on the disk*

$$D(R) = \{z \in \mathbb{C}: |z| < R\}, \qquad R > 0.$$

Suppose $w_0 \in \mathbb{C}$ is a simple zero of the polynomial

$$T^n + c_1(0)T^{n-1} + \cdots + c_n(0) \in \mathbb{C}[T].$$

Then there exist an r, $0 < r \le R$, and a function φ holomorphic on the disk $D(r)$ such that $\varphi(0) = w_0$ and

$$\varphi^n + c_1 \varphi^{n-1} + \cdots + c_n = 0 \text{ on } D(r).$$

PROOF. For $z \in D(R)$ and $w \in \mathbb{C}$ let

$$F(z, w) = w^n + c_1(z)w^{n-1} + \cdots + c_n(z).$$

There exists an $\varepsilon > 0$ such that the function $w \mapsto F(0, w)$ has a unique zero w_0 in the disk $\{w \in \mathbb{C}: |w - w_0| \le \varepsilon\}$. Now because of the continuity of F there exists an r with $0 < r \le R$ such that in the set

$$\{(z, w) \in \mathbb{C}^2: |z| < r, |w - w_0| = \varepsilon\}$$

the function F has no zeros. For fixed $z \in D(r)$ the integral

$$n(z) = \frac{1}{2\pi i} \int_{|w - w_0| = \varepsilon} \frac{F_w(z, w)}{F(z, w)} \, dw, \qquad \left(F_w := \frac{\partial F}{\partial w}\right)$$

gives the number of zeros of the function $w \mapsto F(z, w)$ in the disk with radius ε and center w_0. Since $n(0) = 1$ and n depends continuously on z, one has $n(z) = 1$ for every $z \in D(r)$. By the Residue Theorem the zero of $w \mapsto F(z, w)$ in the disk $|w - w_0| < \varepsilon$ is equal to

$$\varphi(z) = \frac{1}{2\pi i} \int_{|w - w_0| = \varepsilon} w \frac{F_w(z, w)}{F(z, w)} \, dw.$$

Since the integral depends holomorphically on z, the function $z \mapsto \varphi(z)$ is holomorphic on $D(r)$ and $F(z, \varphi(z)) = 0$ for every $z \in D(r)$. □

8.8. Corollary. *Let $\mathcal{O}_x$ be the ring of holomorphic function germs at a point x of a Riemann surface and let*

$$P(T) = T^n + c_1 T^{n-1} + \cdots + c_n \in \mathcal{O}_x[T].$$

Suppose that the polynomial

$$p(T) := T^n + c_1(x)T^{n-1} + \cdots + c_n(x) \in \mathbb{C}[T]$$

has n distinct zeros $w_1, \ldots, w_n$. Then there exist elements $\varphi_1, \ldots, \varphi_n \in \mathcal{O}_x$ such that $\varphi_\nu(x) = w_\nu$ and

$$P(T) = \prod_{\nu=1}^{n} (T - \varphi_\nu).$$

8.9. Theorem. *Suppose X is a Riemann surface and*

$$P(T) = T^n + c_1 T^{n-1} + \cdots + c_n \in \mathcal{M}(X)[T]$$

is an irreducible polynomial of degree n. Then there exist a Riemann surface Y, a branched holomorphic n-sheeted covering $\pi: Y \to X$ and a meromorphic *function $F \in \mathcal{M}(Y)$ such that $(\pi^*P)(F) = 0$. The triple (Y, π, F) is uniquely determined in the following sense. If (Z, τ, G) has the corresponding properties, then there exists exactly one fiber-preserving biholomorphic mapping $\sigma: Z \to Y$ such that $G = \sigma^*F$.*

To simplify the terminology (Y, π, F) is called the *algebraic function* defined by the polynomial $P(T)$.

Remark. The classical case is when X is the Riemann sphere $\mathbb{P}^1$. Then by (2.9) the coefficients c_ν of the polynomial $P(T)$ are rational functions in one variable. Since $\mathbb{P}^1$ is compact and $\pi: Y \to \mathbb{P}^1$ is proper, Y is also compact.

PROOF. Let $\Delta \in \mathcal{M}(X)$ be the discriminant of the polynomial $P(T)$. (Δ is a certain polynomial in the coefficients of P.) The discriminant can not vanish identically, for otherwise P would be reducible. There exists a closed discrete subset $A \subset X$ such that at every point $x \in X' := X \setminus A$ all the functions $c_1, \ldots, c_n$ are holomorphic and $\Delta(x) \neq 0$. Then for every $x \in X'$ the polynomial

$$p_x(T) := T^n + c_1(x)T^{n-1} + \cdots + c_n(x) \in \mathbb{C}[T]$$

has n distinct zeros. Now we will use the topological space $|\mathcal{O}| \to X$ associated to the sheaf $\mathcal{O}$, cf. (6.7). Let $Y' \subset |\mathcal{O}|$ be the set of all the function germs $\varphi \in \mathcal{O}_x$, $x \in X'$, which satisfy the equation $P(\varphi) = 0$ and let $\pi': Y' \to X'$ be the canonical projection. By Corollary (8.8) for every point $x \in X'$ there exist an open neighborhood $U \subset X'$ and holomorphic functions $f_1, \ldots, f_n \in \mathcal{O}(U)$ such that

$$P(T) = \prod_{\nu=1}^{n} (T - f_\nu) \quad \text{on } U.$$

Then $\pi'^{-1}(U) = \bigcup_{\nu=1}^{n} [U, f_\nu]$. The $[U, f_\nu]$ are disjoint and $\pi' | [U, f_\nu] \to U$ is a homeomorphism. This shows that $Y' \to X'$ is a covering map. The connected components of Y' are Riemann surfaces which also admit covering maps over X'. Let $f: Y' \to \mathbb{C}$ be defined by $f(\varphi) := \varphi(\pi'(\varphi))$. Then f is holomorphic and by construction

$$f(y)^n + c_1(\pi'(y))f(y)^{n-1} + \cdots + c_n(\pi'(y)) = 0$$

for every $y \in Y'$. By Theorem (8.4) the covering $\pi': Y' \to X'$ may be continued to a proper holomorphic covering $\pi: Y \to X$, where we identify Y' with $\pi^{-1}(X')$. By Theorem (8.2) f may be extended to a meromorphic function $F \in \mathcal{M}(Y)$, for which

$$(\pi^*P)(F) = F^n + (\pi^*c_1)F^{n-1} + \cdots + \pi^*c_n = 0.$$

Now we will show that Y is connected and thus a Riemann surface. Suppose this is not the case. Then Y has finitely many connected components $Y_1, \ldots, Y_k$ and $\pi | Y_i \to X$ is a proper holomorphic n_i-sheeted covering, where

$\Sigma n_i = n$. Using the elementary symmetric functions of $F \mid Y_i$ one gets polynomials $P_i(T) \in \mathcal{M}(X)[T]$ of degree n_i such that

$$P(T) = P_1(T)P_2(T) \cdots P_k(T).$$

But this contradicts the assumption that $P(T)$ is irreducible.

Uniqueness. Suppose (Z, τ, G) is another algebraic function defined by the polynomial $P(T)$. Let $B \subset Z$ be the union of the poles of G and the branch points of τ and let $A' := \tau(B)$. Let

$$X'' := X' \backslash A', \qquad Y'' := \pi^{-1}(X''), \qquad Z'' := \tau^{-1}(X'').$$

Define a fiber-preserving mapping $\sigma''\colon Z'' \to Y''$ in the following way. Let $z \in Z''$, $\tau(z) = x$ and $\varphi \in \mathcal{O}_x$ be the function germ $\varphi := \tau_* G_z$. Then $P(\varphi) = 0$. By the construction of Y' one sees that φ is a point of Y' over x and thus $\varphi \in Y''$. Set $\sigma''(z) = \varphi$. From the definition it follows directly that σ'' is continuous. Since σ'' is fiber-preserving, σ'' is thus holomorphic. Moreover, σ'' is proper since $\pi \mid Y'' \to X''$ is continuous and $\tau \mid Z'' \to X''$ is proper. Hence σ'' is surjective. Because $Y'' \to X''$ and $Z'' \to X''$ have the same number of sheets, $\sigma''\colon Z'' \to Y''$ is biholomorphic. Also from the definition of σ'' one gets $G \mid Z'' = (\sigma'')^*(F \mid Y'')$. By Theorem (8.5) σ'' can be extended to a fiber-preserving biholomorphic mapping $\sigma\colon Z \to Y$ for which one then has $G = \sigma^* F$. The mapping σ is in fact uniquely determined by the property $G = \sigma^* F$. For, otherwise there would exist a covering transformation $\alpha\colon Y \to Y$ different from the identity such that $\alpha^* F = F$. But this is not possible since F assumes distinct values on the fiber $\pi^{-1}(x)$ over every point $x \in X'$. □

8.10. Example. Suppose $f(z) = (z - a_1) \cdots (z - a_n)$ is a polynomial with distinct roots $a_1, \ldots, a_n \in \mathbb{C}$. Consider f as a meromorphic function on the Riemann sphere $\mathbb{P}^1$. The polynomial $P(T) = T^2 - f$ is irreducible over $\mathcal{M}(\mathbb{P}^1)$ and defines an algebraic function which is usually denoted by $\sqrt{f(z)}$. Its Riemann surface $\pi\colon Y \to \mathbb{P}^1$ may be described using the above construction as follows. Let

$$A := \{a_1, \ldots, a_n\} \cup \{\infty\},$$

$X' := \mathbb{P}^1 \backslash A$ and $Y' := \pi^{-1}(X')$. Then $\pi\colon Y' \to X'$ is an unbranched holomorphic two-sheeted covering. This implies that every function germ $\varphi \in \mathcal{O}_x$, where $x \in X'$, such that $\varphi^2 = f$ can be analytically continued along every curve lying in X'. Now consider the covering over neighborhoods of the exceptional points.

(a) For each $j \in \{1, \ldots, n\}$ choose $r_j > 0$ sufficiently small that no other point of A lies in the disk

$$U_j := \{z \in \mathbb{C}\colon |z - a_j| < r_j\}.$$

Since the function $g(z) = \prod_{k \neq j} (z - a_k)$ has no zeros in U_j and U_j is simply connected, there exists a holomorphic function $h: U_j \to \mathbb{C}$ such that $h^2 = g$. Thus

$$f(z) = (z - a_j)h(z)^2$$

on U_j. Suppose $0 < \rho < r_j$, $\theta \in \mathbb{R}$ and let $\zeta = a_j + \rho e^{i\theta}$. By Lemma (8.7) there exists a function germ $\varphi_\zeta \in \mathcal{O}_\zeta$ such that $\varphi_\zeta^2 = f$ and

$$\varphi_\zeta(\zeta) = \sqrt{\rho}\, e^{i\theta/2} h(\zeta).$$

If one continues this function germ along the closed curve $\zeta = a_j + \rho e^{i\theta}$, $0 \leq \theta \leq 2\pi$, then one obtains the negative of the original germ. Let $U_j^* := U_j \setminus \{a_j\}$ and $V_j^* := \pi^{-1}(U_j^*)$. Then $\pi: V_j^* \to U_j^*$ is a connected two-sheeted covering as in Theorem (5.10.ii) with $k = 2$. For otherwise $\pi: V_j^* \to U_j^*$ would split into two single-sheeted coverings and the analytic continuation of the function germ φ_ζ along the curve $\zeta = a_j + \rho e^{i\theta}$, $0 \leq \theta \leq 2\pi$, would lead back to the same function germ. Hence the Riemann surface Y has exactly one point over the point a_j.

(b) Suppose $r > \max\{|a_1|, \ldots, |a_n|\}$ and let

$$U^* := \{z \in \mathbb{C}: |z| > r\}.$$

Then $U := U^* \cup \{\infty\}$ is a neighborhood of ∞, which is isomorphic to a disk, and which contains no other points of A. On U one can write $f = z^n F$, where F is a holomorphic function having no zeros in U. Now we distinguish two cases:

(i) *n odd.* Then there exists a meromorphic function h on U such that $f(z) = zh(z)^2$.

(ii) *n even.* Then there exists a meromorphic function h on U such that $f(z) = h(z)^2$.

Let $V^* := \pi^{-1}(U^*)$. Now one shows, the same as above, that in case (i) $\pi: V^* \to U^*$ is a connected two-sheeted covering and thus Y has precisely one point over ∞. But in case (ii) $\pi: V^* \to U^*$ splits into two single-sheeted coverings and thus when n is even Y has two points over ∞.

8.11. If X and Y are Riemann surfaces and $\pi: Y \to X$ is a branched holomorphic covering map, then $\operatorname{Deck}(Y/X)$ has a representation into the automorphism group of the field $\mathcal{M}(Y)$ defined in the following way. For $\sigma \in \operatorname{Deck}(Y/X)$ let $\sigma f := f \circ \sigma^{-1}$. Clearly the correspondence $f \mapsto \sigma f$ is an automorphism of $\mathcal{M}(Y)$. The mapping

$$\operatorname{Deck}(Y/X) \to \operatorname{Aut}(\mathcal{M}(Y))$$

is a group homomorphism. For suppose $\sigma, \tau \in \operatorname{Deck}(Y/X)$. Then for every $f \in \mathcal{M}(Y)$

$$(\sigma\tau)f = f \circ (\sigma\tau)^{-1} = f \circ \tau^{-1} \circ \sigma^{-1} = \sigma(f \circ \tau^{-1}) = \sigma(\tau f).$$

Trivially every such automorphism $f \mapsto \sigma f$ leaves invariant the functions of the subfield $\pi^*\mathcal{M}(X) \subset \mathcal{M}(Y)$ and thus is an element of the Galois group $\mathrm{Aut}(\mathcal{M}(Y)/\pi^*\mathcal{M}(X))$.

8.12. Theorem. *Suppose X is a Riemann surface, $K := \mathcal{M}(X)$ is the field of meromorphic functions on X and $P(T) \in K[T]$ is an irreducible monic polynomial of degree n. Let (Y, π, F) be the algebraic function defined by $P(T)$ and $L = \mathcal{M}(Y)$. By means of the monomorphism $\pi^*: K \to L$ consider K as a subfield of L. Then $L: K$ is a field extension of degree n and $L \cong K[T]/(P(T))$. Every covering transformation $\sigma: Y \to Y$ of Y over X induces an automorphism $f \mapsto \sigma f := f \circ \sigma^{-1}$ of L leaving K fixed and the mapping*

$$\mathrm{Deck}(Y/X) \to \mathrm{Aut}(L/K)$$

which is so defined, is a group isomorphism. The covering $Y \to X$ is Galois precisely if the field extension $L: K$ is Galois.

Proof. The fact that $L: K$ is a field extension of degree n follows from the last statement of Theorem (8.3). Since $P(F) = 0$, there is a homomorphism $K[T]/(P(T)) \to L$. Since both these fields are of degree n over K, this is an isomorphism.

The mapping $\mathrm{Deck}(Y/X) \to \mathrm{Aut}(L/K)$ is injective, because $\sigma F \neq F$ for any covering transformation σ which is not the identity. This mapping is also surjective. For, suppose $\alpha \in \mathrm{Aut}(L/K)$. Then $(Y, \pi, \alpha F)$ is also an algebraic function defined by the polynomial $P(T)$. Thus by the uniqueness statement of Theorem (8.9) there exists a covering transformation $\tau \in \mathrm{Deck}(Y/X)$ such that $\alpha F = \tau^* F$. If $\sigma := \tau^{-1}$, then

$$\sigma F = F \circ \sigma^{-1} = F \circ \tau = \tau^* F = \alpha F.$$

Since L is generated by F over K, the automorphism $f \mapsto \sigma f$ of L coincides with α.

The last statement of the Theorem follows from the fact that Y is Galois over X (resp. L is Galois over K) precisely when $\mathrm{Deck}(Y/X)$ (resp. $\mathrm{Aut}(L/K)$) contains n elements. □

8.13. Puiseux Expansions. Denote by $\mathbb{C}\{\{z\}\}$ the field of all Laurent series with finite principal part

$$\varphi(z) = \sum_{\nu=k}^{\infty} c_\nu z^\nu, \qquad k \in \mathbb{Z}, \qquad c_\nu \in \mathbb{C},$$

converging in some punctured disk $\{0 < |z| < r\}$, where $r > 0$ may depend on the element φ. Then $\mathbb{C}\{\{z\}\}$ is isomorphic to the stalk $\mathcal{M}_0$ of the sheaf $\mathcal{M}$ of meromorphic functions in the complex plane and is the quotient field of $\mathbb{C}\{z\}$.

Consider an irreducible polynomial

$$F(z, w) = w^n + a_1(z)w^{n-1} + \cdots + a_n(z) \in \mathbb{C}\{\{z\}\}[w]$$

of degree n over the field $\mathbb{C}\{\{z\}\}$. For some $r > 0$, all the coefficients a_ν are meromorphic functions on

$$D(r) := \{z \in \mathbb{C}: |z| < r\},$$

and thus F may also be considered as an element of $\mathcal{M}(D(r))[w]$. It is clear that F is also irreducible over the field $\mathcal{M}(D(r))$. Now suppose that r has been chosen so small that for every $a \in D(r)\backslash 0$ the polynomial

$$F(a, w) \in \mathbb{C}[w]$$

has no multiple roots. Let (Y, π, f) be the algebraic function defined by $F(z, w) \in \mathcal{M}(D(r))[w]$ in the sense of Theorem 8.9. Then $\pi: Y \to D(r)$ is an n-sheeted proper holomorphic map which is ramified only over the origin. By Theorem (5.11) there exists an isomorphism

$$\alpha: D(\rho) \to Y, \qquad \rho = \sqrt[n]{r},$$

such that

$$\pi(\alpha(\zeta)) = \zeta^n \quad \text{for every } \zeta \in D(\rho).$$

Since $F(\pi, f) = 0$, it follows that

$$F(\zeta^n, \varphi(\zeta)) = 0, \quad \text{where } \varphi := f \circ \alpha.$$

This proves the following Theorem.

8.14. Theorem (Puiseux). *Let*

$$F(z, w) = w^n + a_1(z)w^{n-1} + \cdots + a_n(z) \in \mathbb{C}\{\{z\}\}[w]$$

be an irreducible polynomial of degree n over the field $\mathbb{C}\{\{z\}\}$. *Then there exists a Laurent series*

$$\varphi(\zeta) = \sum_{\nu=k}^{\infty} c_\nu \zeta^\nu \in \mathbb{C}\{\{\zeta\}\}$$

such that

$$F(\zeta^n, \varphi(\zeta)) = 0$$

as an element of $\mathbb{C}\{\{\zeta\}\}$.

Remarks

(1) If all of the coefficients a_ν are holomorphic, i.e., $a_\nu \in \mathbb{C}\{z\}$, then $\varphi \in \mathbb{C}\{\zeta\}$ as well. This follows from the fact that in this case the function f considered in (8.13) is holomorphic on Y.

(2) Another way of expressing the assertion of the Theorem is to say that the equation

$$F(z, w) = 0$$

can be solved by a *Puiseux series*

$$w = \varphi(\sqrt[n]{z}) = \sum_{\nu=k}^{\infty} c_\nu z^{\nu/n}.$$

(3) We can interpret the Theorem of Puiseux in the following algebraic way. By means of the map

$$\mathbb{C}\{\{z\}\} \to \mathbb{C}\{\{\zeta\}\}, \qquad z \mapsto \zeta^n,$$

$\mathbb{C}\{\{\zeta\}\}$ becomes an extension field of $\mathbb{C}\{\{z\}\}$ of degree n. A basis of $\mathbb{C}\{\{\zeta\}\}$ over $\mathbb{C}\{\{z\}\}$ is given by $1, \zeta, \ldots, \zeta^{n-1}$. The series $\varphi(\zeta)$ is a root of F in this extension field. Let ε be a primitive nth root of unity, e.g. $\varepsilon = e^{2\pi i/n}$. Then for $k = 0, 1, \ldots, n-1$ we have $(\varepsilon^k\zeta)^n = \zeta^n$ and hence

$$F(\zeta^n, \varphi(\varepsilon^k\zeta)) = 0.$$

Thus $\varphi(\varepsilon^k\zeta) \in \mathbb{C}\{\{\zeta\}\}$ is also a root of the polynomial F. It is easy to see that the series $\varphi(\varepsilon^k\zeta)$, $k = 0, 1, \ldots, n-1$, are distinct. Thus $\mathbb{C}\{\{\zeta\}\}$ is a splitting field of the polynomial $F \in \mathbb{C}\{\{z\}\}[w]$.

Exercises (§8)

8.1. Suppose X and Y are compact Riemann surfaces such that $\mathcal{M}(X)$ and $\mathcal{M}(Y)$ are isomorphic as $\mathbb{C}$-algebras. Prove that X and Y are isomorphic.
[*Hint*: Represent X and Y as the Riemann surfaces of algebraic functions defined by one and the same irreducible polynomial $P \in \mathcal{M}(\mathbb{P}^1)[T]$. Also use the fact (proved in Corollary (14.13)) that on a compact Riemann surface the meromorphic functions separate points.]

8.2. Let X and Y be compact Riemann surfaces, $a_1, \ldots, a_n \in X$, $b_1, \ldots, b_m \in Y$ and $X' := X \setminus \{a_1, \ldots, a_n\}$, $Y' := Y \setminus \{b_1, \ldots, b_m\}$. Show that every isomorphism $f: X' \to Y'$ extends to an isomorphism $\tilde{f}: X \to Y$.

8.3. Let $F(z, w) := w^2 - z^3 w + z \in \mathbb{C}\{\{z\}\}[w]$.
(a) Show that F is irreducible over $\mathbb{C}\{\{z\}\}$.
(b) Determine the Puiseux expansion

$$w = \sum_{\nu=0}^{\infty} c_\nu z^{\nu/2}$$

of the algebraic function defined by $F(z, w) = 0$.

§9. Differential Forms

In this section we introduce the notion of differential forms on Riemann surfaces. It is important to consider not only holomorphic and meromorphic forms but also forms which are only differentiable in the real sense.

9.1. Suppose U is an open subset of $\mathbb{C}$. We identify $\mathbb{C}$ with $\mathbb{R}^2$ by writing $z = x + iy$, where x and y are the standard real coordinates on $\mathbb{R}^2$. Denote by $\mathscr{E}(U)$ the $\mathbb{C}$-algebra of all those functions $f\colon U \to \mathbb{C}$ which are infinitely differentiable with respect to the real coordinates x and y. Besides the partial derivatives $(\partial/\partial x)$ and $(\partial/\partial y)$, we also consider the differential operators

$$\frac{\partial}{\partial z} := \frac{1}{2}\left(\frac{\partial}{\partial x} - i\frac{\partial}{\partial y}\right), \qquad \frac{\partial}{\partial \bar{z}} := \frac{1}{2}\left(\frac{\partial}{\partial x} + i\frac{\partial}{\partial y}\right).$$

As is well-known, the Cauchy–Riemann equations say that the vector space $\mathcal{O}(U)$ of holomorphic functions on U is the kernel of the mapping $(\partial/\partial\bar{z})\colon \mathscr{E}(U) \to \mathscr{E}(U)$.

9.2. By means of the complex charts one can define the notion of differentiable function on any Riemann surface X. For any open subset $Y \subset X$, let $\mathscr{E}(Y)$ consist of all functions $f\colon Y \to \mathbb{C}$ such that for every chart $z\colon U \to V \subset \mathbb{C}$ on X with $U \subset Y$ there exists a function $\tilde{f} \in \mathscr{E}(V)$ with $f \mid U = \tilde{f} \circ z$. Clearly the function $\tilde{f}$ is uniquely determined by f, for $\tilde{f} = f \circ \psi$, where $\psi\colon V \to U$ is the inverse of $z\colon U \to V$.

Together with the natural restriction mappings one gets the sheaf $\mathscr{E}$ of differentiable functions on the Riemann surface X. In the following differentiable will always mean infinitely differentiable.

If (U, z), where $z = x + iy$, is a coordinate neighborhood on X, then the differential operators

$$\frac{\partial}{\partial x}, \frac{\partial}{\partial y}, \frac{\partial}{\partial z}, \frac{\partial}{\partial \bar{z}} : \mathscr{E}(U) \to \mathscr{E}(U)$$

can be defined in the obvious way.

Suppose a is a point in X. Then the stalk $\mathscr{E}_a$ consists of all the germs of differentiable functions at the point a. Denote by $\mathfrak{m}_a \subset \mathscr{E}_a$ the vector subspace of all function germs which vanish at a and by $\mathfrak{m}_a^2 \subset \mathfrak{m}_a$ the vector subspace of those function germs which vanish to second order. A function germ $\varphi \in \mathfrak{m}_a$ is said to vanish to second order if it can be represented by a function f such that, with respect to a coordinate neighborhood $(U, z = x + iy)$ of a, one has

$$\frac{\partial f}{\partial x}(a) = \frac{\partial f}{\partial y}(a) = 0.$$

This definition is independent of the choice of the local coordinate z.

9.3. Definition. The quotient vector space

$$T_a^{(1)} := \frac{\mathfrak{m}_a}{\mathfrak{m}_a^2}$$

is called the *cotangent space* of X at the point a. If U is an open neighborhood of a and $f \in \mathcal{E}(U)$, then the *differential* $d_a f \in T_a^{(1)}$ of f at a is the element

$$d_a f := (f - f(a)) \bmod \mathfrak{m}_a^2.$$

Note that the function $f - f(a)$ vanishes at the point a and thus represents an element of $\mathfrak{m}_a$. By definition its equivalence class modulo $\mathfrak{m}_a^2$ is $d_a f$.

9.4. Theorem. *Suppose X is a Riemann surface, $a \in X$ and (U, z) is a coordinate neighborhood of a, where $z = x + iy$ is the decomposition of z into its real and imaginary parts. Then the elements $d_a x$ and $d_a y$ form a basis of the cotangent space $T_a^{(1)}$. As well $(d_a z, d_a \bar{z})$ is a basis of $T_a^{(1)}$. If f is a function which is differentiable in a neighborhood of a, then*

$$\begin{aligned} d_a f &= \frac{\partial f}{\partial x}(a)\, d_a x + \frac{\partial f}{\partial y}(a)\, d_a y \\ &= \frac{\partial f}{\partial z}(a)\, d_a z + \frac{\partial f}{\partial \bar{z}}(a)\, d_a \bar{z}. \end{aligned}$$

PROOF

(a) First we will show that $d_a x$ and $d_a y$ span $T_a^{(1)}$. Let $t \in T_a^{(1)}$ and suppose $\varphi \in \mathfrak{m}_a$ is a representative of t. Expanding φ in a Taylor series about a yields

$$\varphi = c_1(x - x(a)) + c_2(y - y(a)) + \psi,$$

where $c_1, c_2 \in \mathbb{C}$ and $\psi \in \mathfrak{m}_a^2$. Taking equivalence classes modulo $\mathfrak{m}_a^2$, we get

$$t = c_1\, d_a x + c_2\, d_a y.$$

(b) Now we claim $d_a x$ and $d_a y$ are linearly independent. For, $c_1\, d_a x + c_2\, d_a y = 0$ implies

$$c_1(x - x(a)) + c_2(y - y(a)) \in \mathfrak{m}_a^2.$$

Then taking partial derivatives with respect to x and y, one has $c_1 = c_2 = 0$.

(c) Suppose f is differentiable in a neighborhood of a. Then

$$f - f(a) = \frac{\partial f}{\partial x}(a)(x - x(a)) + \frac{\partial f}{\partial y}(a)(y - y(a)) + g,$$

where g vanishes at a to second order. Thus

$$d_a f = \frac{\partial f}{\partial x}(a)\, d_a x + \frac{\partial f}{\partial y}(a)\, d_a y.$$

Similarly, one can prove the corresponding results for $(d_a z, d_a \bar{z})$. □

9.5. Cotangent Vectors of Type (1, 0) and (0, 1). Suppose (U, z) and (U', z') are two coordinate neighborhoods of $a \in X$. Then

$$\frac{\partial z'}{\partial z}(a) =: c \in \mathbb{C}^*, \qquad \frac{\partial \bar{z}'}{\partial \bar{z}}(a) = \bar{c},$$

and

$$\frac{\partial z'}{\partial \bar{z}}(a) = \frac{\partial \bar{z}'}{\partial z}(a) = 0.$$

This implies $d_a z' = c\, d_a z$ and $d_a \bar{z}' = \bar{c}\, d_a \bar{z}$.

Thus the one-dimensional vector subspaces of $T_a^{(1)}$, which are spanned by $d_a z$ and $d_a \bar{z}$, are independent of the choice of local coordinate (U, z) about a. Introduce the following notation:

$$T_a^{1,0} := \mathbb{C}\, d_a z, \qquad T_a^{0,1} := \mathbb{C}\, d_a \bar{z}.$$

By construction $T_a^{(1)} = T_a^{1,0} \oplus T_a^{0,1}$. The elements of $T_a^{1,0}$ (resp. $T_a^{0,1}$) are called cotangent vectors of type (1, 0) (resp. (0, 1)).

If f is differentiable in a neighborhood of a, define $d'_a f$ and $d''_a f$ by

$$d_a f = d'_a f + d''_a f, \qquad d'_a f \in T_a^{1,0}, \qquad d''_a f \in T_a^{0,1}.$$

Then

$$d'_a f = \frac{\partial f}{\partial z}(a)\, d_a z, \qquad d''_a f = \frac{\partial f}{\partial \bar{z}}(a)\, d_a \bar{z}.$$

9.6. Definition. Suppose Y is an open subset of the Riemann surface X. By a *differential form of degree one*, or simply a *1-form*, on Y we mean a mapping

$$\omega\colon Y \to \bigcup_{a \in Y} T_a^{(1)}$$

with $\omega(a) \in T_a^{(1)}$ for every $a \in Y$. If $\omega(a) \in T_a^{1,0}$ (resp. $\omega(a) \in T_a^{0,1}$) for every $a \in Y$, then ω is said to be of type (1, 0) (resp. of type (0, 1)).

9.7. Examples

(a) Suppose $f \in \mathcal{E}(Y)$. Then the mappings df, $d'f$, $d''f$, which are defined by

$$(df)(a) := d_a f, \qquad (d'f)(a) := d'_a f, \qquad (d''f)(a) := d''_a f,$$

for every $a \in Y$, are 1-forms. Clearly a function f is holomorphic precisely if $d''f = 0$.

(b) Suppose ω is a 1-form on Y and $f\colon Y \to \mathbb{C}$ is a function. Then the mapping $f\omega$ defined by $(f\omega)(a) := f(a)\omega(a)$ is also a 1-form on Y.

Remark. If (U, z) is a complex chart with $z = x + iy$, then every 1-form on U may be written

$$\omega = f\, dx + g\, dy = \varphi\, dz + \psi\, d\bar{z},$$

where the functions $f, g, \varphi, \psi\colon U \to \mathbb{C}$ are not necessarily continuous in general.

9.8. Definition. Suppose Y is an open subset of a Riemann surface X. A 1-form ω on Y is called *differentiable* (resp. *holomorphic*) if, with respect to every chart (U, z), ω may be written

$$\omega = f\,dz + g\,d\bar{z} \text{ on } U \cap Y, \quad \text{where } f, g \in \mathcal{E}(U \cap Y),$$

resp.

$$\omega = f\,dz \text{ on } U \cap Y, \quad \text{where } f \in \mathcal{O}(U \cap Y).$$

Notation. For any open subset U of a Riemann surface X we will denote by $\mathcal{E}^{(1)}(U)$ the vector space of differentiable 1-forms on U, by $\mathcal{E}^{1,0}(U)$ (resp. $\mathcal{E}^{0,1}(U)$) the subspace of $\mathcal{E}^{(1)}(U)$ of differential forms of type (1, 0) (resp. (0, 1)) and by $\Omega(U)$ the vector space of holomorphic 1-forms. Together with the natural restriction mappings $\mathcal{E}^{(1)}$, $\mathcal{E}^{1,0}$, $\mathcal{E}^{0,1}$ and Ω are sheaves of vector spaces over X.

9.9. The Residue. Suppose Y is an open subset of a Riemann surface, $a \in Y$ and ω is a holomorphic 1-form on $Y \backslash \{a\}$. Let (U, z) be a coordinate neighborhood of a such that $U \subset Y$ and $z(a) = 0$. Then on $U \backslash \{a\}$ one may write $\omega = f\,dz$, where $f \in \mathcal{O}(U \backslash \{a\})$. Let

$$f = \sum_{n=-\infty}^{\infty} c_n z^n$$

be the Laurent series expansion of f about a with respect to the coordinate z. If $c_n = 0$ for every $n < 0$, then ω may be holomorphically continued to all of Y. In this case a is called a removable singularity of ω. If there exists $k < 0$ such that $c_k \neq 0$ and $c_n = 0$ for every $n < k$, then ω has a *pole* of kth order at a. If there are infinitely many $n < 0$ with $c_n \neq 0$, then ω has an essential singularity at a.

The coefficient c_{-1} is called the *residue* of ω at a and is denoted by

$$c_{-1} = \operatorname{Res}_a(\omega).$$

The next lemma shows that this definition makes sense.

Lemma. *The residue is independent of the choice of chart* (U, z).

PROOF. Suppose V is an open neighborhood of a.

Claim (a) If g is holomorphic on $V \backslash \{a\}$, then the residue of dg at a equals zero and is thus independent of the choice of chart.

PROOF. Let (U, z) be any coordinate neighborhood of a with $z(a) = 0$ and suppose

$$g = \sum_{n=-\infty}^{\infty} c_n z^n$$

is the Laurent series expansion of g about a. Then

$$dg = \left(\sum_{n=-\infty}^{\infty} n c_n z^{n-1} \right) dz$$

and thus the coefficient of $z^{-1}\, dz$ is zero.

Claim (*b*) If φ is a holomorphic function on V which has a zero of first order at a, then $\operatorname{Res}_a(\varphi^{-1}\, d\varphi) = 1$ and is thus independent of the choice of chart.

PROOF. Suppose (U, z) is a chart at a with $z(a) = 0$. Then $\varphi = zh$, where h is holomorphic at a and does not vanish there. Thus $d\varphi = h\, dz + z\, dh$ and

$$\frac{d\varphi}{\varphi} = \frac{h\, dz + z\, dh}{zh} = \frac{dz}{z} + \frac{dh}{h}.$$

Since $h(a) \neq 0$, the differential form $h^{-1}\, dh$ is holomorphic at a and thus has residue zero. This implies

$$\operatorname{Res}_a\left(\frac{d\varphi}{\varphi}\right) = \operatorname{Res}_a\left(\frac{dz}{z}\right) = 1.$$

Now using (*a*) and (*b*) one can easily finish the proof. With respect to a chart (U, z) with $z(a) = 0$ let $\omega = f\, dz$, where

$$f = \sum_{n=-\infty}^{\infty} c_n z^n.$$

Let

$$g := \sum_{n=-\infty}^{-2} \frac{c_n}{n+1} z^{n+1} + \sum_{n=0}^{\infty} \frac{c_n}{n+1} z^{n+1}.$$

Then $\omega = dg + c_{-1} z^{-1}\, dz$. From (*a*) and (*b*) one has $\operatorname{Res}_a(\omega) = c_{-1}$, which is independent of the chart. □

9.10. Meromorphic Differential Forms. A 1-form ω on an open subset Y of a Riemann surface is said to be a meromorphic differential form on Y if there exists an open subset $Y' \subset Y$ such that the following hold:

(i) ω is a holomorphic 1-form on Y',
(ii) $Y \backslash Y'$ consists of only isolated points,
(iii) ω has a pole at every point $a \in Y \backslash Y'$.

Let $\mathscr{M}^{(1)}(Y)$ denote the set of all meromorphic 1-forms on Y. With the natural algebraic operations and the usual restriction mappings $\mathscr{M}^{(1)}$ is a sheaf of vector spaces over X. The meromorphic 1-forms on X are also called *abelian differentials.* As well an abelian differential is said to be of the first kind if it is holomorphic everywhere, of the second kind if its residue is zero at every one of its poles and of the third kind otherwise.

9.11. The Exterior Product. In order to be able to define differential forms of degree two, we have to recall some properties of the exterior product of a vector space with itself. Let V be a vector space over $\mathbb{C}$. Then $\Lambda^2 V$ is the vector space over $\mathbb{C}$ whose elements are finite sums of elements of the form $v_1 \wedge v_2$ for $v_1, v_2 \in V$. One has the following rules

$$(v_1 + v_2) \wedge v_3 = v_1 \wedge v_3 + v_2 \wedge v_3$$

$$(\lambda v_1) \wedge v_2 = \lambda(v_1 \wedge v_2)$$

$$v_1 \wedge v_2 = -v_2 \wedge v_1$$

for $v_1, v_2, v_3 \in V$ and $\lambda \in \mathbb{C}$. If $(e_1, \ldots, e_n)$ is a basis of V, then the elements $e_i \wedge e_j$, for $i < j$, form a basis of $\Lambda^2 V$. In fact these properties completely characterize $\Lambda^2 V$.

Now we will apply this to the cotangent space $T_a^{(1)}$ of a Riemann surface X at a point a. Set

$$T_a^{(2)} := \Lambda^2 T_a^{(1)}.$$

Let (U, z) be a coordinate neighborhood of a, where $z = x + iy$. Then, it follows from what was just said, that $d_a x \wedge d_a y$ is a basis of $T_a^{(2)}$. Another basis is $d_a z \wedge d_a \bar{z} = -2i\, d_a x \wedge d_a y$. Thus $T_a^{(2)}$ has dimension one.

9.12. Definition. Suppose Y is an open subset of a Riemann surface X. A *2-form* on Y is a mapping

$$\omega\colon Y \to \bigcup_{a \in Y} T_a^{(2)},$$

where $\omega(a) \in T_a^{(2)}$ for every $a \in Y$. The form ω is called differentiable on Y if, with respect to every complex chart (U, z) on X, it can be written

$$\omega = f\, dz \wedge d\bar{z} \quad \text{with } f \in \mathscr{E}(U \cap Y),$$

where $\omega = f\, dz \wedge d\bar{z}$ means that $\omega(a) = f(a)\, d_a z \wedge d_a \bar{z}$ for every $a \in U \cap Y$. Denote by $\mathscr{E}^{(2)}(Y)$ the vector space of all differentiable 2-forms on Y.

Examples If $\omega_1, \omega_2 \in \mathscr{E}^{(1)}(Y)$ are 1-forms, then one can define a 2-form $\omega_1 \wedge \omega_2 \in \mathscr{E}^{(2)}(Y)$ by letting

$$(\omega_1 \wedge \omega_2)(a) := \omega_1(a) \wedge \omega_2(a)$$

for every $a \in Y$. For $f \in \mathscr{E}(Y)$ and $\omega \in \mathscr{E}^{(2)}(Y)$ one gets a new 2-form $f\omega \in \mathscr{E}^{(2)}(Y)$ by defining $(f\omega)(a) = f(a)\omega(a)$ for every $a \in Y$.

9.13. Exterior Differentiation of Forms. We now define derivations d, d', d'': $\mathscr{E}^{(1)}(U) \to \mathscr{E}^{(2)}(U)$, where U is an open subset of a Riemann surface. Locally a differentiable 1-form may be written as a finite sum

$$\omega = \sum f_k \, dg_k,$$

where the f_k and g_k are differentiable functions, e.g., $\omega = f_1 \, dz + f_2 \, d\bar{z}$ where z is a local coordinate. Set

$$d\omega := \sum df_k \wedge dg_k,$$

$$d'\omega := \sum d'f_k \wedge dg_k,$$

$$d''\omega := \sum d''f_k \wedge dg_k.$$

Now one has to show that this definition is independent of the representation $\omega = \sum f_k \, dg_k$. We will do this for the operator d, the other cases being similar.

Suppose $\omega = \sum f_k \, dg_k = \sum \tilde{f}_j \, d\tilde{g}_j$. Choose a particular coordinate neighborhood (U, z), where $z = x + iy$. One has to show that $\sum df_k \wedge dg_k = \sum d\tilde{f}_j \wedge d\tilde{g}_j$. Because

$$dg_k = \frac{\partial g_k}{\partial x} dx + \frac{\partial g_k}{\partial y} dy,$$

with a corresponding expression for $d\tilde{g}_j$, one has by assumption

$$\sum f_k \frac{\partial g_k}{\partial x} = \sum \tilde{f}_j \frac{\partial \tilde{g}_j}{\partial x}, \qquad \sum f_k \frac{\partial g_k}{\partial y} = \sum \tilde{f}_j \frac{\partial \tilde{g}_j}{\partial y}.$$

Taking appropriate partial derivatives with respect to x and y and subtracting yields

$$\sum \left(\frac{\partial f_k}{\partial y} \frac{\partial g_k}{\partial x} - \frac{\partial f_k}{\partial x} \frac{\partial g_k}{\partial y} \right) = \sum \left(\frac{\partial \tilde{f}_j}{\partial y} \frac{\partial \tilde{g}_j}{\partial x} - \frac{\partial \tilde{f}_j}{\partial x} \frac{\partial \tilde{g}_j}{\partial y} \right).$$

On the other hand

$$\sum df_k \wedge dg_k = \sum \left(\frac{\partial f_k}{\partial x} \frac{\partial g_k}{\partial y} - \frac{\partial f_k}{\partial y} \frac{\partial g_k}{\partial x} \right) dx \wedge dy,$$

with a corresponding formula for $\sum d\tilde{f}_j \wedge d\tilde{g}_j$. The result follows immediately.

9.14. Elementary Properties. Suppose U is an open subset of a Riemann surface, $f \in \mathscr{E}(U)$ and $\omega \in \mathscr{E}^{(1)}(U)$. Then

(i) $ddf = d'd'f = d''d''f = 0$.
(ii) $d\omega = d'\omega + d''\omega$,
(iii) $d(f\omega) = df \wedge \omega + f \, d\omega$ with similar rules for d' and d''.

These rules are straightforward consequences of the definitions; e.g., $ddf = d(1 \cdot df) = d1 \wedge df = 0$.

From (i) and (ii) one gets

$$d'd''f = -d''d'f,$$

since $0 = (d' + d'')(d' + d'')f = d'd''f + d''d'f$.

With respect to a local chart (U, z), where $z = x + iy$, one has

$$d'd''f = \frac{\partial^2 f}{\partial z\, \partial \bar{z}}\, dz \wedge d\bar{z} = \frac{1}{2i}\left(\frac{\partial^2 f}{\partial x^2} + \frac{\partial^2 f}{\partial y^2}\right) dx \wedge dy.$$

Hence a differentiable function f, defined on an open subset of a Riemann surface, is called *harmonic* if $d'd''f = 0$.

9.15. Definition. Suppose Y is an open subset of a Riemann surface. A differentiable 1-form $\omega \in \mathscr{E}^{(1)}(Y)$ is called *closed* if $d\omega = 0$ and *exact* if there exists $f \in \mathscr{E}(Y)$ such that $\omega = df$.

Remark. Because $ddf = 0$, every exact form is closed. However the converse is not true in general. We shall look at this question in more detail in the next section.

9.16. Theorem. *Suppose Y is an open subset of a Riemann surface. Then the following hold:*

(a) *Every holomorphic 1-form $\omega \in \Omega(Y)$ is closed.*
(b) *Every closed 1-form $\omega \in \mathscr{E}^{1,0}(Y)$ is holomorphic.*

PROOF. Suppose ω is a differentiable 1-form of type $(1, 0)$. With respect to a coordinate neighborhood (U, z) one may write $\omega = f\,dz$ for some differentiable function f. Then

$$d\omega = df \wedge dz = \left(\frac{\partial f}{\partial z}\, dz + \frac{\partial f}{\partial \bar{z}}\, d\bar{z}\right) \wedge dz = -\frac{\partial f}{\partial \bar{z}}\, dz \wedge d\bar{z}.$$

Thus $d\omega = 0$ is equivalent to $(\partial f/\partial \bar{z}) = 0$ and the results follow. □

Consequence. If u is a harmonic function, then $d'u$ is a holomorphic 1-form. For, $dd'u = d''d'u = 0$.

9.17. The Pull-Back of Differential Forms. Suppose $F: X \to Y$ is a holomorphic mapping between two Riemann surfaces. For every open set $U \subset Y$ the map F induces a homomorphism

$$F^*: \mathscr{E}(U) \to \mathscr{E}(F^{-1}(U)), \qquad F^*(f) := f \circ F.$$

Generalizing this one can define corresponding mappings for differential forms

$$F^*: \mathscr{E}^{(k)}(U) \to \mathscr{E}^{(k)}(F^{-1}(U)), \qquad k = 1, 2.$$

(Using the same symbol F^* should cause no confusion.) Locally a 1-form (resp. 2-form) may be written as a finite sum $\sum f_j\, dg_j$ (resp. $\sum f_j\, dg_j \wedge dh_j$), where the functions f_j, g_j, h_j are differentiable. Set

$$F^*\left(\sum f_j\, dg_j\right) = \sum (F^*f_j)\, d(F^*g_j),$$

$$F^*\left(\sum f_j\, dg_j \wedge dh_j\right) = \sum (F^*f_j)\, d(F^*g_j) \wedge d(F^*h_j).$$

It is easy to check that these definitions are independent of the local representations chosen and hence piece together to give unique global vector space homomorphisms $F^*\colon \mathscr{E}^{(k)}(U) \to \mathscr{E}^{(k)}(F^{-1}(U))$. For $f \in \mathscr{E}(U)$ and $\omega \in \mathscr{E}^{(1)}(U)$ one has

(i) $F^*(df) = d(F^*f)$, $\quad F^*(d\omega) = d(F^*\omega)$,
(ii) $F^*(d'f) = d'(F^*f)$, $\quad F^*(d'\omega) = d'(F^*\omega)$,

with corresponding formulas for d''.

Consequence. If $f \in \mathscr{E}(U)$ is harmonic, then $F^*f = f \circ F \in \mathscr{E}(F^{-1}(U))$ is also harmonic. For, $d'd''(F^*f) = d'(F^*d''f) = F^*(d'd''f) = 0$.

EXERCISES (§9)

9.1. Suppose $p := \exp\colon \mathbb{C} \to \mathbb{C}^*$ is the universal covering of $\mathbb{C}^*$ and ω is the holomorphic 1-form dz/z on $\mathbb{C}^*$. Find $p^*\omega$.

9.2. Prove that the holomorphic 1-form

$$\frac{dz}{1+z^2},$$

which is defined on $\mathbb{C}\backslash\{\pm i\}$, can be extended to a holomorphic 1-form ω on $\mathbb{P}^1\backslash\{\pm i\}$. Let

$$p := \tan\colon \mathbb{C} \to \mathbb{P}^1\backslash\{\pm i\}$$

(cf. Ex. 4.4) and find $p^*\omega$.

9.3. Suppose $p\colon Y \to X$ is a holomorphic mapping of Riemann surfaces, $a \in X$, $b \in p^{-1}(a)$ and k is the multiplicity of p at b. Given any holomorphic 1-form ω on $X\backslash\{a\}$ show that

$$\operatorname{Res}_b(p^*\omega) = k \operatorname{Res}_a(\omega).$$

§10. The Integration of Differential Forms

Differential 1-forms can be integrated along curves. If the form is closed, then the integral only depends on the homotopy class of the curve. Thus on any simply connected surface X the indefinite integral of a closed 1-form, where the integration takes place along a curve with fixed initial point and variable end point, is a well-defined function on X. In general the integration of closed forms yields multi-valued functions. But these functions display a very special kind of multi-valued behavior. This will be looked at more closely in this section. As well we consider the integration of 2-forms. This will be useful in transforming line integrals into surface integrals and will also be needed to prove the Residue Theorem.

A. Differential 1-Forms

10.1. Suppose X is a Riemann surface and $\omega \in \mathscr{E}^{(1)}(X)$. Further suppose that a piece-wise continuously differentiable curve in X is given. This means there is a continuous mapping

$$c: [0, 1] \to X$$

for which there exists a partition

$$0 = t_0 < t_1 < \cdots < t_n = 1$$

of the interval $[0, 1]$ and charts (U_k, z_k), $z_k = x_k + iy_k$, $k = 1, \ldots, n$, such that $c([t_{k-1}, t_k]) \subset U_k$ and the functions

$$x_k \circ c: [t_{k-1}, t_k] \to \mathbb{R}, \qquad y_k \circ c: [t_{k-1}, t_k] \to \mathbb{R}$$

have continuous first order derivatives. The integral of ω along the curve c is defined in the following way. On U_k one may write ω as $\omega = f_k\, dx_k + g_k\, dy_k$, where the functions f_k, g_k are differentiable. Set

$$\int_c \omega := \sum_{k=1}^{n} \int_{t_{k-1}}^{t_k} \left(f_k(c(t)) \frac{dx_k(c(t))}{dt} + g_k(c(t)) \frac{dy_k(c(t))}{dt} \right) dt.$$

One can easily check that this definition is independent of the choice of partition and charts.

10.2. Theorem. *Suppose X is a Riemann surface, $c: [0, 1] \to X$ is a piece-wise continuously differentiable curve and $F \in \mathscr{E}(X)$. Then*

$$\int_c dF = F(c(1)) - F(c(0)).$$

PROOF. Choose a partition $0 = t_0 < t_1 < \cdots < t_n = 1$ and charts (U_k, z_k) as above. On U_k one has

$$dF = \frac{\partial F}{\partial x_k}\, dx_k + \frac{\partial F}{\partial y_k}\, dy_k.$$

Thus

$$\begin{aligned}\int_c dF &= \sum_{k=1}^{n} \int_{t_{k-1}}^{t_k} \left(\frac{\partial F}{\partial x_k}(c(t))\frac{dx_k(c(t))}{dt} + \frac{\partial F}{\partial y_k}(c(t))\frac{dy_k(c(t))}{dt}\right) dt \\ &= \sum_{k=1}^{n} \int_{t_{k-1}}^{t_k} \left(\frac{d}{dt} F(c(t))\right) dt \\ &= \sum_{k=1}^{n} (F(c(t_k)) - F(c(t_{k-1}))) = F(c(1)) - F(c(0)).\end{aligned}$$

□

10.3. Definition. Suppose X is a Riemann surface and $\omega \in \mathscr{E}^{(1)}(X)$. A function $F \in \mathscr{E}(X)$ is called a *primitive* of ω if $dF = \omega$.

By (9.15) any differential form which has a primitive is necessarily closed. But the primitive of a differential form is not unique. If F is a primitive of ω and $c \in \mathbb{C}$, then $F + c$ is also a primitive of ω. Conversely any two primitives differ by a constant. For, if $dF = 0$, it follows, for example using Theorem (10.2), that F is a constant.

Using Theorem (10.2) one can easily compute any line integral of a differential form if one knows one of its primitives. And it also follows from the Theorem that the integral of an exact differential form along a curve depends only on the initial and end points of the curve.

10.4. The Local Existence of Primitives. Suppose $U := \{z \in \mathbb{C} : |z| < r\}$, where $r > 0$, is an open disk about zero in $\mathbb{C}$ and $\omega \in \mathscr{E}^{(1)}(U)$. The differential form ω may be written

$$\omega = f\,dx + g\,dy, \qquad f, g \in \mathscr{E}(U),$$

where x, y are the usual real coordinates on $\mathbb{R}^2 \cong \mathbb{C}$. Assume that ω is *closed*, i.e., $d\omega = 0$. Since

$$d\omega = df \wedge dx + dg \wedge dy = \left(\frac{\partial g}{\partial x} - \frac{\partial f}{\partial y}\right) dx \wedge dy,$$

this is equivalent to $(\partial g/\partial x) = (\partial f/\partial y)$. We will prove that ω has a primitive F which is given by the integral

$$F(x, y) := \int_0^1 (f(tx, ty)x + g(tx, ty)y)\,dt, \quad \text{for } (x, y) \in U.$$

One sees directly that F is infinitely differentiable. One has only to verify that $dF = \omega$, i.e., $(\partial F/\partial x) = f$ and $(\partial F/\partial y) = g$. Differentiating under the integral sign, we get

$$\frac{\partial F(x, y)}{\partial x} = \int_0^1 \left(\frac{\partial f}{\partial x}(tx, ty)tx + \frac{\partial g}{\partial x}(tx, ty)ty + f(tx, ty)\right) dt.$$

Since

$$\frac{\partial g}{\partial x} = \frac{\partial f}{\partial y} \quad \text{and} \quad \frac{d}{dt} f(tx, ty) = \frac{\partial f}{\partial x}(tx, ty)x + \frac{\partial f}{\partial y}(tx, ty)y,$$

one then has

$$\begin{aligned}\frac{\partial F(x, y)}{\partial x} &= \int_0^1 \left(t \frac{d}{dt} f(tx, ty) + f(tx, ty) \right) dt \\ &= \int_0^1 \frac{d}{dt}(tf(tx, ty))\, dt = f(x, y).\end{aligned}$$

Similarly, $(\partial F/\partial y) = g$. This proves that $dF = \omega$.

In the special case that ω is holomorphic, the proof of the existence of a primitive on the disk U is much easier. Namely, in this case one has

$$\omega = f\, dz \text{ with } f \in \mathcal{O}(U).$$

Let

$$f(z) = \sum_{n=0}^{\infty} c_n z^n$$

be the Taylor series expansion of f. Then defining

$$F(z) := \sum_{n=0}^{\infty} \frac{c_n}{n+1} z^{n+1}$$

gives us a function $F \in \mathcal{O}(U)$ such that $dF = \omega$.

Globally a primitive of a closed differential form exists in general only as a multi-valued function. This is made precise in the next theorem.

10.5. Theorem. *Suppose X is a Riemann surface and $\omega \in \mathcal{E}^{(1)}(X)$ is a closed differential form. Then there exist a covering map $p: \hat{X} \to X$ with $\hat{X}$ connected, and a primitive $F \in \mathcal{E}(\hat{X})$ of the differential form $p^*\omega$.*

Proof. Let $\mathcal{F}$ be the sheaf of primitives of ω. This is defined as follows. For an open set $U \subset X$ let $\mathcal{F}(U)$ consist of all functions $f \in \mathcal{E}(U)$ such that $df = \omega$ on U. The sheaf $\mathcal{F}$ satisfies the Identity Theorem (cf. Definition (6.9)), since any two elements $f_1, f_2 \in \mathcal{F}(U)$, where U is a domain in X, differ by a constant. Consider the associated space $p: |\mathcal{F}| \to X$. By Theorem (6.10) the space $|\mathcal{F}|$ is Hausdorff. Now we will show that $p: |\mathcal{F}| \to X$ is a covering map. For every point $a \in X$ there exist by (10.4) a connected open neighborhood U and a primitive $f \in \mathcal{F}(U)$ of ω. Then $f + c$, for $c \in \mathbb{C}$, are all the primitives of ω on U. Hence

$$p^{-1}(U) = \bigcup_{c \in \mathbb{C}} [U, f + c].$$

The sets $[U, f+c]$ are pairwise disjoint and all the mappings $p|[U, f+c] \to U$ are homeomorphisms. This proves that $p\colon |\mathscr{F}| \to X$ is a covering map. Let $\hat{X} \subset |\mathscr{F}|$ be a connected component. Then $p|\hat{X} \to X$ is also a covering map. Since $\hat{X}$ is a set of function germs, a function $F\colon \hat{X} \to \mathbb{C}$ is defined in a natural way by $F(\varphi) := \varphi(p(\varphi))$. It then follows directly from the definitions that F is a primitive of $p^*\omega$. □

10.6. Corollary. *Suppose X is a Riemann surface, $\pi\colon \tilde{X} \to X$ its universal covering and $\omega \in \mathscr{E}^{(1)}(X)$ a closed differential form. Then there exists a primitive $f \in \mathscr{E}(\tilde{X})$ of $\pi^*\omega$.*

PROOF. Let $p\colon \hat{X} \to X$ be the covering map construction in (10.5) and let $F \in \mathscr{E}(\hat{X})$ be a primitive of $p^*\omega$. Since $\pi\colon \tilde{X} \to X$ is the universal covering, there exists a holomorphic fiber-preserving mapping $\tau\colon \tilde{X} \to \hat{X}$. Let $f := \tau^*F \in \mathscr{E}(\tilde{X})$. Then f is a primitive of $\tau^*(p^*\omega) = \pi^*\omega$. □

10.7. Corollary. *On a simply connected Riemann surface X every closed differential form $\omega \in \mathscr{E}^{(1)}(X)$ has a primitive $F \in \mathscr{E}(X)$.*

This follows from (10.6) since $\mathrm{id}\colon X \to X$ is the universal covering.

10.8. Theorem. *Suppose X is a Riemann surface and $p\colon \tilde{X} \to X$ is its universal covering. Suppose $\omega \in \mathscr{E}^{(1)}(X)$ is a closed differential form and $F \in \mathscr{E}(\tilde{X})$ is a primitive of $p^*\omega$. If $c\colon [0, 1] \to X$ is a piece-wise continuously differentiable curve and $\hat{c}\colon [0, 1] \to \tilde{X}$ is a lifting of c, then*

$$\int_c \omega = F(\hat{c}(1)) - F(\hat{c}(0)).$$

PROOF. For every piece-wise continuously differentiable curve $v\colon [0, 1] \to \tilde{X}$ and every differential form $\omega \in \mathscr{E}^{(1)}(X)$ one has

$$\int_v p^*\omega = \int_{p \circ v} \omega.$$

This follows directly from the definitions. The theorem then follows from Theorem (10.2). □

10.9. Remark. Theorem (10.8) now gives a way to define the integral of a closed differential form along an arbitrary (continuous) curve $c\colon [0, 1] \to X$, namely by the given formula. This definition is independent of the choice of the primitive F of $p^*\omega$, for any two primitives only differ by a constant and taking the difference kills this. The definition is also independent of the lifting of the curve c. For suppose u and v are two liftings of c. Since the covering $p\colon \tilde{X} \to X$ is Galois (cf. 5.6), there is a covering transformation σ

such that $v = \sigma \circ u$. Since $p \circ \sigma = p$, one has $\sigma^*(p^*\omega) = p^*\omega$. Thus σ^*F is also a primitive of $p^*\omega$ and so $\sigma^*F - F = \text{const}$. Hence

$$F(v(1)) - F(v(0)) = \sigma^*F(u(1)) - \sigma^*F(u(0)) = F(u(1)) - F(u(0))$$

and thus the value of the integral is the same for both liftings.

10.10. Theorem. *Suppose X is a Riemann surface and $\omega \in \mathcal{E}^{(1)}(X)$ is a closed differential form.*

(a) *If $a, b \in X$ are two points and $u, v\colon [0, 1] \to X$ are two homotopic curves from a to b, then*

$$\int_u \omega = \int_v \omega.$$

(b) *If $u, v\colon [0, 1] \to X$ are two closed curves which are free homotopic, then*

$$\int_u \omega = \int_v \omega.$$

PROOF

(a) Let $p\colon \tilde{X} \to X$ be the universal covering and suppose $\hat{u}, \hat{v}\colon [0, 1] \to \tilde{X}$ are liftings of u and v resp. with the same initial point. By Theorem (4.10) $\hat{u}$ and $\hat{v}$ also have the same end point. Hence the result follows from Theorem (10.8).

(b) Suppose the curve u has initial and end point x_0 and the curve v has initial and end point x_1. Then there exists a curve w from x_0 to x_1 such that u is homotopic to $w \cdot v \cdot w^-$, cf. (3.13). Hence by (a) one has

$$\int_u \omega = \int_{w \cdot v \cdot w^-} \omega = \int_w \omega + \int_v \omega - \int_w \omega = \int_v \omega. \qquad \square$$

10.11. Periods. Suppose X is a Riemann surface and $\omega \in \mathcal{E}^{(1)}(X)$ is a closed differential form. Then by Theorem (10.10) one can define the integral

$$a_\sigma := \int_\sigma \omega, \qquad \sigma \in \pi_1(X),$$

by choosing any curve representing the homotopy class σ and integrating along that curve. These integrals are called the *periods* of ω. Clearly

$$\int_{\sigma \cdot \tau} \omega = \int_\sigma \omega + \int_\tau \omega \quad \text{for } \sigma, \tau \in \pi_1(X).$$

Thus one gets a homomorphism $\pi_1(X) \to \mathbb{C}$ of the fundamental group of X into the additive group $\mathbb{C}$. This homomorphism is called the *period homomorphism* associated to the closed differential form ω.

Example. Suppose $X = \mathbb{C}^*$. By (5.7.a) $\pi_1(\mathbb{C}^*) \cong \mathbb{Z}$. A generator of $\pi_1(\mathbb{C}^*)$ is represented by the curve u: $[0, 1] \to \mathbb{C}^*$, $u(t) = e^{2\pi i t}$. Let $\omega := (dz/z)$, where z is the canonical coordinate. Then

$$\int_u \omega = \int_u \frac{dz}{z} = 2\pi i.$$

Hence the period homomorphism of ω is

$$\mathbb{Z} \to \mathbb{C}, \qquad n \mapsto 2\pi i n,$$

where we have explicitly realized the isomorphism $\mathbb{Z} \cong \pi_1(\mathbb{C}^*)$ by the correspondence $n \mapsto \mathrm{cl}(u^n)$.

10.12. Summands of Automorphy. Suppose X is a Riemann surface and p: $\tilde{X} \to X$ is its universal covering. The group $G := \mathrm{Deck}(\tilde{X}/X)$ of covering transformations of the universal covering, as was observed in (5.6), is isomorphic to the fundamental group of X. If $\sigma \in G$ and f: $\tilde{X} \to \mathbb{C}$ is a function, then we can define a function σf: $\tilde{X} \to \mathbb{C}$ by $\sigma f := f \circ \sigma^{-1}$. If g: $\tilde{X} \to \mathbb{C}$ is another function, then $\sigma(f + g) = \sigma f + \sigma g$ and $\sigma(fg) = (\sigma f)(\sigma g)$. Also for $\sigma, \tau \in G$ one has $(\sigma\tau)f = \sigma(\tau f)$.

A function f: $\tilde{X} \to \mathbb{C}$ is called *additively automorphic* with constant summands of automorphy, if there exist constants $a_\sigma \in \mathbb{C}$, $\sigma \in G$, such that

$$f - \sigma f = a_\sigma \quad \text{for every } \sigma \in G.$$

The constants a_σ, which are uniquely determined by f, are called the *summands of automorphy* of f. Then $\sigma f - \sigma\tau f = a_\tau$ for any $\sigma, \tau \in G$, since $f - \tau f = a_\tau$. Thus

$$a_{\sigma\tau} = f - \sigma\tau f = (f - \sigma f) + (\sigma f - \sigma\tau f) = a_\sigma + a_\tau.$$

Hence the correspondence $\sigma \mapsto a_\sigma$ is a group homomorphism $\mathrm{Deck}(\tilde{X}/X) \to \mathbb{C}$.

Any function f: $\tilde{X} \to \mathbb{C}$ which is invariant under covering transformations, i.e., $\sigma f = f$ for every $\sigma \in G$, is an example of an additively automorphic function. In particular its summands of automorphy are all zero. For any such function there exists a function f_0: $X \to \mathbb{C}$ such that $f = p^* f_0$. If f is differentiable (resp. holomorphic) then f_0 is differentiable (resp. holomorphic) as well.

10.13. Theorem. *Suppose X is a Riemann surface and p: $\tilde{X} \to X$ is its universal covering.*

(a) *If $\omega \in \mathcal{E}^{(1)}(X)$ is a closed differential form and $F \in \mathcal{E}(\tilde{X})$ is a primitive of $p^*\omega$, then F is additively automorphic with constant summands of automorphy. Its summands of automorphy a_σ, $\sigma \in \mathrm{Deck}(\tilde{X}/X)$, are, with respect to the isomorphism $\pi_1(X) \cong \mathrm{Deck}(\tilde{X}/X)$, exactly the periods of ω.*

(b) *Conversely suppose $F \in \mathcal{E}(\tilde{X})$ is an additive automorphic function with constant summands of automorphy. Then there exists precisely one closed differential form $\omega \in \mathcal{E}^{(1)}(X)$ such that $dF = p^*\omega$.*

PROOF

(a) If σ is any deck transformation, then because $p \circ \sigma^{-1} = p$ the function σF is also a primitive of $p^*\omega$. Thus

$$-a_\sigma := \sigma F - F$$

is a constant. Suppose $x_0 \in X$ and $z_0 \in \tilde{X}$ is a point with $p(z_0) = x_0$. Suppose $\sigma \in \text{Deck}(\tilde{X}/X)$. By (5.6) the element $\bar{\sigma} \in \pi_1(X, x_0)$ which is associated to σ can be represented as follows. Choose a curve $v\colon [0, 1] \to \tilde{X}$ with $v(0) := y_0 := \sigma^{-1}(z_0)$ and $v(1) := z_0 = \sigma(y_0)$. Then $u := p \circ v$ is a closed curve in X and $\bar{\sigma} = \text{cl}(u)$. By Theorem (10.8) the periods of ω with respect to $\bar{\sigma}$ are given by

$$\int_u \omega = F(v(1)) - F(v(0)) = F(z_0) - F(\sigma^{-1}(z_0)) = -a_\sigma.$$

(b) If F has summands of automorphy $a_\sigma \in \mathbb{C}$, then for every $\sigma \in \text{Deck}(\tilde{X}/X)$ one has

$$\sigma^*(dF) = d\sigma^*F = d(F + a_\sigma) = dF.$$

Thus the closed differential form dF is invariant under covering transformations. Since $p\colon \tilde{X} \to X$ is locally biholomorphic, there exists $\omega \in \mathcal{E}^{(1)}(X)$ such that $dF = p^*\omega$. Clearly ω is uniquely determined and is closed. □

10.14. Example. Suppose $\Gamma = \mathbb{Z}\gamma_1 + \mathbb{Z}\gamma_2$, where $\gamma_1, \gamma_2 \in \mathbb{C}$ are linearly independent over $\mathbb{R}$, is a lattice in $\mathbb{C}$. Let $X := \mathbb{C}/\Gamma$.

The canonical quotient mapping $\pi\colon \mathbb{C} \to X$ is also the universal covering map and $\text{Deck}(\mathbb{C}/X)$ is the group of all translations by vectors $\gamma \in \Gamma$, cf. (5.7.c). Consider the identity map $z\colon \mathbb{C} \to \mathbb{C}$. Then the function z is additively automorphic under the action of $\text{Deck}(\mathbb{C}/X)$ with summands of automorphy $a_\gamma = \gamma$, $\gamma \in \Gamma$. Hence dz is invariant under covering transformations. Thus there exists a holomorphic differential form $\omega \in \Omega(X)$ such that $p^*\omega = dz$ and whose periods are exactly the elements of the lattice Γ.

10.15. Theorem. *Suppose X is a Riemann surface. A closed differential form $\omega \in \mathcal{E}^{(1)}(X)$ has a primitive $f \in \mathcal{E}(X)$ if and only if all the periods of ω are zero.*

PROOF. If ω has a primitive, then by (10.2) all its periods are zero.

Conversely, suppose that all the periods of ω are zero. By Corollary (10.6) there exists, on the universal covering $p\colon \tilde{X} \to X$, a primitive $F \in \mathcal{E}(\tilde{X})$ of $p^*\omega$. By (10.3), F has summands of automorphy 0. Thus there is an $f \in \mathcal{E}(X)$

such that $F = p^*f$. Then this function is a primitive of ω, since $p^*\omega = dF = d(p^*f) = p^*(df)$ implies $\omega = df$. □

Remark. If all the periods of ω vanish, then by Theorem (10.2) one gets a special primitive of ω from the integral

$$f(x) := \int_{x_0}^{x} \omega.$$

Here $x_0 \in X$ is a fixed arbitrary point and the integral is along any curve from x_0 to x (the integral is in this case independent of the choice of curve).

10.16. Corollary. *Suppose X is a compact Riemann surface and $\omega_1, \omega_2 \in \Omega(X)$ are two holomorphic differential forms which define the same period homomorphism $\pi_1(X) \to \mathbb{C}$. Then $\omega_1 = \omega_2$.*

PROOF. The difference $\omega := \omega_1 - \omega_2$ has zero periods and thus has a primitive $f \in \mathcal{O}(X)$. Since X is compact, f is constant and thus $\omega = df = 0$. □

B. Differential 2-Forms

10.17. Next we look at integration of differential 2-forms in the complex plane. Suppose $U \subset \mathbb{C}$ is open and $\omega \in \mathcal{E}^{(2)}(U)$. Then ω may be written

$$\omega = f\,dx \wedge dy = \frac{i}{2} f\,dz \wedge d\bar{z}, \quad \text{where } f \in \mathcal{E}(U).$$

Assume that f vanishes outside of a compact subset of U. Then define

$$\iint_U \omega := \iint_U f(x, y)\,dx\,dy,$$

where the right-hand side is the usual double integral.

Now suppose V is another open subset of $\mathbb{C}$ and $\varphi\colon V \to U$ is a biholomorphic mapping. If $\varphi = u + iv$ is the splitting of φ into its real and imaginary parts, then by the Cauchy–Riemann equations the Jacobian determinant of the mapping φ is

$$\frac{\partial(u, v)}{\partial(x, y)} = \frac{\partial u}{\partial x}\frac{\partial v}{\partial y} - \frac{\partial u}{\partial y}\frac{\partial v}{\partial x} = |\varphi'|^2.$$

Thus the transformation formula for the integral becomes

$$\iint_U f\,dx\,dy = \iint_V (f \circ \varphi)|\varphi'|^2\,dx\,dy.$$

On the other hand,

$$\varphi^*(dz \wedge d\bar{z}) = d\varphi \wedge d\bar{\varphi} = (\varphi' \, dz) \wedge (\bar{\varphi}' \, d\bar{z}) = |\varphi'|^2 \, dz \wedge d\bar{z}$$

and thus $\varphi^*\omega = (f \circ \varphi)|\varphi'|^2 \, dx \wedge dy$. Hence

$$\iint_U \omega = \iint_V \varphi^*\omega.$$

10.18. Now suppose X is a Riemann surface. By the *support* of a differential form ω on X we mean the closed set

$$\operatorname{Supp}(\omega) := \overline{\{a \in X : \omega(a) \neq 0\}}.$$

The support $\operatorname{Supp}(f)$ of a function $f: X \to \mathbb{C}$ is defined analogously.

(a) Suppose $\varphi: U \to V$ is a chart on X and $\omega \in \mathscr{E}^{(2)}(X)$ is a differential form whose support is compact and contained in U. Then $(\varphi^{-1})^*\omega$ is a differential form with compact support in $V \subset \mathbb{C}$ and thus one can define

$$\iint_X \omega := \iint_U \omega := \iint_V (\varphi^{-1})^*\omega.$$

This definition is independent of the choice of chart. For, suppose $\varphi_1: U_1 \to V_1$ is another chart with $\operatorname{Supp}(\omega) \subset U_1$. Without loss of generality we may assume $U = U_1$ (otherwise take the intersection). Then

$$\psi := \varphi_1 \circ \varphi^{-1}: V \to V_1$$

is a biholomorphic mapping. Since

$$(\varphi^{-1})^*\omega = (\varphi_1^{-1} \circ \psi)^*\omega = \psi^*((\varphi_1^{-1})^*\omega),$$

by (10.17) one has

$$\iint_V (\varphi^{-1})^*\omega = \iint_{V_1} (\varphi_1^{-1})^*\omega.$$

Thus $\iint_X \omega$ is defined independently of the choice of chart.

(b) Now suppose $\omega \in \mathscr{E}^{(2)}(X)$ is an arbitrary differential form with compact support. Then there exist finitely many charts $\varphi_k: U_k \to V_k$, $k = 1, \ldots, n$ such that

$$\operatorname{Supp}(\omega) \subset \bigcup_{k=1}^{n} U_k.$$

Then one can find functions $f_k \in \mathscr{E}(X)$ with the following properties (a so-called "partition of unity," cf. Appendix A):

(i) $\operatorname{Supp}(f_k) \subset U_k$,
(ii) $\sum_{k=1}^{n} f_k(x) = 1$ for every $x \in \operatorname{Supp}(\omega)$.

Then $f_k \omega$ is a differential form with $\mathrm{Supp}(f_k \omega) \Subset U_k$ and

$$\omega = \sum_{k=1}^{n} f_k \omega.$$

Define

$$\iint_X \omega := \sum_{k=1}^{n} \iint_X f_k \omega.$$

Here the right-hand side is defined by (a). Again it is straightforward to check that the definition is independent of the choice of charts and functions f_k.

10.19. Later on we want to use a special case of *Stokes' Theorem* in the plane. Suppose $U \subset \mathbb{C}$ is open and $A \subset U$ is a compact subset with smooth boundary ∂A. Then for every differential form $\omega \in \mathscr{E}^{(1)}(U)$

$$\iint_A d\omega = \int_{\partial A} \omega.$$

Here the boundary is oriented so that the outward pointing normal of A and the tangent vector to ∂A in this order determine a positively oriented basis of the plane.

We will need the theorem only in the case that A is a disk or an annulus

$$A = \{z \in \mathbb{C} : \varepsilon \le |z| \le R\}, \qquad 0 < \varepsilon < R.$$

In the second case, ∂A consists of the positively oriented circle $|z| = R$ and the negatively oriented circle $|z| = \varepsilon$. Then Stokes' Theorem for $\omega = f\,dx + g\,dy$ says

$$\iint_{\varepsilon \le |z| \le R} \left(\frac{\partial g}{\partial x} - \frac{\partial f}{\partial y}\right) dx\,dy = \int_{|z|=R} (f\,dx + g\,dy) - \int_{|z|=\varepsilon} (f\,dx + g\,dy).$$

We would now like to prove this formula directly by introducing polar coordinates $z = re^{i\theta}$, i.e.,

$$x = r\cos\theta, \qquad y = r\sin\theta.$$

First we look at the case $\omega = g\,dy$. Thus $d\omega = (\partial g/\partial x)\,dx \wedge dy$. Noting that

$$\frac{\partial}{\partial x} = \cos\theta \frac{\partial}{\partial r} - \frac{\sin\theta}{r}\frac{\partial}{\partial \theta},$$

and letting $\tilde{g}(r, \theta) := g(re^{i\theta})$, one gets

$$\iint_A d\omega = \iint_{\varepsilon \le |z| \le R} \frac{\partial g}{\partial x}\, dx\, dy = \iint_{\substack{\varepsilon \le r \le R \\ 0 \le \theta \le 2\pi}} \left(\cos\theta \frac{\partial \tilde{g}}{\partial r} - \frac{\sin\theta}{r}\frac{\partial \tilde{g}}{\partial \theta}\right) r\, dr\, d\theta$$

$$= \iint_{\substack{\varepsilon \le r \le R \\ 0 \le \theta \le 2\pi}} \left(\cos\theta \frac{\partial}{\partial r}(r\tilde{g}) - \frac{\partial}{\partial \theta}(\sin\theta\tilde{g})\right) dr\, d\theta.$$

Now for every fixed $r \in [\varepsilon, R]$

$$\int_0^{2\pi} \frac{\partial}{\partial \theta}(\sin\theta\,\tilde{g})\, d\theta = \sin\theta\,\tilde{g}(r, \theta)\Big|_{\theta=0}^{\theta=2\pi} = 0.$$

Then

$$\iint_A d\omega = \int_0^{2\pi} \cos\theta\left(\int_\varepsilon^R \frac{\partial}{\partial r}(r\tilde{g})\, dr\right) d\theta$$

$$= \int_0^{2\pi} \tilde{g}(R, \theta) R \cos\theta\, d\theta - \int_0^{2\pi} \tilde{g}(\varepsilon, \theta)\, \varepsilon \cos\theta\, d\theta$$

$$= \int_{|z|=R} g\, dy - \int_{|z|=\varepsilon} g\, dy = \int_{\partial A} \omega.$$

The case $\omega = f\, dx$ is reduced to the case just considered by making the change of coordinates $(x, y) \mapsto (y, -x)$ and noting that this transformation has Jacobian determinant 1. This proves Stokes' Theorem for an annulus. The case of the disk is obtained by letting $\varepsilon \to 0$.

10.20. Theorem. *Suppose X is a Riemann surface and $\omega \in \mathscr{E}^{(1)}(X)$ is a differential form with compact support. Then*

$$\iint_X d\omega = 0.$$

Proof. By multiplying by a partition of unity as in (10.18.b) we may write ω as a sum $\omega = \omega_1 + \cdots + \omega_n$, where each ω_k has compact support which lies entirely in one chart.

Without loss of generality we may thus assume $X = \mathbb{C}$.

Choose $R > 0$ so large that

$$\operatorname{Supp}(\omega) \subset \{z \in \mathbb{C} : |z| < R\}.$$

Then

$$\iint_{\mathbb{C}} d\omega = \iint_{|z| \le R} d\omega = \int_{|z|=R} \omega = \int_{|z|=R} 0 = 0. \qquad \square$$

10.21. The Residue Theorem. *Suppose X is a compact Riemann surface and $a_1, \ldots, a_n$ are distinct points in X. Let $X' := X \backslash \{a_1, \ldots, a_n\}$. Then for every holomorphic 1-form $\omega \in \Omega(X')$, one has*

$$\sum_{k=1}^{n} \operatorname{Res}_{a_k}(\omega) = 0.$$

PROOF. Choose coordinate neighborhoods (U_k, z_k) of the a_k such that $U_j \cap U_k = \emptyset$ if $j \neq k$. Also we may assume that $z_k(a_k) = 0$ and $z_k(U_k) \subset \mathbb{C}$ is a disk. For every $k = 1, \ldots, n$ choose a function f_k with compact support $\operatorname{Supp}(f_k) \subset U_k$ such that there exists an open neighborhood $U'_k \subset U_k$ of a_k with $f_k \mid U'_k = 1$. Set $g := 1 - (f_1 + \cdots + f_n)$. Then $g \mid U'_k = 0$. Thus $g\omega$ may be continued to the point a_k by assigning it the value zero, and may thus be considered as an element of $\mathscr{E}^{(1)}(X)$. By (10.20)

$$\iint_X d(g\omega) = 0.$$

Since ω is holomorphic, $d\omega = 0$ on X'. On $U'_k \cap X'$ one has $f_k\omega = \omega$ and thus $d(f_k\omega) = 0$. Hence $d(f_k\omega)$ may be considered to be an element of $\mathscr{E}^{(2)}(X)$ whose support is a compact subset of $U_k \backslash \{a_k\}$. Now $d(g\omega) = -\sum d(f_k\omega)$ implies

$$\sum_{k=1}^{n} \iint_X d(f_k\omega) = 0.$$

Hence the proof will be complete once we show

$$\iint_X d(f_k\omega) = -2\pi i \operatorname{Res}_{a_k}(\omega).$$

Since the support of $d(f_k\omega)$ is contained in U_k, we only have to integrate over U_k. Using the coordinate z_k we may identify U_k with the unit disk. There exist $0 < \varepsilon < R < 1$ such that

$$\operatorname{Supp}(f_k) \subset \{|z_k| < R\} \quad \text{and} \quad f_k | \{|z_k| \leq \varepsilon\} = 1.$$

But then

$$\iint_X d(f_k\omega) = \iint_{\varepsilon \leq |z_k| \leq R} d(f_k\omega) = \int_{|z_k|=R} f_k\omega - \int_{|z_k|=\varepsilon} f_k\omega$$

$$= -\int_{|z_k|=\varepsilon} \omega = -2\pi i \operatorname{Res}_{a_k}(\omega)$$

by the Residue Theorem in the complex plane. □

10.22. Corollary. *Any non-constant meromorphic function f on a compact Riemann surface X has, counting multiplicities, as many zeros as poles.*

PROOF. The differential form $\omega := df/f$ is holomorphic except at the zeros and poles of f. If $a \in X$ is a zero (resp. pole) of mth order of f, then $\text{Res}_a(\omega) = m$ (resp. $\text{Res}_a(\omega) = -m$). Hence the result follows from the Residue Theorem. □

Remark. We already proved this Corollary in (4.25) using coverings.

EXERCISES (§10)

10.1. Let X be a Riemann surface and ω be a holomorphic 1-form on X. Suppose φ is a primitive of ω on a neighborhood of a point $a \in X$ and (Y, p, f, b) is a maximal analytic continuation of φ. Prove

(a) $p\colon Y \to X$ is a covering map
(b) f is a primitive of $p^*\omega$
(c) The covering $p\colon Y \to X$ is Galois and $\text{Deck}(Y/X)$ is abelian.

10.2. Let $X = \mathbb{C}/\Gamma$ be a torus. Given any homomorphism

$$a\colon \pi_1(X) \to \mathbb{C}$$

show that there exists a closed 1-form $\omega \in \mathscr{E}^{(1)}(X)$ whose period homomorphism is equal to a.

10.3. Suppose X is a Riemann surface and $\omega \in \mathscr{M}^{(1)}(X)$ is a meromorphic 1-form on X which has residue zero at every pole. Show that there is a covering $p\colon Y \to X$ and a meromorphic function $F \in \mathscr{M}(X)$ such that $dF = p^*\omega$.

10.4. Let $\Gamma \subset \mathbb{C}$ be a lattice. Use the Residue Theorem to show that there is no meromorphic function $f \in \mathscr{M}(\mathbb{C}/\Gamma)$ having a single pole of order 1.

§11. Linear Differential Equations

In this section we consider linear differential equations of the form $w' = A(z)w$, where $A(z)$ is a given $n \times n$ matrix which depends holomorphically on z. A vector-valued function $w = w(z)$ is sought which satisfies the differential equation. Locally, for any given initial condition $w(z_0) = w_0$, there always exists a unique holomorphic solution. This solution may be continued along every curve in the domain of definition of A. However this continuation is, in general, no longer a single-valued function. It turns out that closer consideration of this multi-valued behavior gives a good insight into the structure of the solutions.

11.1. Notation. Denote by $M(n \times m, \mathbb{C})$ the vector space of all $n \times m$ matrices with coefficients in $\mathbb{C}$ and by $\text{GL}(n, \mathbb{C})$ the group of all invertible $n \times n$ matrices with complex coefficients. If X is a Riemann surface, then a mapping

$$A\colon X \to M(n \times m, \mathbb{C})$$

is called holomorphic if all the coefficients $a_{ij}: X \to \mathbb{C}$ are holomorphic. The set of all holomorphic mappings $A: X \to M(n \times m, \mathbb{C})$ will be denoted by $M(n \times m, \mathcal{O}(X))$. One can define $\mathrm{GL}(n, \mathcal{O}(X))$ similarly.

11.2. Theorem. *Suppose $A \in M(n \times n, \mathcal{O}(D))$ is a holomorphic $n \times n$-matrix on the disk*

$$D := \{z \in \mathbb{C}: |z| < R\}, \quad \text{where } 0 < R \leq \infty.$$

Then for every $w_0 \in \mathbb{C}^n$ there exists precisely one holomorphic function $w: D \to \mathbb{C}^n$ such that

(1) $w'(z) = A(z)w(z)$ *for every* $z \in D$,
(2) $w(0) = w_0$.

(Here we are identifying $\mathbb{C}^n$ with the space $M(n \times 1, \mathbb{C})$ of column vectors.)

PROOF

(a) The matrix A can be expanded in a Taylor series

$$A(z) = \sum_{\nu=0}^{\infty} A_\nu z^\nu, \qquad A_\nu = (a_{ij\nu}) \in M(n \times n, \mathbb{C})$$

in D. (This is to be understood as a system of n^2 equations for the entries of $A(z)$.) Now suppose that the solution w has the form

$$w(z) = \sum_{\nu=0}^{\infty} c_\nu z^\nu, \qquad c_\nu = (c_{i\nu}) \in \mathbb{C}^n.$$

If this series converges in D, then (1) is equivalent to

$$\sum_{k=1}^{\infty} k c_k z^{k-1} = \left(\sum_{\mu=0}^{\infty} A_\mu z^\mu\right)\left(\sum_{\nu=0}^{\infty} c_\nu z^\nu\right) = \sum_{k=0}^{\infty} \left(\sum_{\mu+\nu=k} A_\mu c_\nu\right) z^k,$$

i.e.,

(3) $(k+1)c_{k+1} = \sum_{\nu=0}^{k} A_{k-\nu} c_\nu$ for every $k \in \mathbb{N}$.

The initial condition (2) is equivalent to $c_0 = w_0$. Hence by (3) one can recursively compute all the coefficients c_k. This shows the uniqueness of the solution.

(b) In order to prove the existence of a solution we have to show that the series for w, having the coefficients computed in (3), does in fact converge in D. To do this we will use the majorant method of Cauchy.

For an arbitrary r with $0 < r < R$ the series

$$\sum_{\nu=0}^{\infty} |a_{ij\nu}| r^\nu$$

converges. Hence there exists $N \in \mathbb{N}$ such that

(4) $|a_{ij\nu}| \leq N r^{-\nu-1}$ for every $\nu \in \mathbb{N}$ and $1 \leq i, j \leq n$.

Define an $n \times n$ matrix $B = (b_{ij})$ which is holomorphic in $|z| < r$ by letting

(5) $b_{ij}(z) := \frac{N}{r}\left(1 - \frac{z}{r}\right)^{-1} = \frac{N}{r} \sum_{\nu=0}^{\infty} \frac{z^\nu}{r^\nu}$ for all i, j.

Let $w_0 = (w_{10}, \ldots, w_{n0})$ and $K := \max(|w_{10}|, \ldots, |w_{n0}|)$. Now we can find a solution of the differential equation

$$v'(z) = B(z)v(z)$$

in the disk $|z| < r$ which satisfies the initial condition $v(0) = (K, \ldots, K)$. By (a) the solution is unique and is given by

$$v(z) = (\psi(z), \ldots, \psi(z)),$$

where

$$\psi(z) = K\left(1 - \frac{z}{r}\right)^{-nN}$$

The function ψ is a solution because

$$\psi'(z) = \frac{KnN}{r}\left(1 - \frac{z}{r}\right)^{-nN-1} = n\frac{N}{r}\left(1 - \frac{z}{r}\right)^{-1} \psi(z).$$

On the other hand, the differential equation $v' = Bv$ can be solved using power series. If

$$B(z) = \sum_{\nu=0}^{\infty} B_\nu z^\nu, \qquad B_\nu = (b_{ij\nu}) \in M(n \times n, \mathbb{C})$$

and

$$v(z) = \sum_{\nu=0}^{\infty} \gamma_\nu z^\nu, \qquad \gamma_\nu = (\gamma_{i\nu}) \in \mathbb{C}^n$$

are the appropriate power series, then analogous to (a) one has

(6) $(k + 1)\gamma_{k+1} = \sum_{\nu=0}^{k} B_{k-\nu}\gamma_\nu$.

Then from (4) and (5), it follows that

$$|a_{ij\nu}| \leq b_{ij\nu} \quad \text{for every } i, j, \nu.$$

Since $|c_{i0}| = |w_{i0}| \leq K = \gamma_{i0}$ for $i = 1, \ldots, n$, comparison of (3) and (6) and induction on k implies

$$|c_{ij}| \leq \gamma_{ik} \quad \text{for every } k \in \mathbb{N} \text{ and } i = 1, \ldots, n.$$

Since the series $\sum_k \gamma_{ik} z^k = \psi(z)$ converges for $|z| < r$, one has that $\sum_k c_k z^k = w(z)$ converges as well.

Since $r < R$ is arbitrary, the series converges on all of $D = \{|z| < R\}$.

□

11.3. On a Riemann surface X a linear differential equation for an unknown holomorphic function $w\colon X \to \mathbb{C}^n$ may be written in the form

$$dw = Aw,$$

where $A = (a_{ij}) \in M(n \times n, \Omega(X))$ is a given $n \times n$ matrix of holomorphic 1-forms $a_{ij} \in \Omega(X)$. For any local chart (U, z) on X one has $A = F\,dz$, where $F \in M(n \times n, \mathcal{O}(U))$ and the differential equation becomes

$$\frac{dw}{dz} = F \cdot w.$$

But this is just the form of equation studied in (11.2).

11.4. Theorem. *Suppose X is a simply connected Riemann surface, $A \in M(n \times n, \Omega(X))$ and $x_0 \in X$. Then for every $c \in \mathbb{C}^n$ there exists a unique solution $w \in \mathcal{O}(X)^n$ of the differential equation*

$$dw = Aw$$

satisfying $w(x_0) = c$.

PROOF

(a) By Theorem (11.2) there exists a connected open neighborhood U_0 of x_0 and a solution $f \in \mathcal{O}(U_0)^n$ of the differential equation $df = Af$ with $f(x_0) = c$. Now we will show that f may be analytically continued along any curve $\alpha\colon [0, 1] \to X$ having initial point x_0. Then by Corollary (7.4) these continuations will piece together to form a global function $w \in \mathcal{O}(X)^n$ which, because of the Identity Theorem, satisfies the differential equation $dw = Aw$ on all of X.

(b) By Theorem (11.2) there exists a partition

$$0 = t_0 < t_1 < \cdots < t_k = 1$$

of the interval $[0, 1]$ and domains U_j, $j = 1, \ldots, k - 1$, with the following properties:

(i) $\alpha([t_j, t_{j+1}]) \subset U_j$ for $j = 0, \ldots, k - 1$, where U_0 is the neighborhood of x_0 mentioned above.

(ii) For any initial value $c_j \in \mathbb{C}^n$ there exists $f_j \in \mathcal{O}(U_j)^n$ with $df_j = Af_j$ and $f_j(\alpha(t_j)) = c_j$, $j = 1, \ldots, k - 1$.

Now, starting with the solution $f_0 := f$ on U_0 found in (a) and using induction on j one can construct solutions f_j on U_j satisfying

$$f_j(\alpha(t_j)) = f_{j-1}(\alpha(t_j)).$$

From the uniqueness proved in Theorem (11.2) and the Identity Theorem it follows that f_{j-1} and f_j agree on the connected component of $U_{j-1} \cap U_j$ containing $\alpha(t_j)$. This proves that f can be analytically continued along α.

□

11.5. Corollary. *Suppose X is a Riemann surface, $p\colon \tilde{X} \to X$ is its universal covering, $x_0 \in X$ is a point and $y_0 \in \tilde{X}$ is a point such that $p(y_0) = x_0$. Suppose $A \in M(n \times n, \Omega(X))$ and $c \in \mathbb{C}^n$. Then there exists a unique solution $w \in \mathcal{O}(\tilde{X})^n$ on the universal covering $\tilde{X}$ of X of the differential equation*

$$dw = (p^*A)w$$

satisfying $w(y_0) = c$.

11.6. Factors of Automorphy. Suppose X is a Riemann surface and $A \in M(n \times n, \Omega(X))$. On the universal covering $p\colon \tilde{X} \to X$ let L_A be the set of all solutions $w \in \mathcal{O}(\tilde{X})^n$ of the differential equation

$$dw = (p^*A)w.$$

Just as in the theory of real linear differential equations one can show that L_A is an n-dimensional vector space over $\mathbb{C}$ and that $w_1, \ldots, w_n \in L_A$ are linearly independent precisely if for an arbitrary point $a \in \tilde{X}$ the vectors $w_1(a), \ldots, w_n(a) \in \mathbb{C}^n$ are linearly independent. Therefore a basis $w_1, \ldots, w_n$ of L_A defines an invertible matrix

$$\Phi := (w_1, \ldots, w_n) \in \mathrm{GL}(n, \mathcal{O}(\tilde{X}))$$

such that $d\Phi = (p^*A)\Phi$. Such a matrix is called a *fundamental system of solutions* of the differential equation $dw = Aw$. Let $G := \mathrm{Deck}(\tilde{X}/X) \cong \pi_1(X)$ be the group of covering transformations of $p\colon \tilde{X} \to X$. Analogous to (10.12), for $\sigma \in G$ we can set $\sigma\Phi := \Phi \circ \sigma^{-1}$. Then $\sigma\Phi$ as well as Φ satisfies the differential equation $d(\sigma\Phi) = (p^*A)(\sigma\Phi)$ and thus is another fundamental system of solutions. Hence there exists a constant matrix $T_\sigma \in \mathrm{GL}(n, \mathbb{C})$ such that

$$\sigma\Phi = \Phi T_\sigma.$$

If τ is another covering transformation, then

$$\Phi T_{\tau\sigma} = \tau\sigma\Phi = \tau(\Phi T_\sigma) = (\tau\Phi)T_\sigma = \Phi T_\tau T_\sigma,$$

i.e., $T_{\tau\sigma} = T_\tau T_\sigma$. Hence the correspondence $\sigma \mapsto T_\sigma$ defines a group homomorphism

$$\pi_1(X) \cong \mathrm{Deck}(\tilde{X}/X) \to \mathrm{GL}(n, \mathbb{C}).$$

The matrices T_σ are called the *factors of automorphy* of Φ. Now conversely suppose a homomorphism

$$T\colon \mathrm{Deck}(\tilde{X}/X) \to \mathrm{GL}(n, \mathbb{C}), \qquad \sigma \mapsto T_\sigma$$

and a holomorphic mapping

$$\Phi\colon \tilde{X} \to \mathrm{GL}(n, \mathbb{C})$$

are given such that

$$\sigma\Phi = \Phi T_\sigma \quad \text{for every } \sigma \in \mathrm{Deck}(\tilde{X}/X).$$

The matrix $(d\Phi)\Phi^{-1} \in M(n \times n, \Omega(\tilde{X}))$ is then invariant under covering transformations, for

$$\sigma(d\Phi \cdot \Phi^{-1}) = (d\Phi \cdot T_\sigma)(\Phi T_\sigma)^{-1} = d\Phi \cdot \Phi^{-1}.$$

Hence there is a matrix $A \in M(n \times n, \Omega(X))$ such that $p^*A = d\Phi \cdot \Phi^{-1}$ and Φ is a fundamental system of solutions of the differential equation $dw = Aw$.

11.7. Now consider the special case

$$X := \{z \in \mathbb{C} : 0 < |z| < R\}, \quad \text{where } 0 < R \leq \infty.$$

Then by (5.7.b) the group of covering transformations of the universal covering $p: \tilde{X} \to X$ is $\mathbb{Z}$. Let σ be one of the generators of $\text{Deck}(\tilde{X}/X)$. On $\tilde{X}$ the logarithm of the coordinate function on X exists, i.e., there exists a holomorphic function

$$\log: \tilde{X} \to \mathbb{C}$$

such that $\exp \circ \log = p$. Now we may assume that σ is chosen so that

$$\sigma \log = \log + 2\pi i.$$

Suppose $A \in M(n \times n, \Omega(X))$ and $\Phi \in \text{GL}(n, \mathcal{O}(\tilde{X}))$ is a fundamental system of solutions of the differential equation $dw = Aw$. Since $\text{Deck}(\tilde{X}/X) = \{\sigma^n : n \in \mathbb{Z}\}$, the behavior of Φ as an automorphic function is determined by the matrix $T \in \text{GL}(n, \mathbb{C})$ which satisfies

$$\sigma\Phi = \Phi T.$$

If $\Psi \in \text{GL}(n, \mathcal{O}(\tilde{X}))$ is another fundamental system of solutions of $dw = Aw$, then there exists a matrix $S \in \text{GL}(n, \mathbb{C})$ with $\Psi = \Phi S$. Thus

$$\sigma\Psi = \Psi S^{-1} T S = \Psi \tilde{T},$$

where $\tilde{T} := S^{-1}TS$. Hence by a suitable choice of the fundamental system Ψ one can in fact arrange it so that the factor of automorphy T has Jordan normal form.

11.8. The Exponential of Matrices. For a matrix $A \in M(n \times n, \mathbb{C})$ define the exponential of A by

$$\exp A = \sum_{k=0}^{\infty} \frac{1}{k!} A^k.$$

Then each entry of the matrix converges absolutely. If $A, B \in M(n \times n, \mathbb{C})$ are matrices which commute with each other, i.e., $AB = BA$, then

$$\exp(A + B) = \exp(A)\exp(B).$$

One proves this in the same way that one proves the comparable result for the exponential of complex numbers, namely by multiplying together the series for $\exp(A)$ and $\exp(B)$ to get the series for $\exp(A + B)$. In particular if $B = -A$, then $\exp(A)\exp(-A) = I$, i.e., $\exp(A) \in \mathrm{GL}(n, \mathbb{C})$.

If $S \in \mathrm{GL}(n, \mathbb{C})$ and $A \in M(n \times n, \mathbb{C})$, then

$$\exp(S^{-1}AS) = S^{-1}(\exp A)S. \tag{*}$$

Now for every matrix $B \in \mathrm{GL}(n, \mathbb{C})$ there exists a matrix $A \in M(n \times n, \mathbb{C})$ such that

$$\exp A = B.$$

Because of (*), it suffices to prove this in the case that B has Jordan normal form. If B is a diagonal matrix with entries $\lambda_1, \ldots, \lambda_n \in \mathbb{C}^*$, then one can simply choose A to be the diagonal matrix with entries $\mu_1, \ldots, \mu_n$, where $\exp(\mu_j) = \lambda_j$. A general matrix in Jordan normal form is made up of Jordan blocks of the form

$$B_1 := \begin{pmatrix} \lambda & 1 & & & & 0 \\ & \lambda & 1 & & & \\ & & \ddots & \ddots & & \\ & & & \ddots & \ddots & \\ & & & & \lambda & 1 \\ 0 & & & & & \lambda \end{pmatrix} = \lambda(E + (1/\lambda)N),$$

$$\text{where } N = \begin{pmatrix} 0 & 1 & & & & 0 \\ & 0 & 1 & & & \\ & & \ddots & \ddots & & \\ & & & \ddots & \ddots & \\ & & & & 0 & 1 \\ 0 & & & & & 0 \end{pmatrix}.$$

A matrix A_1 such that $\exp(A_1) = B_1$ is given by

$$A_1 = \mu E + M,$$

where $\exp(\mu) = \lambda$ and

$$M = \log\left(E + \frac{1}{\lambda}N\right) := \sum_{k=1}^{\infty} (-1)^{k-1} \frac{1}{k\lambda^k} N^k.$$

The series contains only finitely many non-zero terms since N is nilpotent.

11.9. Suppose A is an $n \times n$ matrix whose coefficients are holomorphic functions on a Riemann surface X. Then the coefficients of the matrix $\exp A$ are also holomorphic on X, since the series converges uniformly on compact subsets of X.

If $A \in M(n \times n, \mathcal{O}(X))$ is a matrix such that

$$A \cdot dA = dA \cdot A,$$

then

$$d(\exp A) = dA \cdot \exp A = \exp A \cdot dA.$$

This follows immediately when one differentiates the exponential series term by term.

11.10. Theorem. *Suppose $T \in \mathrm{GL}(n, \mathbb{C})$ is a given matrix and $B \in M(n \times n, \mathbb{C})$ is a matrix such that*

$$\exp(2\pi i B) = T.$$

Now consider the differential equation

$$w' = \frac{1}{z} Bw$$

on $X := \{z \in \mathbb{C} : 0 < |z| < R\}$. Then

$$\Phi_0 := \exp(B \log)$$

is a fundamental system of solutions of $w' = \frac{1}{z} Bw$ on the universal covering $p: \tilde{X} \to X$ which has T as its factors of automorphy, i.e.,

$$\sigma\Phi_0 = \Phi_0 T.$$

Here σ is defined the same as in (11.7).

Proof. From the remark in (11.9) it follows that $\Phi_0' = (1/z)B\Phi_0$. Moreover,

$$\begin{aligned} \sigma\Phi_0 &= \sigma \exp(B \log) = \exp(B\sigma \log) \\ &= \exp(B(\log + 2\pi i)) = \exp(B \log)\exp(2\pi i B) = \Phi_0 T. \end{aligned}$$ □

Remark. The theorem shows that given any punctured disk X and prescribed factor of automorphy one can always find a differential equation whose solution has this as its factor of automorphy. We will look at the same problem on an arbitrary non-compact Riemann surface X in §31.

11.11. Theorem. *Suppose the notation is the same as in Theorem* (11.10) *and $A \in M(n \times n, \mathcal{O}(X))$. Then the differential equation*

$$w' = Aw$$

has a fundamental system of solutions $\Phi \in \mathrm{GL}(n, \mathcal{O}(\tilde{X}))$ of the form

$$\Phi = \Psi\Phi_0,$$

where $\Phi_0 = \exp(B \log)$ for a constant matrix $B \in M(n \times n, \mathbb{C})$ and Ψ is invariant under covering transformations, i.e., Ψ defines an element in $\mathrm{GL}(n, \mathcal{O}(X))$.

PROOF. Suppose $\Phi \in GL(n, \mathcal{O}(\tilde{X}))$ is a fundamental system of solutions of $w' = Aw$ and

$$\sigma\Phi = \Phi T, \quad \text{where } T \in GL(n, \mathbb{C}).$$

By (11.10) one can find $\Phi_0 = \exp(B \log) \in GL(n, \mathcal{O}(\tilde{X}))$ such that

$$\sigma\Phi_0 = \Phi_0 T.$$

Then for $\Psi := \Phi\Phi_0^{-1}$ one has $\sigma\Psi = \Psi$. □

11.12. By Theorem (11.11) any fundamental system of solutions of a differential equation $w' = Aw$ on the punctured disk $X = \{0 < |z| < R\}$ may be represented as the product of a very special kind of multi-valued function $\Phi_0 = \exp(B \log)$ and a single-valued (matrix-valued) function Ψ. Now this function Ψ can be expanded in a Laurent series on X. The origin is called a *regular singular point* or a singularity of *Fuchsian type* of the differential equation $w' = Aw$ if Ψ has at most a pole at the origin, i.e., if only finitely many terms with negative exponent occur in the Laurent series.

11.13. Theorem. *Let* $X := \{z \in \mathbb{C}: 0 < |z| < R\}$. *If the matrix* $A \in M(n \times n, \mathcal{O}(X))$ *has at most a pole of first order at the origin, then the origin is a regular singular point of the differential equation* $w' = Aw$.

PROOF. The proof requires two lemmas.

(1) *Suppose* $K \geq 0$ *is a constant and* $F\colon]0, r_0] \to \mathbb{R}_+^*$ *is a continuously differentiable positive-valued function, whose derivative satisfies the inequality*

$$|F'(r)| \leq \frac{K}{r} F(r) \quad \textit{for every } r \in]0, r_0].$$

Then

$$F(r) \leq F(r_0)\left(\frac{r}{r_0}\right)^{-K} \quad \textit{for every } r \in]0, r_0].$$

Proof of (1). From the assumption one gets

$$\frac{d}{dr} \log F(r) = \frac{F'(r)}{F(r)} \geq -\frac{K}{r}.$$

By integrating over the interval $[r, r_0]$, one obtains

$$\log \frac{F(r_0)}{F(r)} \geq -K \log \frac{r_0}{r}.$$

Thus

$$F(r) \leq F(r_0)\left(\frac{r}{r_0}\right)^{-K}.$$

(2) *Suppose f is a holomorphic function on X. Then*

$$\left|\frac{\partial}{\partial r}|f|^2\right| \le 2|f||f'|.$$

Here $(\partial/\partial r)$ denotes the radial derivative with respect to polar coordinates $z = re^{i\theta}$.

Proof of (2). Since f is complex differentiable,

$$f' = \frac{df}{dz} = \frac{1}{e^{i\theta}}\frac{\partial f}{\partial r} \quad \text{and thus} \quad \left|\frac{\partial f}{\partial r}\right| = |f'|.$$

Moreover

$$\frac{\partial \bar{f}}{\partial r} = \overline{\left(\frac{\partial f}{\partial r}\right)} \quad \text{and thus} \quad \left|\frac{\partial \bar{f}}{\partial r}\right| = |f'|.$$

This implies

$$\left|\frac{\partial}{\partial r}|f|^2\right| = \left|\bar{f}\frac{\partial f}{\partial r} + f\frac{\partial \bar{f}}{\partial r}\right| \le 2|f||f'|.$$

Now the actual proof of Theorem 11.13! By (11.11) there exists a fundamental system of solutions of $w' = Aw$ which may be written $\Phi = \Psi\Phi_0$, where $\Psi \in \mathrm{GL}(n, \mathcal{O}(X))$ and $\Phi_0 = \exp(B \log)$ with $B \in M(n \times n, \mathbb{C})$. Then

$$\Phi' = A\Phi = \Psi'\Phi_0 + \Psi\Phi_0' = \Psi'\Phi_0 + \Psi \cdot \frac{1}{z}B\Phi_0.$$

Multiplication by Φ_0^{-1} on the right yields $A\Psi = \Psi' + (1/z)\Psi B$, i.e.,

$$\Psi' = A\Psi - \frac{1}{z}\Psi B. \tag{*}$$

Since the matrix A has at most a pole of first order at 0, there exists a matrix A_1 which is holomorphic on the whole disk $|z| < R$ such that $A = (1/z)A_1$. Define the norm of a matrix $C = (c_{jk})$ by

$$\|C\| := \left(\sum_{j,k}|c_{jk}|^2\right)^{1/2}.$$

Then from (*) it follows that given any $r_0 \in \,]0, R[$ there exists a constant $M \ge 0$ such that

$$\|\Psi'(z)\| \le \frac{M}{r}\|\Psi(z)\| \quad \text{for } 0 < |z| = r \le r_0.$$

Suppose ψ_{jk} are the components of the matrix Ψ. For fixed $\theta \in \mathbb{R}$ let

$$F(r) := \|\Psi(re^{i\theta})\|^2 = \sum_{j,k}|\psi_{jk}(re^{i\theta})|^2.$$

By means of (2) one gets

$$\begin{aligned}|F'(r)| &\le 2\sum_{j,k} |\psi_{jk}(re^{i\theta})|\cdot|\psi'_{jk}(re^{i\theta})|\\ &\le 2\|\Psi(re^{i\theta})\|\cdot\|\Psi'(re^{i\theta})\| \le \frac{2M}{r}\|\Psi(re^{i\theta})\|^2,\end{aligned}$$

i.e., $|F'(r)| \le (2M/r)F(r)$. Now from (1) it follows that

$$F(r) \le F(r_0)\left(\frac{r}{r_0}\right)^{-2M},$$

i.e.,

$$\|\Psi(re^{i\theta})\| \le \|\Psi(r_0 e^{i\theta})\|\left(\frac{r}{r_0}\right)^{-M}.$$

Hence Ψ can have a pole of order at most M at the origin. □

We are now going to use Theorem (11.13) to determine the form of the solutions of certain linear second order differential equations which arise quite often in practice.

11.14. Theorem. *Suppose $D = \{z \in \mathbb{C}: |z| < R\}$ and $X := D\setminus\{0\}$. Suppose a, $b \in \mathcal{O}(D)$. Then the differential equation*

$$w'' + \frac{a(z)}{z}w' + \frac{b(z)}{z^2}w = 0 \tag{1}$$

has, on the universal covering $p: \tilde{X} \to X$, a fundamental system (φ_1, φ_2) of solutions which has one of the following forms: Either

$$\begin{cases}\varphi_1(z) = z^{\rho_1}\psi_1(z),\\ \varphi_2(z) = z^{\rho_2}\psi_2(z),\end{cases}$$

or

$$\begin{cases}\varphi_1(z) = z^{\rho}\psi_1(z),\\ \varphi_2(z) = z^{\rho}(\psi_1(z)\log z + \psi_2(z)).\end{cases}$$

Here ρ, ρ_1, ρ_2 denote complex numbers and $\psi_1, \psi_2 \in \mathcal{O}(D)$.

Remark. $\log z$ and $z^\rho = e^{\rho \log z}$ are single-valued holomorphic functions on $\tilde{X}$. Holomorphic functions on D will be interpreted as functions on $\tilde{X}$ which are invariant under covering transformations.

PROOF. We reduce the differential equation to a system of two differential equations of first order by setting

$$w_1 := w, \qquad w_2 := zw'.$$

Since $w_2' = zw'' + w'$, equation (1) is equivalent to the system

$$\begin{cases} w_1' = \dfrac{1}{z} w_2, \\ w_2' = \dfrac{-b(z)}{z} w_1 + \dfrac{1 - a(z)}{z} w_2. \end{cases} \tag{2}$$

Theorem (11.13) may be applied to the system (2). Thus it has a fundamental system of solutions of the form

$$\Phi(z) = z^n \Psi(z) \exp(B \log z),$$

where $n \in \mathbb{Z}$, $\Psi \in GL(2, \mathcal{O}(D))$, $B \in M(2 \times 2, \mathbb{C})$. By a change of basis one may even assume that B has Jordan normal form.

Case 1: B is a diagonal matrix, i.e.,

$$B = \begin{pmatrix} \alpha_1 & 0 \\ 0 & \alpha_2 \end{pmatrix}.$$

Then

$$\exp(B \log z) = \begin{pmatrix} z^{\alpha_1} & 0 \\ 0 & z^{\alpha_2} \end{pmatrix}$$

$$\Phi(z) = \begin{pmatrix} \varphi_1(z) & \varphi_2(z) \\ z\varphi_1'(z) & z\varphi_2'(z) \end{pmatrix} = z^n \begin{pmatrix} \psi_1(z) & \psi_2(z) \\ \psi_3(z) & \psi_4(z) \end{pmatrix} \begin{pmatrix} z^{\alpha_1} & 0 \\ 0 & z^{\alpha_2} \end{pmatrix}.$$

Thus $\varphi_1(z) = z^{n+\alpha_1}\psi_1(z)$, $\varphi_2(z) = z^{n+\alpha_2}\psi_2(z)$.

Case 2: B is a Jordan block, i.e.,

$$B = \begin{pmatrix} \alpha & 1 \\ 0 & \alpha \end{pmatrix}.$$

Then

$$\exp(B \log z) = z^{\alpha} \begin{pmatrix} 1 & \log z \\ 0 & 1 \end{pmatrix}$$

and this yields

$$\varphi_1(z) = z^{n+\alpha}\psi_1(z), \qquad \varphi_2(z) = z^{n+\alpha}(\psi_1(z)\log z + \psi_2(z)). \qquad \square$$

11.15. Bessel's Equation. As an example consider the differential equation, which is known as Bessel's equation, on $\mathbb{C}^*$

$$w'' + \frac{1}{z} w' + \left(1 - \frac{p^2}{z^2}\right) w = 0. \tag{1}$$

Here p is an arbitrary complex number. By Theorem (11.14) the equation has at least one solution of the form

$$\varphi(z) = z^{\rho} \sum_{n=0}^{\infty} c_n z^n, \qquad \rho \in \mathbb{C}, \qquad c_0 \neq 0. \tag{2}$$

Differentiation of the series gives

$$\varphi'(z) = z^{\rho} \sum_{n=0}^{\infty} (\rho + n) c_n z^{n-1},$$

$$\varphi''(z) = z^{\rho} \sum_{n=0}^{\infty} (\rho + n)(\rho + n - 1) c_n z^{n-2}.$$

Substituting this into the differential equation and collecting together the resulting coefficient of $z^{\rho+n-2}$, one sees that the differential equation is satisfied precisely if

(i) $(\rho^2 - p^2)c_0 = 0,$
(ii) $((\rho + 1)^2 - p^2)c_1 = 0,$
(iii) $((\rho + n)^2 - p^2)c_n + c_{n-2} = 0$ for every $n \geq 2$.

Since $c_0 \neq 0$, (i) implies $\rho = \pm p$. For $n = 2k$ even, (iii) becomes

(iii)′ $2^2 k(\rho + k)c_{2k} + c_{2k-2} = 0.$

We now distinguish two cases.

Case 1. $p \in \mathbb{C} \backslash \mathbb{Z}$. A solution to the system of equations (i)–(iii) is given by

$$c_{2k+1} = 0$$

$$c_{2k} = (-1)^k (\tfrac{1}{2})^{2k} \frac{c_0}{k!\,(\rho + 1) \cdots (\rho + k)}$$

for every $k \in \mathbb{N}$ and arbitrary c_0. Since

$$\Gamma(\rho + k + 1) = (\rho + k)(\rho + k - 1) \cdots (\rho + 1)\Gamma(\rho + 1),$$

for the special choice $c_0 = 1/\Gamma(\rho + 1)$ this gives

$$c_{2k} = (-1)^k (\tfrac{1}{2})^{2k} \frac{1}{\Gamma(k + 1)\Gamma(\rho + k + 1)}.$$

The *Bessel function* of order ρ is defined to be

$$J_\rho(z) := \left(\frac{z}{2}\right)^{\rho} \sum_{k=0}^{\infty} \frac{(-1)^k}{\Gamma(k + 1)\Gamma(\rho + k + 1)} \left(\frac{z}{2}\right)^{2k}.$$

What we have done shows that J_p and J_{-p} are two linearly independent solutions of the differential equation (1).

Case 2: $p \in \mathbb{Z}$. We may assume $p \geq 0$. In this case equation (2) with $\rho = p$ necessarily leads to the solution $\varphi(z) = \text{const} \cdot J_p(z)$. If $p \neq 0$ and $\rho = -p$, then using (iii)' and the fact that $c_0 \neq 0$ first gives $c_{2k} \neq 0$ for all $k < p$ and then for $k = p$ the contradiction $0 \cdot c_{2p} + c_{2p-2} = 0$. Thus equation (2) for p an integer can only give us one linearly independent solution. By Theorem (11.14) equation (1) has a second solution which is linearly independent of J_p and has the form

$$\psi(z) = J_p(z)\log z + g(z),$$

where g is a function holomorphic on $\mathbb{C}^*$ having at most a pole at 0. Differentiation gives

$$\psi'(z) = J_p'(z)\log z + \frac{1}{z} J_p(z) + g'(z),$$

$$\psi''(z) = J_p''(z)\log z + \frac{2}{z} J_p'(z) - \frac{1}{z^2} J_p(z) + g''(z).$$

Substituting this into the differential equation and using the fact that $w = J_p(z)$ is already a solution of (1), one gets that ψ is a solution of (1) precisely if

$$g''(z) + \frac{1}{z} g'(z) + \left(1 - \frac{p^2}{z^2}\right) g(z) = -\frac{2}{z} J_p'(z).$$

This equation can be solved by a power series of the form

$$g(z) = z^{-p} \sum_{n=0}^{\infty} a_n z^n.$$

This solution is then uniquely determined up to the addition of a constant multiple of J_p. This solution, when properly normalized, is the so-called *Neumann Function* N_p (or *Bessel function of the second kind*) and together with J_p forms a fundamental system of solutions of *Bessel's* equation (1), cf. [52], [58].

Exercises (§11)

11.1. Show that for every $A \in M(n \times n, \mathbb{C})$

$$\det(\exp A) = \exp(\operatorname{trace} A).$$

11.2. Calculate $\exp(A_j)$ for

$$A_1 = \begin{pmatrix} 2 & 1 \\ -1 & 2 \end{pmatrix}, \quad A_2 = \begin{pmatrix} 2 & 2 & 0 \\ 0 & 2 & 1 \\ 0 & 0 & 1 \end{pmatrix}, \quad A_3 = \begin{pmatrix} 0 & 1 & 1 \\ 1 & 1 & 0 \\ 1 & 0 & 0 \end{pmatrix}.$$

11.3. Let $X = \{z \in \mathbb{C} : 0 < |z| < R\}$, $p: \tilde{X} \to X$ be its universal covering and σ be a generator of $\operatorname{Deck}(\tilde{X}/X)$. Find a holomorphic map

$$\Phi: \tilde{X} \to \mathrm{GL}(3, \mathbb{C})$$

such that

$$\sigma\Phi = \Phi \cdot \begin{pmatrix} i & 1 & 0 \\ 0 & i & 1 \\ 0 & 0 & i \end{pmatrix}.$$

CHAPTER 2
Compact Riemann Surfaces

Amongst all Riemann surfaces the compact ones are especially important. They arise, for example, as those covering surfaces of the Riemann sphere defined by algebraic functions. As well their function theory is subject to interesting restrictions, like the Riemann–Roch Theorem and Abel's Theorem. More recently the theory of Riemann surfaces has been generalized to an extensive theory for complex manifolds of higher dimension. And the methods developed for this are very well suited to proving the classical theorems. One such method is sheaf cohomology and we give a short introduction to this in the present chapter.

To a large extent Chapter 2 is independent of Chapter 1. Essentially only §1 (the definition of Riemann surfaces), the first half of §6 (the definition of sheaves) and §§9 and 10 (differential forms) will be needed.

§12. Cohomology Groups

The goal of this section is to define the cohomology groups $H^1(X, \mathscr{F})$, where $\mathscr{F}$ is a sheaf of abelian groups on a topological space X. In our further study of Riemann surfaces, these cohomology groups play a very decided role.

12.1. Cochains, Cocycles, Coboundaries. Suppose X is a topological space and $\mathscr{F}$ is a sheaf of abelian groups on X. Also suppose that an open covering of X is given, i.e., a family $\mathfrak{U} = (U_i)_{i \in I}$ of open subsets of X such that $\bigcup_{i \in I} U_i = X$. For $q = 0, 1, 2, \ldots$ define the *qth cochain group* of $\mathscr{F}$, with respect to $\mathfrak{U}$, as

$$C^q(\mathfrak{U}, \mathscr{F}) := \prod_{(i_0, \ldots, i_q) \in I^{q+1}} \mathscr{F}(U_{i_0} \cap \cdots \cap U_{i_q}).$$

The elements of $C^q(\mathfrak{U}, \mathcal{F})$ are called *q*-cochains. Thus a *q*-cochain is a family

$$(f_{i_0\ldots i_q})_{i_0,\ldots,i_q \in I^{q+1}} \quad \text{such that } f_{i_0\ldots i_q} \in \mathcal{F}(U_{i_0} \cap \cdots \cap U_{i_q})$$

for all $(i_0, \ldots, i_q) \in I^{q+1}$. The addition of two cochains is defined component-wise. Now define *coboundary operators*

$$\delta\colon C^0(\mathfrak{U}, \mathcal{F}) \to C^1(\mathfrak{U}, \mathcal{F})$$

$$\delta\colon C^1(\mathfrak{U}, \mathcal{F}) \to C^2(\mathfrak{U}, \mathcal{F})$$

as follows:

(i) For $(f_i)_{i \in I} \in C^0(\mathfrak{U}, \mathcal{F})$ let $\delta((f_i)_{i \in I}) = (g_{ij})_{i,j \in I}$ where

$$g_{ij} := f_j - f_i \in \mathcal{F}(U_i \cap U_j).$$

Here it is understood that one restricts f_i and f_j to the intersection $U_i \cap U_j$ and then takes their difference.

(ii) For $(f_{ij})_{i,j \in I} \in C^1(\mathfrak{U}, \mathcal{F})$ let $\delta((f_{ij})) = (g_{ijk})$ where

$$g_{ijk} := f_{jk} - f_{ik} + f_{ij} \in \mathcal{F}(U_i \cap U_j \cap U_k).$$

Again the terms on the right are restricted to their common domain of definition $U_i \cap U_j \cap U_k$.

These coboundary operators are group homomorphisms. Let

$$Z^1(\mathfrak{U}, \mathcal{F}) := \operatorname{Ker}(C^1(\mathfrak{U}, \mathcal{F}) \xrightarrow{\delta} C^2(\mathfrak{U}, \mathcal{F})),$$

$$B^1(\mathfrak{U}, \mathcal{F}) := \operatorname{Im}(C^0(\mathfrak{U}, \mathcal{F}) \xrightarrow{\delta} C^1(\mathfrak{U}, \mathcal{F})).$$

The elements of $Z^1(\mathfrak{U}, \mathcal{F})$ are called *1-cocycles*. Thus by definition a 1-cochain $(f_{ij}) \in C^1(\mathfrak{U}, \mathcal{F})$ is a cocycle precisely if

$$f_{ik} = f_{ij} + f_{jk} \quad \text{on } U_i \cap U_j \cap U_k \tag{*}$$

for all $i, j, k \in I$. One calls (*) the *cocycle relation* and it implies

$$f_{ii} = 0, \qquad f_{ij} = -f_{ji}.$$

One obtains these from (*) by letting $i = j = k$ for the first and $i = k$ for the second.

The elements of $B^1(\mathfrak{U}, \mathcal{F})$ are called *1-coboundaries*. In particular every coboundary is a cocycle. A coboundary is also called a *splitting cocycle*. Thus a 1-cocycle $(f_{ij}) \in Z^1(\mathfrak{U}, \mathcal{F})$ splits if and only if there is a 0-cochain $(g_i) \in C^0(\mathfrak{U}, \mathcal{F})$ such that

$$f_{ij} = g_i - g_j \quad \text{on } U_i \cap U_j \quad \text{for every } i, j \in I.$$

12.2. Definition. The quotient group

$$H^1(\mathfrak{U}, \mathcal{F}) := Z^1(\mathfrak{U}, \mathcal{F})/B^1(\mathfrak{U}, \mathcal{F})$$

is called the *1st cohomology group* with coefficients in $\mathscr{F}$ with respect to the covering $\mathfrak{U}$. Its elements are called cohomology classes and two cocycles which belong to the same cohomology class are called *cohomologous*. Thus two cocycles are cohomologous precisely if their difference is a coboundary.

The groups $H^1(\mathfrak{U}, \mathscr{F})$ depend on the covering $\mathfrak{U}$. In order to have cohomology groups which depend only on X and $\mathscr{F}$, one has to use finer and finer coverings and then take a limit. We shall now make this idea precise.

An open covering $\mathfrak{V} = (V_k)_{k \in K}$ is called *finer* than the covering $\mathfrak{U} = (U_i)_{i \in I}$, denoted $\mathfrak{V} < \mathfrak{U}$, if every V_k is contained in at least one U_i. Thus there is a mapping $\tau: K \to I$ such that

$$V_k \subset U_{\tau k} \quad \text{for every } k \in K.$$

By means of the mapping τ we can define a mapping

$$t_{\mathfrak{V}}^{\mathfrak{U}}: Z^1(\mathfrak{U}, \mathscr{F}) \to Z^1(\mathfrak{V}, \mathscr{F})$$

in the following way. For $(f_{ij}) \in Z^1(\mathfrak{U}, \mathscr{F})$ let $t_{\mathfrak{V}}^{\mathfrak{U}}((f_{ij})) = (g_{kl})$ where

$$g_{kl} := f_{\tau k, \tau l} | V_k \cap V_l \quad \text{for every } k, l \in K.$$

This mapping takes coboundaries into coboundaries and thus induces a homomorphism of the cohomology groups $H^1(\mathfrak{U}, \mathscr{F}) \to H^1(\mathfrak{V}, \mathscr{F})$, which we also denote by $t_{\mathfrak{V}}^{\mathfrak{U}}$.

12.3. Lemma. *The mapping*

$$t_{\mathfrak{V}}^{\mathfrak{U}}: H^1(\mathfrak{U}, \mathscr{F}) \to H^1(\mathfrak{V}, \mathscr{F})$$

is independent of the choice of the refining mapping $\tau: K \to I$.

PROOF. Suppose $\tilde{\tau}: K \to I$ is another mapping such that $V_k \subset U_{\tilde{\tau} k}$ for every $k \in K$. Suppose $(f_{ij}) \in Z^1(\mathfrak{U}, \mathscr{F})$ and let

$$g_{kl} := f_{\tau k, \tau l} | V_k \cap V_l \quad \text{and} \quad \tilde{g}_{kl} := f_{\tilde{\tau} k, \tilde{\tau} l} | V_k \cap V_l.$$

We have to show that the cocycles (g_{kl}) and $(\tilde{g}_{kl})$ are cohomologous. Since $V_k \subset U_{\tau k} \cap U_{\tilde{\tau} k}$, one can define

$$h_k := f_{\tau k, \tilde{\tau} k} | V_k \in \mathscr{F}(V_k).$$

On $V_k \cap V_l$ one has

$$\begin{aligned} g_{kl} - \tilde{g}_{kl} &= f_{\tau k, \tau l} - f_{\tilde{\tau} k, \tilde{\tau} l} \\ &= f_{\tau k, \tau l} + f_{\tau l, \tilde{\tau} k} - f_{\tau l, \tilde{\tau} k} - f_{\tilde{\tau} k, \tilde{\tau} l} \\ &= f_{\tau k, \tilde{\tau} k} - f_{\tau l, \tilde{\tau} l} = h_k - h_l. \end{aligned}$$

Thus the cocycle $(g_{kl}) - (\tilde{g}_{kl})$ is a coboundary. □

12.4. Lemma. *The mapping*

$$t_{\mathfrak{B}}^{\mathfrak{U}}: H^1(\mathfrak{U}, \mathscr{F}) \to H^1(\mathfrak{B}, \mathscr{F})$$

is injective.

PROOF. Suppose $(f_{ij}) \in Z^1(\mathfrak{U}, \mathscr{F})$ is a cocycle whose image in $Z^1(\mathfrak{B}, \mathscr{F})$ splits. One has to show that (f_{ij}) itself splits.

Now suppose $f_{\tau k, \tau l} = g_k - g_l$ on $V_k \cap V_l$, where $g_k \in \mathscr{F}(V_k)$. Then on $U_i \cap V_k \cap V_l$ one has

$$g_k - g_l = f_{\tau k, \tau l} = f_{\tau k, i} + f_{i, \tau l} = f_{i, \tau l} - f_{i, \tau k},$$

and thus $f_{i, \tau k} + g_k = f_{i, \tau l} + g_l$. Applying sheaf axiom II (see Definition (6.3)) to the family of open sets $(U_i \cap V_k)_{k \in K}$, one obtains $h_i \in \mathscr{F}(U_i)$ such that

$$h_i = f_{i, \tau k} + g_k \quad \text{on } U_i \cap V_k.$$

With the elements h_i found in this way, on $U_i \cap U_j \cap V_k$ one has

$$f_{ij} = f_{i, \tau k} + f_{\tau k, j} = f_{i, \tau k} + g_k - f_{j, \tau k} - g_k = h_i - h_j.$$

Since k is arbitrary, it follows from sheaf axiom I that this equation is valid over $U_i \cap U_j$, i.e., the cocycle (f_{ij}) splits with respect to the covering $\mathfrak{U}$. □

12.5. The definition of $H^1(X, \mathscr{F})$. If one has three open coverings such that $\mathfrak{W} < \mathfrak{B} < \mathfrak{U}$, then

$$t_{\mathfrak{W}}^{\mathfrak{B}} \circ t_{\mathfrak{B}}^{\mathfrak{U}} = t_{\mathfrak{W}}^{\mathfrak{U}}.$$

Thus one can define the following equivalence relation $\sim$ on the disjoint union of the $H^1(\mathfrak{U}, \mathscr{F})$, where $\mathfrak{U}$ runs through all open coverings of X. Two cohomology classes $\xi \in H^1(\mathfrak{U}, \mathscr{F})$ and $\eta \in H^1(\mathfrak{U}', \mathscr{F})$ are defined to be equivalent, denoted $\xi \sim \eta$, if there exists an open covering $\mathfrak{B}$ with $\mathfrak{B} < \mathfrak{U}$ and $\mathfrak{B} < \mathfrak{U}'$ such that $t_{\mathfrak{B}}^{\mathfrak{U}}(\xi) = t_{\mathfrak{B}}^{\mathfrak{U}'}(\eta)$. The set of equivalence classes is the so-called *inductive limit* of the cohomology groups $H^1(\mathfrak{U}, \mathscr{F})$ and is called the 1st cohomology group of X with coefficients in the sheaf $\mathscr{F}$. In symbols

$$H^1(X, \mathscr{F}) = \varinjlim_{\mathfrak{U}} H^1(\mathfrak{U}, \mathscr{F}) = \left(\dot{\bigcup_{\mathfrak{U}}} H^1(\mathfrak{U}, \mathscr{F}) \right) \Big/ \sim.$$

Addition in $H^1(X, \mathscr{F})$ is defined by means of representatives as follows. Suppose the elements $x, y \in H^1(X, \mathscr{F})$ are represented by $\xi \in H^1(\mathfrak{U}, \mathscr{F})$ resp. $\eta \in H^1(\mathfrak{U}', \mathscr{F})$. Let $\mathfrak{B}$ be a common refinement of $\mathfrak{U}$ and $\mathfrak{U}'$. Then $x + y \in H^1(X, \mathscr{F})$ is defined to be the equivalence class of $t_{\mathfrak{B}}^{\mathfrak{U}}(\xi) + t_{\mathfrak{B}}^{\mathfrak{U}'}(\eta) \in H^1(\mathfrak{B}, \mathscr{F})$. One can easily check that this definition is independent of the various choices made and makes $H^1(X, \mathscr{F})$ into an abelian group. If $\mathscr{F}$ is a sheaf of vector spaces, then in a natural way $H^1(\mathfrak{U}, \mathscr{F})$ and $H^1(X, \mathscr{F})$ are also vector spaces.

From Lemma (12.4) it follows that for any open covering of X the canonical mapping

$$H^1(\mathfrak{U}, \mathcal{F}) \to H^1(X, \mathcal{F})$$

is injective. In particular this implies that $H^1(X, \mathcal{F}) = 0$ precisely if $H^1(\mathfrak{U}, \mathcal{F}) = 0$ for every open covering $\mathfrak{U}$ of X.

12.6. Theorem. *Suppose X is a Riemann surface and $\mathcal{E}$ is the sheaf of differentiable functions on X. Then $H^1(X, \mathcal{E}) = 0$.*

PROOF. We give the proof under the assumption that X has a countable topology. However this assumption is always valid, see §23.

Suppose $\mathfrak{U} = (U_i)_{i \in I}$ is an arbitrary open covering of X. Then there is a partition of unity subordinate to $\mathfrak{U}$, i.e. a family $(\psi_i)_{i \in I}$ of functions $\psi_i \in \mathcal{E}(X)$ with the following properties (cf. the Appendix):

(i) $\operatorname{Supp}(\psi_i) \subset U_i$.

(ii) Every point of X has a neighborhood meeting only finitely many of the sets $\operatorname{Supp}(\psi_i)$.

(iii) $\sum_{i \in I} \psi_i = 1$.

We will show that $H^1(\mathfrak{U}, \mathcal{E}) = 0$, i.e., every cocycle $(f_{ij}) \in Z^1(\mathfrak{U}, \mathcal{E})$ splits.

The function $\psi_j f_{ij}$, which is defined on $U_i \cap U_j$, may be differentiably extended to all of U_i by assigning it the value zero outside its support. Thus it may be considered as an element of $\mathcal{E}(U_i)$. Set

$$g_i := \sum_{j \in I} \psi_j f_{ij}.$$

Because of (ii), in a neighborhood of any point in U_i, this sum has only finitely many terms which are not zero and thus defines an element $g_i \in \mathcal{E}(U_i)$. For $i, j \in I$

$$g_i - g_j = \sum_{k \in I} \psi_k f_{ik} - \sum_{k \in I} \psi_k f_{jk} = \sum_k \psi_k (f_{ik} - f_{jk})$$
$$= \sum_k \psi_k (f_{ik} + f_{kj}) = \sum_k \psi_k f_{ij} = f_{ij}$$

on $U_i \cap U_j$ and thus (f_{ij}) is a coboundary. □

Remark. In exactly the same way one can show that on a Riemann surface X the 1st cohomology groups with coefficients in the sheaves $\mathcal{E}^{(1)}, \mathcal{E}^{1,0}, \mathcal{E}^{0,1}$ and $\mathcal{E}^{(2)}$ also vanish.

12.7. Theorem. *Suppose X is a simply connected Riemann surface. Then*

(a) $H^1(X, \mathbb{C}) = 0$,

(b) $H^1(X, \mathbb{Z}) = 0$.

Here $\mathbb{C}$ (resp. $\mathbb{Z}$) denotes the sheaf of locally constant functions with values in the complex numbers (resp. integers), cf. (6.4.e).

PROOF

(a) Suppose $\mathfrak{U}$ is an open covering of X and $(c_{ij}) \in Z^1(\mathfrak{U}, \mathbb{C})$. Since $Z^1(\mathfrak{U}, \mathbb{C}) \subset Z^1(\mathfrak{U}, \mathscr{E})$ and $H^1(\mathfrak{U}, \mathscr{E}) = 0$, there exists a cochain $(f_i) \in C^0(\mathfrak{U}, \mathscr{E})$ such that

$$c_{ij} = f_i - f_j \quad \text{on } U_i \cap U_j.$$

But $dc_{ij} = 0$ implies $df_i = df_j$ on $U_i \cap U_j$, and thus there exists a global differential form $\omega \in \mathscr{E}^{(1)}(X)$ such that $\omega \mid U_i = df_i$. Since $ddf_i = 0$, it follows that ω is closed. Because X is simply connected, by (10.7) there exists $f \in \mathscr{E}(X)$ such that $df = \omega$. Set

$$c_i := f_i - f \mid U_i.$$

Since $dc_i = df_i - df = \omega - \omega = 0$ on U_i, c_i is locally constant, i.e., $(c_i) \in C^0(\mathfrak{U}, \mathbb{C})$. On $U_i \cap U_j$ one has

$$c_{ij} = f_i - f_j = (f_i - f) - (f_j - f) = c_i - c_j,$$

and thus the cocycle (c_{ij}) splits.

(b) Suppose $(a_{jk}) \in Z^1(\mathfrak{U}, \mathbb{Z})$. By (a) there exists a cochain $(c_j) \in C^0(\mathfrak{U}, \mathbb{C})$ such that

$$a_{jk} = c_j - c_k \quad \text{on } U_j \cap U_k.$$

Since $\exp(2\pi i a_{jk}) = 1$, one has $\exp(2\pi i c_j) = \exp(2\pi i c_k)$ on the intersection $U_j \cap U_k$. Since X is connected, there exists a constant $b \in \mathbb{C}^*$ such that

$$b = \exp(2\pi i c_j) \quad \text{for every } j \in I.$$

Choose $c \in \mathbb{C}$ such that $\exp(2\pi i c) = b$ and let

$$a_j := c_j - c.$$

Since $\exp(2\pi i a_j) = \exp(2\pi i c_j)\exp(-2\pi i c) = 1$, it follows that a_j is an integer, i.e., $(a_j) \in C^0(\mathfrak{U}, \mathbb{Z})$. Moreover

$$a_{jk} = c_j - c_k = (c_j - c) - (c_k - c) = a_j - a_k,$$

i.e., the cocycle (a_{jk}) lies in $B^1(\mathfrak{U}, \mathbb{Z})$. □

The next theorem shows that in certain cases one can calculate $H^1(X, \mathscr{F})$ using only a single covering of X.

12.8 Theorem (Leray). *Suppose $\mathscr{F}$ is a sheaf of abelian groups on the topological space X and $\mathfrak{U} = (U_i)_{i \in I}$ is an open covering of X such that $H^1(U_i, \mathscr{F}) = 0$ for every $i \in I$. Then*

$$H^1(X, \mathscr{F}) \cong H^1(\mathfrak{U}, \mathscr{F}).$$

Such a $\mathfrak{U}$ is called a *Leray covering* (of 1st order) for the sheaf $\mathscr{F}$.

PROOF. It suffices to show that, for every open covering $\mathfrak{V} = (V_\alpha)_{\alpha \in A}$, with $\mathfrak{V} < \mathfrak{U}$, the mapping $t^{\mathfrak{U}}_{\mathfrak{V}}: H^1(\mathfrak{U}, \mathscr{F}) \to H^1(\mathfrak{V}, \mathscr{F})$ is an isomorphism. From (12.4) this mapping is injective.

Suppose $\tau: A \to I$ is a refining mapping with $V_\alpha \subset U_{\tau\alpha}$ for every $\alpha \in A$. To prove the surjectivity of $t^{\mathfrak{U}}_{\mathfrak{V}}$, we must show that given any cocycle $(f_{\alpha\beta}) \in Z^1(\mathfrak{V}, \mathscr{F})$, there exists a cocycle $(F_{ij}) \in Z^1(\mathfrak{U}, \mathscr{F})$ such that the cocycle

$$(F_{\tau\alpha, \tau\beta}) - (f_{\alpha\beta})$$

is cohomologous to zero relative to the covering $\mathfrak{V}$. Now the family $(U_i \cap V_\alpha)_{\alpha \in A}$ is an open covering of U_i which we denote by $U_i \cap \mathfrak{V}$. By assumption $H^1(U_i \cap \mathfrak{V}, \mathscr{F}) = 0$, i.e., there exist $g_{i\alpha} \in \mathscr{F}(U_i \cap V_\alpha)$ such that

$$f_{\alpha\beta} = g_{i\alpha} - g_{i\beta} \quad \text{on } U_i \cap V_\alpha \cap V_\beta .$$

Now on the intersection $U_i \cap U_j \cap V_\alpha \cap V_\beta$ one has

$$g_{j\alpha} - g_{i\alpha} = g_{j\beta} - g_{i\beta}$$

and thus by sheaf axiom II there exist elements $F_{ij} \in \mathscr{F}(U_i \cap U_j)$ such that

$$F_{ij} = g_{j\alpha} - g_{i\alpha} \quad \text{on } U_i \cap U_j \cap V_\alpha .$$

Clearly, (F_{ij}) satisfies the cocycle relation and thus lies in $Z^1(\mathfrak{U}, \mathscr{F})$. Let $h_\alpha := g_{\tau\alpha, \alpha} | V_\alpha \in \mathscr{F}(V_\alpha)$. Then on $V_\alpha \cap V_\beta$ one has

$$\begin{aligned} F_{\tau\alpha, \tau\beta} - f_{\alpha\beta} &= (g_{\tau\beta, \alpha} - g_{\tau\alpha, \alpha}) - (g_{\tau\beta, \alpha} - g_{\tau\beta, \beta}) \\ &= g_{\tau\beta, \beta} - g_{\tau\alpha, \alpha} = h_\beta - h_\alpha , \end{aligned}$$

and thus $(F_{\tau\alpha, \tau\beta}) - (f_{\alpha\beta})$ splits. □

12.9. Example. As an application of Leray's Theorem, we will show

$$H^1(\mathbb{C}^*, \mathbb{Z}) = \mathbb{Z}.$$

Let $U_1 := \mathbb{C}^* \backslash \mathbb{R}_-$ and $U_2 := \mathbb{C}^* \backslash \mathbb{R}_+$, where $\mathbb{R}_+$ and $\mathbb{R}_-$ denote the positive and negative real axes respectively. Then $\mathfrak{U} = (U_1, U_2)$ is an open covering of $\mathbb{C}^*$. By (12.7) $H^1(U_i, \mathbb{Z}) = 0$ since U_i is star-shaped and thus simply connected. Thus $H^1(\mathbb{C}^*, \mathbb{Z}) = H^1(\mathfrak{U}, \mathbb{Z})$.

Since any cocycle $(a_{ij}) \in Z^1(\mathfrak{U}, \mathbb{Z})$ is alternating, i.e., $a_{ii} = 0$ and $a_{ij} = -a_{ji}$, it is completely determined by a_{12} and thus $Z^1(\mathfrak{U}, \mathbb{Z}) \cong \mathbb{Z}(U_1 \cap U_2)$. But the intersection $U_1 \cap U_2$ has two connected components, namely the upper and lower half planes, and thus $\mathbb{Z}(U_1 \cap U_2) \cong \mathbb{Z} \times \mathbb{Z}$. Since U_i is connected, $\mathbb{Z}(U_i) \cong \mathbb{Z}$ and hence $C^0(\mathfrak{U}, \mathbb{Z}) \cong \mathbb{Z} \times \mathbb{Z}$. The coboundary operator $\delta: C^0(\mathfrak{U}, \mathbb{Z}) \to Z^1(\mathfrak{U}, \mathbb{Z})$ is given with respect to these isomorphisms by

$$\mathbb{Z} \times \mathbb{Z} \to \mathbb{Z} \times \mathbb{Z}, \qquad (b_1, b_2) \mapsto (b_2 - b_1, b_2 - b_1).$$

Thus the coboundaries are exactly the subgroup $B \subset \mathbb{Z} \times \mathbb{Z}$ of those elements (a_1, a_2) with $a_1 = a_2$. Hence $H^1(\mathfrak{U}, \mathbb{Z}) \cong \mathbb{Z} \times \mathbb{Z}/B \cong \mathbb{Z}$.

Similarly one can show $H^1(\mathbb{C}^*, \mathbb{C}) \cong \mathbb{C}$.

12.10. The Zeroth Cohomology Group. Suppose $\mathcal{F}$ is a sheaf of abelian groups on the topological space X and $\mathfrak{U} = (U_i)_{i \in I}$ is an open covering of X. Set

$$Z^0(\mathfrak{U}, \mathcal{F}) := \operatorname{Ker}(C^0(\mathfrak{U}, \mathcal{F}) \xrightarrow{\delta} C^1(\mathfrak{U}, \mathcal{F})),$$

$$B^0(\mathfrak{U}, \mathcal{F}) := 0,$$

$$H^0(\mathfrak{U}, \mathcal{F}) := Z^0(\mathfrak{U}, \mathcal{F})/B^0(\mathfrak{U}, \mathcal{F}) = Z^0(\mathfrak{U}, \mathcal{F}).$$

From the definition of δ it follows that a 0-cochain $(f_i) \in C^0(\mathfrak{U}, \mathcal{F})$ belongs to $Z^0(\mathfrak{U}, \mathcal{F})$ precisely if $f_i \mid U_i \cap U_j = f_j \mid U_i \cap U_j$ for every $i, j \in I$. By sheaf axiom II the elements f_i piece together to give a global element $f \in \mathcal{F}(X)$ and there is a natural isomorphism

$$H^0(\mathfrak{U}, \mathcal{F}) = Z^0(\mathfrak{U}, \mathcal{F}) \cong \mathcal{F}(X).$$

Thus the groups $H^0(\mathfrak{U}, \mathcal{F})$ are entirely independent of the covering $\mathfrak{U}$ and one can define

$$H^0(X, \mathcal{F}) := \mathcal{F}(X).$$

Exercises (§12)

12.1. Suppose $p_1, \ldots, p_n$ are distinct points of $\mathbb{C}$ and let

$$X := \mathbb{C} \setminus \{p_1, \ldots, p_n\}.$$

Prove

$$H^1(X, \mathbb{Z}) \cong \mathbb{Z}^n.$$

[*Hint*: Construct a covering $\mathfrak{U} = (U_1, U_2)$ of X such that U_1 and U_2 are connected and simply connected and $U_1 \cap U_2$ has $n+1$ connected components.]

12.2. (a) Let X be a manifold, $U \subset X$ open and $V \Subset U$. Show that V meets only a finite number of connected components of U.

(b) Let X be a compact manifold and $\mathfrak{U} = (U_i)_{i \in I}$, $\mathfrak{V} = (V_i)_{i \in I}$ be two finite open coverings of X such that $V_i \Subset U_i$ for every $i \in I$. Prove that

$$\operatorname{Im}(Z^1(\mathfrak{U}, \mathbb{C}) \to Z^1(\mathfrak{V}, \mathbb{C}))$$

is a finite-dimensional vector space.

(c) Let X be a compact Riemann surface. Prove that $H^1(X, \mathbb{C})$ is a finite-dimensional vector space.

[*Hint*: Use finite coverings $\mathfrak{U} = (U_i)$, $\mathfrak{V} = (V_i)$ of X with $V_i \Subset U_i$, such that all the U_i and V_i are isomorphic to disks.]

12.3. (a) Let X be a compact Riemann surface. Prove that the map

$$H^1(X, \mathbb{Z}) \to H^1(X, \mathbb{C}),$$

induced by the inclusion $\mathbb{Z} \subset \mathbb{C}$, is injective.

(b) Let X be a compact Riemann surface. Show that $H^1(X, \mathbb{Z})$ is a finitely generated free $\mathbb{Z}$-module.

[*Hint*: Show first, as in Ex. 12.2.c), that $H^1(X, \mathbb{Z})$ is finitely generated and then use 12.3.a) to prove that $H^1(X, \mathbb{Z})$ is free.]

§13. Dolbeault's Lemma

In this section we solve the inhomogeneous Cauchy–Riemann differential equation $(\partial f/\partial \bar{z}) = g$, where g is a given differentiable function on the disk X. This is then used to show that the cohomology group $H^1(X, \mathcal{O})$ vanishes.

13.1. Lemma. *Suppose $g \in \mathcal{E}(\mathbb{C})$ has compact support. Then there exists a function $f \in \mathcal{E}(\mathbb{C})$ such that*

$$\frac{\partial f}{\partial \bar{z}} = g.$$

PROOF. Define the function $f: \mathbb{C} \to \mathbb{C}$ by

$$f(\zeta) := \frac{1}{2\pi i} \iint_{\mathbb{C}} \frac{g(z)}{z - \zeta}\, dz \wedge d\bar{z}.$$

Since the integrand has a singularity when $z = \zeta$, one has to show that the integral exists and depends differentiably on ζ. The simplest way to do this is to change variables by translation and then introduce polar coordinates r, θ, namely let

$$z = \zeta + re^{i\theta}.$$

With regard to the integration ζ is a constant and one has

$$dz \wedge d\bar{z} = -2i\, dx \wedge dy = -2ir\, dr \wedge d\theta.$$

Thus

$$f(\zeta) = -\frac{1}{\pi} \iint \frac{g(\zeta + re^{i\theta})}{re^{i\theta}}\, r\, dr\, d\theta$$

$$= -\frac{1}{\pi} \iint g(\zeta + re^{i\theta}) e^{-i\theta}\, dr\, d\theta.$$

Since g has compact support, one has only to integrate over a rectangle $0 \le r \le R$, $0 \le \theta \le 2\pi$, provided R is chosen sufficiently large. One may differentiate under the integral sign, i.e., $f \in \mathcal{E}(\mathbb{C})$ and

$$\frac{\partial f}{\partial \bar{\zeta}}(\zeta) = -\frac{1}{\pi} \iint \frac{\partial g(\zeta + re^{i\theta})}{\partial \bar{\zeta}} e^{-i\theta}\, dr\, d\theta.$$

Changing back to the original coordinates, one has

$$\frac{\partial f}{\partial \bar{\zeta}}(\zeta) = \frac{1}{2\pi i} \lim_{\varepsilon \to 0} \iint_{B_\varepsilon} \frac{\partial g(\zeta + z)}{\partial \bar{\zeta}} \frac{1}{z}\, dz \wedge d\bar{z},$$

where $B_\varepsilon := \{z \in \mathbb{C}: \varepsilon \leq |z| \leq R\}$. Since

$$\frac{\partial g(\zeta + z)}{\partial \bar{\zeta}} \frac{1}{z} = \frac{\partial g(\zeta + z)}{\partial \bar{z}} \frac{1}{z} = \frac{\partial}{\partial \bar{z}} \left(\frac{g(\zeta + z)}{z} \right)$$

for $z \neq 0$, one has

$$\frac{\partial f}{\partial \bar{\zeta}}(\zeta) = \frac{1}{2\pi i} \lim_{\varepsilon \to 0} \iint_{B_\varepsilon} \frac{\partial}{\partial \bar{z}} \left(\frac{g(\zeta + z)}{z} \right) dz \wedge d\bar{z} = - \lim_{\varepsilon \to 0} \iint_{B_\varepsilon} d\omega,$$

where the differential form ω is given by

$$\omega(z) = \frac{1}{2\pi i} \frac{g(\zeta + z)}{z}\, dz$$

(here one considers z as a variable and ζ as a constant). By Stokes' Theorem

$$\frac{\partial f}{\partial \bar{\zeta}}(\zeta) = - \lim_{\varepsilon \to 0} \iint_{B_\varepsilon} d\omega = - \lim_{\varepsilon \to 0} \int_{\partial B_\varepsilon} \omega = \lim_{\varepsilon \to 0} \int_{|z| = \varepsilon} \omega.$$

Parametrizing the circle $|z| = \varepsilon$ by $z = \varepsilon e^{i\theta}$, $0 \leq \theta \leq 2\pi$, one gets

$$\frac{\partial f}{\partial \bar{\zeta}}(\zeta) = \lim_{\varepsilon \to 0} \frac{1}{2\pi} \int_0^{2\pi} g(\zeta + \varepsilon e^{i\theta})\, d\theta.$$

Now the integral gives the average value of the function g over the circle $\zeta + \varepsilon e^{i\theta}$ for $0 \leq \theta \leq 2\pi$. Since g is continuous, this converges to $g(\zeta)$ as $\varepsilon \to 0$, i.e.,

$$\frac{\partial f}{\partial \bar{\zeta}}(\zeta) = g(\zeta). \qquad \square$$

The next theorem shows that one may drop the assumption that g has compact support.

13.2. Theorem. *Suppose* $X := \{z \in \mathbb{C}: |z| < R\}$, $0 < R \leq \infty$, *and* $g \in \mathscr{E}(X)$. *Then there exists* $f \in \mathscr{E}(X)$ *such that*

$$\frac{\partial f}{\partial \bar{z}} = g.$$

This theorem is a special case of the so-called Dolbeault Lemma in several complex variables, see [32].

Proof. In this case a solution cannot simply be given as an integral as in (13.1), for the integral will not converge in general. For this reason we use an exhaustion process which allows (13.1) to be applied in the present setting.

Suppose $0 < R_0 < R_1 < \cdots < R_n$ is a sequence of radii such that $\lim_{n\to\infty} R_n = R$ and set

$$X_n := \{z \in \mathbb{C}: |z| < R_n\}.$$

There exist functions $\psi_n \in \mathscr{E}(X)$ with compact supports $\operatorname{Supp}(\psi_n) \subset X_{n+1}$ and $\psi_n | X_n = 1$. The functions $\psi_n g$ vanish outside X_{n+1} and thus if one extends them by zero, they become functions on $\mathbb{C}$. By (13.1) there exist functions $f_n \in \mathscr{E}(X)$ such that

$$\bar{\partial} f_n = \psi_n g \quad \text{on } X.$$

Here and in the following we use the abbreviation $\bar{\partial} := (\partial/\partial \bar{z})$.

By induction we alter the sequence (f_n) to another sequence $(\tilde{f}_n)$, which for all $n \geq 1$ satisfies

(i) $\bar{\partial}\tilde{f}_n = g \quad \text{on } X_n$,
(ii) $\|\tilde{f}_{n+1} - \tilde{f}_n\|_{X_{n-1}} \leq 2^{-n}$.

(As usual let $\|f\|_K := \sup_{x\in K} |f(x)|$ denote the supremum norm.) Set $\tilde{f}_1 := f_1$. Suppose $\tilde{f}_1, \ldots, \tilde{f}_n$ are already constructed. Then

$$\bar{\partial}(f_{n+1} - \tilde{f}_n) = 0 \quad \text{on } X_n,$$

and thus $f_{n+1} - \tilde{f}_n$ is holomorphic on X_n. Hence there exists a polynomial P (e.g., a finite number of terms of the Taylor series of $f_{n+1} - \tilde{f}_n$) such that

$$\|f_{n+1} - \tilde{f}_n - P\|_{X_{n-1}} \leq 2^{-n}.$$

If we set $\tilde{f}_{n+1} := f_{n+1} - P$, then (ii) is satisfied. Moreover, on X_{n+1} one has

$$\bar{\partial}\tilde{f}_{n+1} = \bar{\partial} f_{n+1} - \bar{\partial} P = \bar{\partial} f_{n+1} = \psi_{n+1} g = g,$$

i.e., (i) also holds. Since every point $z \in X$ is contained in almost all X_n, the limit

$$f(z) := \lim_{n\to\infty} \tilde{f}_n(z)$$

exists. On X_n one may write

$$f = \tilde{f}_n + \sum_{k=n}^{\infty} (\tilde{f}_{k+1} - \tilde{f}_k).$$

For $k \geq n$, the functions $\tilde{f}_{k+1} - \tilde{f}_k$ are holomorphic on X_n, since $\bar{\partial}(\tilde{f}_{k+1} - \tilde{f}_k) = 0$.

Because of (ii), the series

$$F_n := \sum_{k=n}^{\infty} (\tilde{f}_{k+1} - \tilde{f}_k)$$

converges uniformly on X_n and is thus holomorphic there. Hence $f = \tilde{f}_n + F_n$ is infinitely differentiably on X_n for every n and thus $f \in \mathscr{E}(X)$. As well

$$\bar{\partial} f = \bar{\partial} \tilde{f}_n = g \quad \text{on } X_n$$

for every n and thus $\bar{\partial} f = g$ on all of X. □

Remark. Naturally the solution of the equation $\bar{\partial} f = g$ is not uniquely determined, only up to the addition of an arbitrary holomorphic function.

13.3. Corollary. *Suppose* $X := \{z \in \mathbb{C}: |z| < R\}, 0 < R \le \infty$. *Then given any* $g \in \mathscr{E}(X)$, *there exists* $f \in \mathscr{E}(X)$ *such that* $\Delta f = g$.

Here

$$\Delta = \frac{\partial^2}{\partial x^2} + \frac{\partial^2}{\partial y^2} = 4 \frac{\partial^2}{\partial z\, \partial \bar{z}}$$

is the Laplace operator.

PROOF. Choose $f_1 \in \mathscr{E}(X)$ such that $\bar{\partial} f_1 = g$ and $f_2 \in \mathscr{E}(X)$ such that $\bar{\partial} f_2 = \bar{f}_1$. Then $f := \frac{1}{4}\bar{f}_2$ satisfies $\Delta f = g$, for

$$\Delta f = \frac{\partial^2 \bar{f}_2}{\partial z\, \partial \bar{z}} = \frac{\partial}{\partial \bar{z}}\left(\frac{\partial \bar{f}_2}{\partial z}\right) = \frac{\partial}{\partial \bar{z}}\left(\overline{\frac{\partial f_2}{\partial \bar{z}}}\right) = \frac{\partial f_1}{\partial \bar{z}} = g.$$ □

13.4. Theorem. *Suppose* $X := \{z \in \mathbb{C}: |z| < R\}$, $0 < R \le \infty$. *Then* $H^1(X, \mathcal{O}) = 0$.

PROOF. Suppose $\mathfrak{U} = (U_i)$ is an open covering of X and $(f_{ij}) \in Z^1(\mathfrak{U}, \mathcal{O})$ is a cocycle. Since $Z^1(\mathfrak{U}, \mathcal{O}) \subset Z^1(\mathfrak{U}, \mathscr{E})$ and $H^1(X, \mathscr{E}) = 0$, there exists a cochain $(g_i) \in C^0(\mathfrak{U}, \mathscr{E})$ such that

$$f_{ij} = g_i - g_j \quad \text{on } U_i \cap U_j.$$

Since $\bar{\partial} f_{ij} = 0$, one has $\bar{\partial} g_i = \bar{\partial} g_j$ on $U_i \cap U_j$ and thus there exists a global function $h \in \mathscr{E}(X)$ with $h \mid U_i = \bar{\partial} g_i$. By (13.2) we can find a function $g \in \mathscr{E}(X)$ such that $\bar{\partial} g = h$. Define

$$f_i := g_i - g.$$

Now f_i is holomorphic, since $\bar{\partial} f_i = \bar{\partial} g_i - \bar{\partial} g = 0$, and thus $(f_i) \in C^0(\mathfrak{U}, \mathcal{O})$. As well on $U_i \cap U_j$ one has

$$f_i - f_j = g_i - g_j = f_{ij},$$

i.e., the cocycle (f_{ij}) splits. □

13.5. Theorem. *For the Riemann sphere* $H^1(\mathbb{P}^1, \mathcal{O}) = 0$.

PROOF. Set $U_1 := \mathbb{P}^1 \backslash \infty$ and $U_2 := \mathbb{P}^1 \backslash 0$. Since $U_1 = \mathbb{C}$ and U_2 is biholomorphic to $\mathbb{C}$, it follows from (13.4) that $H^1(U_i, \mathcal{O}) = 0$. Thus $\mathfrak{U} = (U_1, U_2)$ is a Leray covering of $\mathbb{P}^1$ and $H^1(\mathbb{P}^1, \mathcal{O}) = H^1(\mathfrak{U}, \mathcal{O})$ by (12.8). Thus the proof is complete once one shows that every cocycle $(f_{ij}) \in Z^1(\mathfrak{U}, \mathcal{O})$ splits. In order to do this, it is clearly enough to find functions $f_i \in \mathcal{O}(U_i)$ such that

$$f_{12} = f_1 - f_2 \quad \text{on } U_1 \cap U_2 = \mathbb{C}^*.$$

Let

$$f_{12}(z) = \sum_{n=-\infty}^{\infty} c_n z^n$$

be the Laurent expansion of f_{12} on $\mathbb{C}^*$. Set

$$f_1(z) := \sum_{n=0}^{\infty} c_n z^n \quad \text{and} \quad f_2(z) := -\sum_{n=-\infty}^{-1} c_n z^n.$$

Then $f_i \in \mathcal{O}(U_i)$ and $f_1 - f_2 = f_{12}$. □

EXERCISES (§13)

13.1. Let $X = \{z \in \mathbb{C} : |z| < R\}$, where $0 < R \leq \infty$. Denote by $\mathscr{H}$ the sheaf of harmonic functions on X, i.e.

$$\mathscr{H}(U) = \{f : U \to \mathbb{C} : f \text{ is harmonic}\}$$

for $U \subset X$ open. Prove

$$H^1(X, \mathscr{H}) = 0.$$

13.2. (a) Show that $\mathfrak{U} = (\mathbb{P}^1 \backslash \infty, \mathbb{P}^1 \backslash 0)$ is a Leray covering for the sheaf Ω of holomorphic 1-forms on $\mathbb{P}^1$.

(b) Prove that

$$H^1(\mathbb{P}^1, \Omega) \cong H^1(\mathfrak{U}, \Omega) \cong \mathbb{C}$$

and that the cohomology class of

$$\frac{dz}{z} \in \Omega(U_1 \cap U_2) \cong Z^1(\mathfrak{U}, \Omega)$$

is a basis of $H^1(\mathbb{P}^1, \Omega)$.

13.3. Suppose $g \in \mathscr{E}(\mathbb{C})$ is a function with compact support. Prove that there is a solution $f \in \mathscr{E}(\mathbb{C})$ of the equation

$$\frac{\partial f}{\partial \bar{z}} = g$$

having compact support if and only if

$$\iint_{\mathbb{C}} z^n g(z)\, dz \wedge d\bar{z} = 0 \quad \text{for every } n \in \mathbb{N}.$$

§14. A Finiteness Theorem

In this section we prove that for any compact Riemann surface X the cohomology group $H^1(X, \mathcal{O})$ is a finite dimensional complex vector space. Its dimension is called the genus of X. One of the consequences of the finiteness theorem is the existence of non-constant meromorphic functions on every compact Riemann surface. With regard to further applications in Chapter 3 we will do everything not only for compact Riemann surfaces but also for relatively compact subsets of arbitrary Riemann surfaces.

14.1. The L^2-Norm for Holomorphic Functions. Suppose $D \subset \mathbb{C}$ is an open set. Given a holomorphic function $f \in \mathcal{O}(D)$ define its L^2-norm by

$$\|f\|_{L^2(D)} := \left(\iint_D |f(x+iy)|^2 \, dx \, dy\right)^{1/2}.$$

Then $\|f\|_{L^2(D)} \in \mathbb{R}_+ \cup \{\infty\}$. If $\|f\|_{L^2(D)} < \infty$, then f is called square integrable. We denote by $L^2(D, \mathcal{O})$ the vector space of all square integrable holomorphic functions on D. If

$$\operatorname{Vol}(D) := \iint_D dx \, dy < \infty,$$

then for every bounded function $f \in \mathcal{O}(D)$ one has

$$\|f\|_{L^2(D)} \le \sqrt{\operatorname{Vol}(D)} \|f\|_D,$$

where $\|f\|_D := \sup\{|f(z)| : z \in D\}$ denotes the supremum norm.

For $f, g \in L^2(D, \mathcal{O})$ one can define an inner product $\langle f, g \rangle \in \mathbb{C}$ by

$$\langle f, g \rangle := \iint_D f\bar{g} \, dx \, dy.$$

The integral exists because for every $z \in D$

$$|f(z)\overline{g(z)}| \le \tfrac{1}{2}(|f(z)|^2 + |g(z)|^2).$$

With this inner product $L^2(D, \mathcal{O})$ is a unitary vector space and in particular has a well-defined notion of orthogonality. Now suppose $B = B(a, r) := \{z \in \mathbb{C} : |z - a| < r\}$ is the disk with center a and radius $r > 0$. Then the monomials $(\psi_n)_{n \in \mathbb{N}}$ given by

$$\psi_n(z) := (z-a)^n$$

form an orthogonal system in $L^2(B, \mathcal{O})$ and one can easily check using polar coordinates that

$$\|\psi_n\|_{L^2(B)} = \frac{\sqrt{\pi}\, r^{n+1}}{\sqrt{n+1}} \quad \text{for every } n \in \mathbb{N}.$$

If $f \in L^2(B, \mathcal{O})$ and

$$f(z) = \sum_{n=0}^{\infty} c_n(z-a)^n$$

is the Taylor series of f about a, it follows from Pythagoras that

$$\|f\|_{L^2(B)}^2 = \sum_{n=0}^{\infty} \frac{\pi r^{2n+2}}{n+1} |c_n|^2. \tag{*}$$

14.2. Theorem. *Suppose $D \subset \mathbb{C}$ is open, $r > 0$ and*

$$D_r := \{z \in \mathbb{C} : B(z, r) \subset D\}$$

is the set of points in D whose distance from the boundary is greater than or equal to r. Then for every $f \in L^2(D, \mathcal{O})$ one has

$$\|f\|_{D_r} \leq \frac{1}{\sqrt{\pi}\, r} \|f\|_{L^2(D)}.$$

PROOF. Suppose $a \in D_r$ and $f(z) = \sum c_n(z-a)^n$ is the Taylor series of f about a. Using (*) one gets

$$|f(a)| = |c_0| \leq \frac{1}{\sqrt{\pi}\, r} \|f\|_{L^2(B(a,r))} \leq \frac{1}{\sqrt{\pi}\, r} \|f\|_{L^2(D)}.$$

Since $\|f\|_{D_r} = \sup\{|f(a)| : a \in D_r\}$, the result follows. □

In particular, it follows from Theorem (14.2) that if $(f_n)_{n \in \mathbb{N}}$ is a Cauchy sequence in $L^2(D, \mathcal{O})$, then the sequence converges uniformly on every compact subset of D. Thus the limit function is holomorphic. Hence $L^2(D, \mathcal{O})$ is complete and thus is a Hilbert space.

The following lemma may be viewed as a certain generalization of Schwarz' Lemma.

14.3. Lemma. *Suppose $D' \Subset D$ are open subsets of $\mathbb{C}$. Then given any $\varepsilon > 0$, there exists a closed vector subspace $A \subset L^2(D, \mathcal{O})$ of finite codimension such that*

$$\|f\|_{L^2(D')} \leq \varepsilon \|f\|_{L^2(D)} \quad \textit{for every } f \in A.$$

PROOF. Since $\bar{D}'$ is compact and lies in D, there exist $r > 0$ and finitely many points $a_1, \ldots, a_k \in D$ with the following properties:

(i) $B(a_j, r) \subset D$ for $j = 1, \ldots, k$,
(ii) $D' \subset \bigcup_{j=1}^{k} B(a_j, r/2)$.

Choose n so large that $2^{-n-1}k \leq \varepsilon$. Let A be the set of all functions $f \in L^2(D, \mathcal{O})$ which vanish at every point a_j at least to order n. Then A is a closed vector subspace of $L^2(D, \mathcal{O})$ of codimension $\leq kn$.

Let $f \in A$. Then f has a Taylor series about a_j

$$f(z) = \sum_{\nu=n}^{\infty} c_\nu (z - a_j)^\nu.$$

For every $\rho \le r$ one has

$$\|f\|^2_{L^2(B(a_j,\rho))} = \sum_{\nu=n}^{\infty} \frac{\pi \rho^{2n+2}}{\nu+1} |c_\nu|^2,$$

from which it follows that

$$\|f\|_{L^2(B(a_j, r/2))} \le 2^{-n-1} \|f\|_{L^2(B(a_j, r))}.$$

Using (i) and (ii) one has

$$\|f\|_{L^2(B(a_j, r))} \le \|f\|_{L^2(D)}$$

and

$$\|f\|_{L^2(D')} \le \sum_{j=1}^{k} \|f\|_{L^2(B(a_j, r/2))}.$$

Thus

$$\|f\|_{L^2(D')} \le k \cdot 2^{-n-1} \|f\|_{L^2(D)} \le \varepsilon \|f\|_{L^2(D)}. \qquad \square$$

14.4. Square Integrable Cochains. Suppose X is a Riemann surface. Choose a finite family (U_i^*, z_i), $i = 1, \ldots, n$, of charts on X such that every $z_i(U_i^*) \subset \mathbb{C}$ is a disk. Note however that we are not assuming that $\mathfrak{U}^* = (U_i^*)_{1 \le i \le n}$ is a covering of X. Suppose $U_i \subset U_i^*$ are open subsets and set $\mathfrak{U} := (U_i)_{1 \le i \le n}$. We introduce L^2-norms on the cochain groups $C^0(\mathfrak{U}, \mathcal{O})$ and $C^1(\mathfrak{U}, \mathcal{O})$, defined on the space

$$|\mathfrak{U}| := U_1 \cup \cdots \cup U_n,$$

in the following way:

(i) For $\eta = (f_i) \in C^0(\mathfrak{U}, \mathcal{O})$ let

$$\|\eta\|^2_{L^2(\mathfrak{U})} := \sum_i \|f_i\|^2_{L^2(U_i)}.$$

(ii) For $\xi = (f_{ij}) \in C^1(\mathfrak{U}, \mathcal{O})$ let

$$\|\xi\|^2_{L^2(\mathfrak{U})} := \sum_{i,j} \|f_{ij}\|^2_{L^2(U_i \cap U_j)}.$$

Here the norms of f_i, resp. f_{ij}, are calculated with respect to the chart (U_i^*, z_i), i.e.,

$$\|f_i\|_{L^2(U_i)} := \|f_i \circ z_i^{-1}\|_{L^2(z_i(U_i))},$$

$$\|f_{ij}\|_{L^2(U_i \cap U_j)} := \|f_{ij} \circ z_i^{-1}\|_{L^2(z_i(U_i \cap U_j))}.$$

The set of cochains having finite norm is a vector subspace $C^q_{L^2}(\mathfrak{U}, \mathcal{O}) \subset C^q(\mathfrak{U}, \mathcal{O})$, $q = 0, 1$, and these subspaces are Hilbert spaces. The cocycles in $C^1_{L^2}(\mathfrak{U}, \mathcal{O})$ form a closed vector subspace which we denote by $Z^1_{L^2}(\mathfrak{U}, \mathcal{O})$.

14.5. If $V_i \Subset U_i$, $i = 1, \ldots, n$, are relatively compact open subsets and $\mathfrak{V} = (V_i)_{1 \le i \le n}$, then to simplify the notation we will write $\mathfrak{V} \ll \mathfrak{U}$. For any cochain $\xi \in C^q(\mathfrak{U}, \mathcal{O})$ one has $\|\xi\|_{L^2(\mathfrak{V})} < \infty$. It then follows directly from Lemma (14.3) that given any $\varepsilon > 0$, there exists a closed vector subspace $A \subset Z^1_{L^2}(\mathfrak{U}, \mathcal{O})$ of finite codimension such that

$$\|\xi\|_{L^2(\mathfrak{V})} \le \varepsilon \|\xi\|_{L^2(\mathfrak{U})} \quad \text{for every } \xi \in A.$$

14.6. Lemma. *Suppose X is a Riemann surface and $\mathfrak{U}^*$ is a finite family of charts on X as in (14.4). Further suppose that one has $\mathfrak{W} \ll \mathfrak{V} \ll \mathfrak{U} \ll \mathfrak{U}^*$, i.e., fixed shrinkings of $\mathfrak{U}^*$ are given. Then there exists a constant $C > 0$ such that for every $\xi \in Z^1_{L^2}(\mathfrak{V}, \mathcal{O})$ there exist elements $\zeta \in Z^1_{L^2}(\mathfrak{U}, \mathcal{O})$ and $\eta \in C^0_{L^2}(\mathfrak{W}, \mathcal{O})$ with*

$$\zeta = \xi + \delta\eta \quad \text{on } \mathfrak{W}$$

and

$$\max(\|\zeta\|_{L^2(\mathfrak{U})}, \|\eta\|_{L^2(\mathfrak{W})}) \le C\|\xi\|_{L^2(\mathfrak{V})}.$$

PROOF

(a) Suppose $\xi = (f_{ij}) \in Z^1_{L^2}(\mathfrak{V}, \mathcal{O})$ is given. Forgetting for the moment the restriction on the norms, we first construct $\zeta \in Z^1_{L^2}(\mathfrak{U}, \mathcal{O})$ and $\eta \in C^0_{L_2}(\mathfrak{W}, \mathcal{O})$ such that $\zeta = \xi + \delta\eta$ on $\mathfrak{W}$. By Theorem (12.6) there exists a cochain $(g_i) \in C^0(\mathfrak{V}, \mathcal{E})$ such that

$$f_{ij} = g_j - g_i \quad \text{on } V_i \cap V_j.$$

Since $d''f_{ij} = 0$, one has $d''g_i = d''g_j$ on $V_i \cap V_j$, and thus there exists a differential form $\omega \in \mathcal{E}^{0,1}(|\mathfrak{V}|)$ with $\omega | V_i = d''g_i$. Since $|\mathfrak{W}| \Subset |\mathfrak{V}|$, there exists a function $\psi \in \mathcal{E}(X)$ with

$$\operatorname{Supp}(\psi) \subset |\mathfrak{V}| \quad \text{and} \quad \psi \,|\, |\mathfrak{W}| = 1.$$

Hence $\psi\omega$ can be considered as an element of $\mathcal{E}(|\mathfrak{U}^*|)$. By Theorem (13.2) there exist functions $h_i \in \mathcal{E}(U_i^*)$ such that

$$d''h_i = \psi\omega \quad \text{on } U_i^*.$$

Because $d''h_i = d''h_j$ on $U_i^* \cap U_j^*$, it follows that

$$F_{ij} := h_j - h_i \in \mathcal{O}(U_i^* \cap U_j^*).$$

Set $\zeta := (F_{ij}) | \mathfrak{U}$. Since $\mathfrak{U} \ll \mathfrak{U}^*$, one has $\zeta \in Z^1_{L^2}(\mathfrak{U}, \mathcal{O})$. On W_i one has $d''h_i = \psi\omega = \omega = d''g_i$, thus $h_i - g_i$ is holomorphic on W_i. Since $h_i - g_i$ is also bounded on W_i, one has

$$\eta := (h_i - g_i) | \mathfrak{W} \in C^0_{L^2}(\mathfrak{W}, \mathcal{O}).$$

Now $F_{ij} - f_{ij} = (h_j - g_j) - (h_i - g_i)$ on $W_i \cap W_j$ and thus

$$\zeta - \xi = \delta\eta \quad \text{on } \mathfrak{W}.$$

(b) In order to get the desired estimate on the norms, we consider the Hilbert space

$$H := Z^1_{L^2}(\mathfrak{U}, \mathcal{O}) \times Z^1_{L^2}(\mathfrak{V}, \mathcal{O}) \times C^0_{L^2}(\mathfrak{W}, \mathcal{O})$$

with the norm

$$\|(\zeta, \xi, \eta)\|_H := (\|\zeta\|^2_{L^2(\mathfrak{U})} + \|\xi\|^2_{L^2(\mathfrak{V})} + \|\eta\|^2_{L^2(\mathfrak{W})})^{1/2}.$$

Let $L \subset H$ be the subspace

$$L := \{(\zeta, \xi, \eta) \in H : \zeta = \xi + \delta\eta \quad \text{on } \mathfrak{W}\}.$$

Since L is closed in H, it is also a Hilbert space. From part (a) the continuous linear mapping

$$\pi: L \to Z^1_{L^2}(\mathfrak{V}, \mathcal{O}), \qquad (\zeta, \xi, \eta) \mapsto \xi,$$

is surjective. By the Theorem of Banach (cf. Appendix B. 6, 7) the mapping π is open. Thus there exists a constant $C > 0$ such that for every $\xi \in Z^1_{L^2}(\mathfrak{V}, \mathcal{O})$ there exists $x = (\zeta, \xi, \eta) \in L$ with $\pi(x) = \xi$ and $\|x\|_H \leq C\|\xi\|_{L^2(\mathfrak{V})}$. This constant then satisfies the desired conditions. □

14.7. Lemma. *Under the same assumptions as in Lemma* (14.6), *there exists a finite dimensional vector subspace* $S \subset Z^1(\mathfrak{U}, \mathcal{O})$ *with the following property. For every* $\xi \in Z^1(\mathfrak{U}, \mathcal{O})$ *there exist elements* $\sigma \in S$ *and* $\eta \in C^0(\mathfrak{W}, \mathcal{O})$ *such that*

$$\sigma = \xi + \delta\eta \quad \text{on } \mathfrak{W}.$$

Remark. The lemma says that the natural restriction mapping

$$H^1(\mathfrak{U}, \mathcal{O}) \to H^1(\mathfrak{W}, \mathcal{O})$$

has a finite dimensional image.

PROOF. Suppose C is the constant in Lemma (14.6) and set $\varepsilon := (1/2C)$. By (14.5) there exists a finite codimensional closed vector subspace $A \subset Z^1_{L^2}(\mathfrak{U}, \mathcal{O})$ such that

$$\|\xi\|_{L^2(\mathfrak{V})} \leq \varepsilon\|\xi\|_{L^2(\mathfrak{U})} \quad \text{for every } \xi \in A.$$

Let S be the orthogonal complement of A in $Z^1_{L^2}(\mathfrak{U}, \mathcal{O})$, i.e., $A \oplus S = Z^1_{L^2}(\mathfrak{U}, \mathcal{O})$.

Now suppose $\xi \in Z^1(\mathfrak{U}, \mathcal{O})$ is arbitrary. Because $\mathfrak{V} \ll \mathfrak{U}$,

$$\|\xi\|_{L^2(\mathfrak{V})} =: M < \infty.$$

By (14.6) there exist $\zeta_0 \in Z^1_{L^2}(\mathfrak{U}, \mathcal{O})$ and $\eta_0 \in C^0_{L^2}(\mathfrak{W}, \mathcal{O})$ such that

$$\zeta_0 = \xi + \delta\eta_0 \quad \text{on } \mathfrak{W}$$

and $\|\zeta_0\|_{L^2(\mathfrak{U})} \leq CM$, $\|\eta_0\|_{L^2(\mathfrak{W})} \leq CM$. Suppose

$$\zeta_0 = \xi_0 + \sigma_0, \qquad \xi_0 \in A, \qquad \sigma_0 \in S,$$

is the orthogonal decomposition.

We now construct, by induction, elements

$$\zeta_\nu \in Z^1_{L^2}(\mathfrak{U}, \mathcal{O}), \qquad \eta_\nu \in C^0_{L^2}(\mathfrak{W}, \mathcal{O}), \qquad \xi_\nu \in A, \qquad \sigma_\nu \in S$$

with the following properties:

(i) $\zeta_\nu = \xi_{\nu-1} + \delta\eta_\nu$ on $\mathfrak{W}$
(ii) $\zeta_\nu = \xi_\nu + \sigma_\nu$
(iii) $\|\zeta_\nu\|_{L^2(\mathfrak{U})} \leq 2^{-\nu}CM, \qquad \|\eta_\nu\|_{L^2(\mathfrak{W})} \leq 2^{-\nu}CM.$

Consider the induction step from ν to $\nu + 1$. Since $\zeta_\nu = \xi_\nu + \sigma_\nu$ is an orthogonal decomposition, one has

$$\|\xi_\nu\|_{L^2(\mathfrak{U})} \leq \|\zeta_\nu\|_{L^2(\mathfrak{U})} \leq 2^{-\nu}CM.$$

Thus

$$\|\xi_\nu\|_{L^2(\mathfrak{V})} \leq \varepsilon\|\xi_\nu\|_{L^2(\mathfrak{U})} \leq 2^{-\nu}\varepsilon CM \leq 2^{-\nu-1}M.$$

By Lemma (14.6) there exist elements $\zeta_{\nu+1} \in Z^1_{L^2}(\mathfrak{U}, \mathcal{O})$ and $\eta_{\nu+1} \in C^0_{L^2}(\mathfrak{W}, \mathcal{O})$ such that

$$\zeta_{\nu+1} = \xi_\nu + \delta\eta_{\nu+1} \quad \text{on } \mathfrak{W}$$

and

$$\max(\|\zeta_{\nu+1}\|_{L^2(\mathfrak{U})}, \|\eta_{\nu+1}\|_{L^2(\mathfrak{W})}) \leq 2^{-\nu-1}CM.$$

Now one has an orthogonal decomposition $\zeta_{\nu+1} = \xi_{\nu+1} + \sigma_{\nu+1}$, where $\xi_{\nu+1} \in A$ and $\sigma_{\nu+1} \in S$, and thus the induction step is complete.

From the equation $\zeta_0 = \xi + \sigma\eta_0$, together with equations (i) and (ii) up to $\nu = k$, one gets

$$\xi_k + \sum_{\nu=0}^{k} \sigma_\nu = \xi + \delta\left(\sum_{\nu=0}^{k} \eta_\nu\right) \quad \text{on } \mathfrak{W}. \tag{*}$$

From (ii) and (iii) it follows that

$$\max(\|\xi_\nu\|_{L^2(\mathfrak{U})}, \|\sigma_\nu\|_{L^2(\mathfrak{U})}, \|\eta_\nu\|_{L^2(\mathfrak{W})}) \leq 2^{-\nu}CM.$$

Hence $\lim_{k\to\infty} \xi_k = 0$ and the series

$$\sigma := \sum_{\nu=0}^{\infty} \sigma_\nu \in S$$

$$\eta := \sum_{\nu=0}^{\infty} \eta_\nu \in C^0_{L^2}(\mathfrak{W}, \mathcal{O})$$

converge. Finally from (*) one gets $\sigma = \xi + \delta\eta$ on $\mathfrak{W}$. □

Remark. By using more powerful tools from functional analysis one could make the proof shorter, cf. the proof of Theorem (29.13).

14.8. Suppose X is a topological space, $Y \subset X$ is open and $\mathcal{F}$ is a sheaf of abelian groups on X. For every open covering $\mathfrak{U} = (U_i)_{i \in I}$ of X, $\mathfrak{U} \cap Y := (U_i \cap Y)_{i \in I}$ is an open covering of Y and the natural restriction mapping $Z^1(\mathfrak{U}, \mathcal{F}) \to Z^1(\mathfrak{U} \cap Y, \mathcal{F})$ induces a homomorphism $H^1(\mathfrak{U}, \mathcal{F}) \to H^1(\mathfrak{U} \cap Y, \mathcal{F})$. These homomorphisms for all $\mathfrak{U}$ give rise to a restriction homomorphism

$$H^1(X, \mathcal{F}) \to H^1(Y, \mathcal{F}).$$

Clearly, if one has open sets $Y \subset Y' \subset X$, then the homomorphism $H^1(X, \mathcal{F}) \to H^1(Y, \mathcal{F})$ is the composition of the homomorphisms $H^1(X, \mathcal{F}) \to H^1(Y', \mathcal{F})$ and $H^1(Y', \mathcal{F}) \to H^1(Y, \mathcal{F})$.

14.9. Theorem. *Suppose X is a Riemann surface and $Y_1 \Subset Y_2 \subset X$ are open subsets. Then the restriction homomorphism*

$$H^1(Y_2, \mathcal{O}) \to H^1(Y_1, \mathcal{O})$$

has a finite dimensional image.

Proof. There exists a finite family of charts $(U_i^*, z_i)_{1 \le i \le n}$ on X and relatively compact open subsets $W_i \Subset V_i \Subset U_i \Subset U_i^*$ with the following properties:

(i) $Y_1 \subset \bigcup_{i=1}^n W_i =: Y' \Subset Y'' := \bigcup_{i=1}^n U_i \subset Y_2$,
(ii) all $z_i(U_i^*)$, $z_i(U_i)$ and $z_i(W_i)$ are disks in $\mathbb{C}$.

Let $\mathfrak{U} := (U_i)_{1 \le i \le n}$, $\mathfrak{W} := (W_i)_{1 \le i \le n}$. By Lemma (14.7) the restriction mapping $H^1(\mathfrak{U}, \mathcal{O}) \to H^1(\mathfrak{W}, \mathcal{O})$ has a finite dimensional image. By Theorem (13.4), $H^1(U_i, \mathcal{O}) = H^1(W_i, \mathcal{O}) = 0$. Thus by Leray's Theorem (12.8), $H^1(\mathfrak{U}, \mathcal{O}) = H^1(Y'', \mathcal{O})$ and $H^1(\mathfrak{W}, \mathcal{O}) = H^1(Y', \mathcal{O})$. Since the restriction mapping $H^1(Y_2, \mathcal{O}) \to H^1(Y_1, \mathcal{O})$ can be factored as follows

$$H^1(Y_2, \mathcal{O}) \to H^1(Y'', \mathcal{O}) \to H^1(Y', \mathcal{O}) \to H^1(Y_1, \mathcal{O}),$$

the proof of the theorem is complete. □

14.10. Corollary. *Suppose X is a compact Riemann surface. Then*

$$\dim H^1(X, \mathcal{O}) < \infty.$$

Proof. Since X is compact, one can choose $Y_1 = Y_2 = X$ in the previous theorem. □

14.11. Definition. Suppose X is a compact Riemann surface. Then

$$g := \dim H^1(X, \mathcal{O})$$

is called the *genus* of X.

By Theorem (13.5) the Riemann sphere $\mathbb{P}^1$ has genus zero.

14.12. Theorem. *Suppose X is a Riemann surface and $Y \Subset X$ is a relatively compact open subset. Then for every point $a \in Y$ there exists a meromorphic function $f \in \mathcal{M}(Y)$ which has a pole at a and is holomorphic on $Y\backslash\{a\}$.*

PROOF. By Theorem (14.9)

$$k := \dim \operatorname{Im}(H^1(X, \mathcal{O}) \to H^1(Y, \mathcal{O})) < \infty.$$

Suppose (U_1, z) is a coordinate neighborhood of a with $z(a) = 0$. Set $U_2 := X\backslash\{a\}$. Then $\mathfrak{U} = (U_1, U_2)$ is an open covering of X. The functions z^{-j} are holomorphic on $U_1 \cap U_2 = U_1 \backslash\{a\}$ and represent cocycles

$$\zeta_j \in Z^1(\mathfrak{U}, \mathcal{O}), \qquad j = 1, \ldots, k+1.$$

Since $\dim \operatorname{Im}(H^1(\mathfrak{U}, \mathcal{O}) \to H^1(\mathfrak{U} \cap Y, \mathcal{O})) < k+1$, the cocycles

$$\zeta_j \mid Y \in Z^1(\mathfrak{U} \cap Y, \mathcal{O}),$$

$1 \le j \le k+1$, are linearly dependent modulo the coboundaries. Thus there exist complex numbers $c_1, \ldots, c_{k+1}$, not all zero, and a cochain $\eta = (f_1, f_2) \in C^0(\mathfrak{U} \cap Y, \mathcal{O})$ such that

$$c_1\zeta_1 + \cdots + c_{k+1}\zeta_{k+1} = \delta\eta \quad \text{with respect to } \mathfrak{U} \cap Y,$$

i.e.,

$$\sum_{j=1}^{k+1} c_j z^{-j} = f_2 - f_1 \quad \text{on } U_1 \cap U_2 \cap Y.$$

Hence there is a function $f \in \mathcal{M}(Y)$, which coincides with

$$f_1 + \sum_{j=1}^{k+1} c_j z^{-j}$$

on $U_1 \cap Y$ and which is equal to f_2 on $U_2 \cap Y = Y\backslash\{a\}$. This is the desired function. □

14.13. Corollary. *Suppose X is a compact Riemann surface and $a_1, \ldots, a_n$ are distinct points on X. Then for any given complex numbers $c_1, \ldots, c_n \in \mathbb{C}$, there exists a meromorphic function $f \in \mathcal{M}(X)$ such that $f(a_i) = c_i$ for $i = 1, \ldots, n$.*

PROOF. For every pair $i \neq j$, by applying Theorem (14.12) in the case $Y = X$, one gets a function $f_{ij} \in \mathcal{M}(X)$ which has a pole at a_i but is holomorphic at a_j. Choose a constant $\lambda_{ij} \in \mathbb{C}^*$ such that $f_{ij}(a_k) \neq f_{ij}(a_j) - \lambda_{ij}$ for every $k = 1, \ldots, n$. Then the function

$$g_{ij} := \frac{f_{ij} - f_{ij}(a_j)}{f_{ij} - f_{ij}(a_j) + \lambda_{ij}} \in \mathcal{M}(X)$$

is holomorphic at the points a_k, $1 \le k \le n$, and satisfies $g_{ij}(a_i) = 1$ and $g_{ij}(a_j) = 0$. Now the functions

$$h_i := \prod_{j \neq i} g_{ij}, \qquad i = 1, \ldots, n,$$

satisfy $h_i(a_j) = \delta_{ij}$ and thus

$$f := \sum_{i=1}^{n} c_i h_i$$

solves the problem. □

We now note a few consequences of the finiteness theorem for non-compact Riemann surfaces. The reader who is only interested in compact Riemann surfaces may skip over these if he wants.

14.14. Corollary. *Suppose Y is a relatively compact open subset of a non-compact Riemann surface X. Then there exists a holomorphic function $f: Y \to \mathbb{C}$ which is not constant on any connected component of Y.*

PROOF. Choose a domain Y_1 such that $Y \Subset Y_1 \Subset X$ and a point $a \in Y_1 \setminus Y$. (Since X is non-compact and connected, $Y_1 \setminus Y$ is not empty.) Now apply Theorem (14.12) to Y_1 and the point a. □

14.15. Theorem. *Suppose X is a non-compact Riemann surface and $Y \Subset Y' \subset X$ are open subsets. Then*

$$\operatorname{Im}(H^1(Y', \mathcal{O}) \to H^1(Y, \mathcal{O})) = 0.$$

PROOF. By Theorem (14.9) we already know that

$$L := \operatorname{Im}(H^1(Y', \mathcal{O}) \to H^1(Y, \mathcal{O}))$$

is a finite dimensional vector space. Choose cohomology classes $\xi_1, \ldots, \xi_n \in H^1(Y', \mathcal{O})$ such that their restrictions to Y span the vector space L. According to (14.14) we may choose a function $f \in \mathcal{O}(Y')$ which is not constant on any connected component of Y'. Since $H^1(Y', \mathcal{O})$ is in a natural way a module over $\mathcal{O}(Y')$, the products $f\xi_\nu \in H^1(Y', \mathcal{O})$ are defined. By the choice of the ξ_ν there exist constants $c_{\nu\mu} \in \mathbb{C}$ such that

$$f\xi_\nu = \sum_{\mu=1}^{n} c_{\nu\mu}\xi_\mu \quad \text{on } Y \quad \text{for } \nu = 1, \ldots, n. \tag{1}$$

Set

$$F := \det(f\delta_{\nu\mu} - c_{\nu\mu})_{1 \le \nu, \mu \le n}.$$

Then F is a holomorphic function on Y' which is not identically zero on any connected component of Y'. From (1) it follows that

$$F\xi_\nu \mid Y = 0 \quad \text{for } \nu = 1, \ldots, n. \tag{2}$$

An arbitrary cohomology class $\zeta \in H^1(Y', \mathcal{O})$ can be represented by a cocycle $(f_{ij}) \in Z^1(\mathfrak{U}, \mathcal{O})$, where $\mathfrak{U} = (U_i)_{i \in I}$ is an open covering of Y' such that each zero of F is contained in at most one U_i. Thus for $i \neq j$ one has

$F | U_i \cap U_j \in \mathcal{O}^*(U_i \cap U_j)$. Hence there exists a cocycle $(g_{ij}) \in Z^1(\mathfrak{U}, \mathcal{O})$ such that $f_{ij} = Fg_{ij}$. Let $\xi \in H^1(Y', \mathcal{O})$ be the cohomology class of (g_{ij}). Then $\zeta = F\xi$. Hence from (2) one gets $\zeta | Y = F\xi | Y = 0$. □

14.16. Corollary. *Suppose X is a non-compact Riemann surface and $Y \Subset Y' \subset X$ are open subsets. Then for every differential form $\omega \in \mathcal{E}^{0,1}(Y')$ there exists a function $f \in \mathcal{E}(Y)$ such that $d''f = \omega | Y$.*

PROOF. By Theorem (13.2) the problem has a solution locally. Thus there exist an open covering $\mathfrak{U} = (U_i)_{i \in I}$ of Y' and functions $f_i \in \mathcal{E}(U_i)$ such that $d''f_i = \omega | U_i$. The differences $f_i - f_j$ are holomorphic on $U_i \cap U_j$ and thus define a cocycle in $Z^1(\mathfrak{U}, \mathcal{O})$. By (14.15) this cocycle is cohomologous to zero on Y and thus there exist holomorphic functions $g_i \in \mathcal{O}(U_i \cap Y)$ such that

$$f_i - f_j = g_i - g_j \quad \text{on } U_i \cap U_j \cap Y.$$

Hence there exists a function $f \in \mathcal{E}(Y)$ such that

$$f = f_i - g_i \quad \text{on } U_i \cap Y, \quad \text{for every } i \in I.$$

But then the function f satisfies the equation $d''f = \omega | Y$. □

Remark. Theorems (25.6) and (26.1) will extend the results of (14.15) and (14.16).

EXERCISES (§14)

14.1. Let $X = \{z \in \mathbb{C} : r < |z| < R\}$, where $0 < r < R < \infty$. Determine an orthonormal basis of $L^2(X, \mathcal{O})$ consisting of functions of the form

$$\varphi_n(z) = c_n z^n, \qquad n \in \mathbb{Z}.$$

14.2. Let $X \subset \mathbb{C}$ be a bounded open subset, $p_1, \ldots, p_k \in X$ and $X' := X \setminus \{p_1, \ldots, p_k\}$. Show that the restriction map

$$L^2(X, \mathcal{O}) \to L^2(X', \mathcal{O})$$

is an isomorphism.

§15. The Exact Cohomology Sequence

In this section we consider sheaf homomorphisms, exact sequences of sheaves and the long exact cohomology sequence. These tools prove useful in calculating various cohomology groups.

15.1. Definition. Suppose $\mathcal{F}$ and $\mathcal{G}$ are sheaves of abelian groups on the topological space X. A *sheaf homomorphism* $\alpha: \mathcal{F} \to \mathcal{G}$ is a family of group homomorphisms

$$\alpha_U: \mathcal{F}(U) \to \mathcal{G}(U), \quad U \text{ open in } X,$$

which are compatible with the restriction homomorphisms, i.e., for every pair of open sets $U, V \subset X$ with $V \subset U$ the diagram

$$\begin{array}{ccc} \mathscr{F}(U) & \xrightarrow{\alpha_U} & \mathscr{G}(U) \\ \text{restr.}\big\downarrow & & \big\downarrow\text{restr.} \\ \mathscr{F}(V) & \xrightarrow{\alpha_V} & \mathscr{G}(V) \end{array}$$

is commutative. If all the α_U are isomorphisms, then α is called an isomorphism.

Similarly, one can define homomorphisms of sheaves of vector spaces. Often one just writes $\alpha: \mathscr{F}(U) \to \mathscr{G}(U)$ instead of $\alpha_U: \mathscr{F}(U) \to \mathscr{G}(U)$.

15.2. Examples

(a) Suppose $\mathscr{E}$(resp. $\mathscr{E}^{(1)}$, $\mathscr{E}^{(2)}$) are the sheaves of differentiable functions (resp. 1-forms and 2-forms) on a Riemann surface X. The exterior derivative d on functions (resp. differential forms) induces sheaf homomorphisms

$$d: \mathscr{E} \to \mathscr{E}^{(1)}, \qquad d: \mathscr{E}^{(1)} \to \mathscr{E}^{(2)}.$$

Similarly the mappings d' and d'' also induce sheaf homomorphisms.

(b) On a Riemann surface X the natural inclusions $\mathcal{O} \to \mathscr{E}$, $\mathbb{C} \to \mathscr{E}$, $\mathbb{Z} \to \mathcal{O}$, $\Omega \to \mathscr{E}^{1,0}$ etc., are sheaf homomorphisms.

(c) On a Riemann surface X one can define a sheaf homomorphism $\text{ex}: \mathcal{O} \to \mathcal{O}^*$ from the sheaf of holomorphic functions into the multiplicative sheaf of holomorphic functions with values in $\mathbb{C}^*$. For U an open subset of X and $f \in \mathcal{O}(U)$ let $\text{ex}_U(f) := \exp(2\pi i f)$.

15.3. The Kernel of a Sheaf Homomorphism. Suppose $\mathscr{F}$ and $\mathscr{G}$ are sheaves on the topological space X and $\alpha: \mathscr{F} \to \mathscr{G}$ is a sheaf homomorphism. For U open in X let

$$\mathscr{K}(U) := \text{Ker}(\mathscr{F}(U) \xrightarrow{\alpha} \mathscr{G}(U)).$$

One can easily show that the family of groups $\mathscr{K}(U)$, together with the restriction homomorphisms induced from the sheaf $\mathscr{F}$, is again a sheaf. It is called the kernel of α and is denoted by $\mathscr{K} = \text{Ker}\,\alpha$.

Examples. On any Riemann surface one has

(a) $\mathcal{O} = \text{Ker}(\mathscr{E} \xrightarrow{d''} \mathscr{E}^{0,1})$, (see 9.1),

(b) $\Omega = \text{Ker}(\mathscr{E}^{1,0} \xrightarrow{d} \mathscr{E}^{(2)})$ (see 9.16),

(c) $\mathbb{Z} = \text{Ker}(\mathcal{O} \xrightarrow{\text{ex}} \mathcal{O}^*)$ (see 15.2.c).

15.4. Remark. Given a homomorphism $\alpha\colon \mathcal{F} \to \mathcal{G}$ of sheaves on the topological space X one can define

$$\mathcal{B}(U) := \operatorname{Im}(\mathcal{F}(U) \xrightarrow{\alpha} \mathcal{G}(U)) \quad \text{for every open } U \text{ in } X.$$

This defines a presheaf $\mathcal{B}$ which in general does not satisfy sheaf axiom II. As a counterexample consider the sheaf homomorphism

$$\mathrm{ex}\colon \mathcal{O} \to \mathcal{O}^*, \qquad f \mapsto \exp(2\pi i f),$$

on the space $\mathbb{C}^*$. Let $U_1 = \mathbb{C}^* \backslash \mathbb{R}_-$ and $U_2 = \mathbb{C}^* \backslash \mathbb{R}_+$. Define $f_k \in \mathcal{O}^*(U_k)$ by $f_k(z) = z$ for every $z \in U_k$, $k = 1, 2$. Since U_k is simply connected,

$$f_k \in \operatorname{Im}\left(\mathcal{O}(U_k) \xrightarrow{\mathrm{ex}} \mathcal{O}^*(U_k)\right).$$

Moreover, $f_1 \mid U_1 \cap U_2 = f_2 \mid U_1 \cap U_2$. But there is no element

$$f \in \operatorname{Im}\left(\mathcal{O}(\mathbb{C}^*) \xrightarrow{\mathrm{ex}} \mathcal{O}^*(\mathbb{C}^*)\right)$$

with $f \mid U_k = f_k$, since the function $z \mapsto z$ has no single-valued logarithm on all of $\mathbb{C}^*$.

15.5. Exact Sequences. Suppose $\alpha\colon \mathcal{F} \to \mathcal{G}$ is a sheaf homomorphism on the topological space X. Then for each $x \in X$ there is an induced homomorphism of the stalks

$$\alpha_x\colon \mathcal{F}_x \to \mathcal{G}_x.$$

A sequence of sheaf homomorphisms $\mathcal{F} \xrightarrow{\alpha} \mathcal{G} \xrightarrow{\beta} \mathcal{H}$ is called *exact*, if for each $x \in X$ the sequence

$$\mathcal{F}_x \xrightarrow{\alpha_x} \mathcal{G}_x \xrightarrow{\beta_x} \mathcal{H}_x$$

is exact, i.e., $\operatorname{Ker} \beta_x = \operatorname{Im} \alpha_x$. A sequence

$$\mathcal{F}_1 \xrightarrow{\alpha_1} \mathcal{F}_2 \xrightarrow{\alpha_2} \cdots \longrightarrow \mathcal{F}_{n-1} \xrightarrow{\alpha_{n-1}} \mathcal{F}_n, \qquad (n > 3),$$

of sheaf homomorphisms is called exact if the sequence

$$\mathcal{F}_k \xrightarrow{\alpha_k} \mathcal{F}_{k+1} \xrightarrow{\alpha_{k+1}} \mathcal{F}_{k+2}$$

is exact for every $1 \le k \le n-2$. A sheaf homomorphism $\alpha\colon \mathcal{F} \to \mathcal{G}$ is called a *monomorphism* if $0 \to \mathcal{F} \xrightarrow{\alpha} \mathcal{G}$ is exact and an *epimorphism* if $\mathcal{F} \xrightarrow{\alpha} \mathcal{G} \to 0$ is exact. An exact sequence of the form $0 \to \mathcal{F} \to \mathcal{G} \to \mathcal{H} \to 0$ is called a *short exact sequence.*

15.6. Lemma. *Suppose $\alpha\colon \mathcal{F} \to \mathcal{G}$ is a sheaf monomorphism on the topological space X. Then for every open subset $U \subset X$ the mapping $\alpha_U\colon \mathcal{F}(U) \to \mathcal{G}(U)$ is injective.*

PROOF. Suppose $f \in \mathscr{F}(U)$ and $\alpha_U(f) = 0$. Since $\alpha_x : \mathscr{F}_x \to \mathscr{G}_x$ is injective for every $x \in X$, every $x \in U$ has an open neighborhood $V_x \subset U$ such that $f \mid V_x = 0$. From sheaf axiom I it follows that $f = 0$. □

15.7. Remark. If $\alpha : \mathscr{F} \to \mathscr{G}$ is a sheaf epimorphism, it is not necessarily true that for every open set U the mapping $\alpha_U : \mathscr{F}(U) \to \mathscr{G}(U)$ is surjective. This is illustrated by the example ex: $\mathcal{O} \to \mathcal{O}^*$ in (15.4). For every x the map ex: $\mathcal{O}_x \to \mathcal{O}_x^*$ is surjective, since every non-vanishing function locally has a logarithm. But ex: $\mathcal{O}(\mathbb{C}^*) \to \mathcal{O}^*(\mathbb{C}^*)$ is not surjective.

15.8. Lemma. *Suppose $0 \to \mathscr{F} \xrightarrow{\alpha} \mathscr{G} \xrightarrow{\beta} \mathscr{H}$ is an exact sequence of sheaves on the topological space X. Then for every open set $U \subset X$ the sequence*

$$0 \to \mathscr{F}(U) \xrightarrow{\alpha} \mathscr{G}(U) \xrightarrow{\beta} \mathscr{H}(U)$$

is exact.

PROOF

(a) The exactness of $0 \to \mathscr{F}(U) \xrightarrow{\alpha} \mathscr{G}(U)$ was proved in (15.6).

(b) $\operatorname{Im} \alpha \subset \operatorname{Ker} \beta$. Suppose $f \in \mathscr{F}(U)$ and $g := \alpha(f)$. Since the sequence of stalks $\mathscr{F}_x \to \mathscr{G}_x \to \mathscr{H}_x$ is exact for every $x \in U$, it follows that each point $x \in U$ has a neighborhood $V_x \subset U$ such that $\beta(g) \mid V_x = 0$. Hence by sheaf axiom I one has $\beta(g) = 0$.

(c) To prove the inclusion $\operatorname{Ker} \beta \subset \operatorname{Im} \alpha$ suppose $g \in \mathscr{G}(U)$ with $\beta(g) = 0$. Since for every $x \in U$ one has $\operatorname{Ker} \beta_x = \operatorname{Im} \alpha_x$, there is an open covering $(V_i)_{i \in I}$ of U and elements $f_i \in \mathscr{F}(V_i)$ such that $\alpha(f_i) = g \mid V_i$ for every $i \in I$. On the intersection $V_i \cap V_j$ one then has $\alpha(f_i - f_j) = 0$. Hence by (15.6) it follows that $f_i = f_j$ on $V_i \cap V_j$. Now by sheaf axiom II there exists an $f \in \mathscr{F}(U)$ with $f \mid V_i = f_i$ for every $i \in I$. Since $\alpha(f) \mid V_i = \alpha(f \mid V_i) = g \mid V_i$, it follows from sheaf axiom I, applied to the sheaf $\mathscr{G}$, that $\alpha(f) = g$. □

15.9. Examples. We now give several examples of short exact sequences of sheaves

$$0 \to \mathscr{F} \to \mathscr{G} \to \mathscr{H} \to 0$$

on a Riemann surface X.

(a) $0 \to \mathcal{O} \to \mathscr{E} \xrightarrow{d''} \mathscr{E}^{0,1} \to 0$.

Here $\mathcal{O} \to \mathscr{E}$ is the natural inclusion. The exactness follows from the Dolbeault Lemma (13.2).

(b) Let

$$\mathscr{Z} := \operatorname{Ker}\left(\mathscr{E}^{(1)} \xrightarrow{d} \mathscr{E}^{(2)}\right)$$

be the sheaf of closed differential forms. The sequence

$$0 \to \mathbb{C} \to \mathscr{E} \xrightarrow{d} \mathscr{Z} \to 0$$

is exact. That $d: \mathscr{E} \to \mathscr{Z}$ is an epimorphism follows from the fact that locally every closed differential form is exact, see (10.4).

(c) $0 \to \mathbb{C} \to \mathscr{O} \xrightarrow{d} \Omega \to 0.$
This exact sequence is the holomorphic analogue of (b).

(d) Since

$$\Omega = \operatorname{Ker}\left(\mathscr{E}^{1,0} \xrightarrow{d} \mathscr{E}^{(2)}\right),$$

in order to prove the exactness of

$$0 \to \Omega \to \mathscr{E}^{1,0} \xrightarrow{d} \mathscr{E}^{(2)} \to 0,$$

one has only to show that $d: \mathscr{E}^{1,0} \to \mathscr{E}^{(2)}$ is onto. With respect to a local chart (U, z), one has

$$d(f\,dz) = \frac{\partial f}{\partial \bar{z}}\, d\bar{z} \wedge dz.$$

Thus for every open set $V \subset U$ such that $z(V) \subset \mathbb{C}$ is a disk one sees by using the Dolbeault Lemma that $d: \mathscr{E}^{1,0}(V) \to \mathscr{E}^{(2)}(V)$ is surjective. Hence $d: \mathscr{E}_a^{1,0} \to \mathscr{E}_a^{(2)}$ is surjective for every point $a \in X$.

(e) The exactness of the sequence

$$0 \to \mathbb{Z} \to \mathscr{O} \xrightarrow{\text{ex}} \mathscr{O}^* \to 0$$

follows from (15.3.c) and the remark (15.7).

15.10. Any homomorphism $\alpha: \mathscr{F} \to \mathscr{G}$ of sheaves on the topological space X induces homomorphisms

$$\alpha^0: H^0(X, \mathscr{F}) \to H^0(X, \mathscr{G}),$$
$$\alpha^1: H^1(X, \mathscr{F}) \to H^1(X, \mathscr{G}).$$

The homomorphism α^0 is nothing but the mapping $\alpha_X: \mathscr{F}(X) \to \mathscr{G}(X)$. The homomorphism α^1 is constructed as follows. Let $\mathfrak{U} = (U_i)_{i \in I}$ be an open covering of X. Consider the mapping

$$\alpha_{\mathfrak{U}}: C^1(\mathfrak{U}, \mathscr{F}) \to C^1(\mathfrak{U}, \mathscr{G})$$

which assigns to each cochain $\xi = (f_{ij}) \in C^1(\mathfrak{U}, \mathscr{F})$ the cochain

$$\alpha_{\mathfrak{U}}(\xi) := (\alpha(f_{ij})) \in C^1(\mathfrak{U}, \mathscr{G}).$$

This mapping takes cocycles to cocycles and coboundaries to coboundaries and thus induces a homomorphism

$$\tilde{\alpha}_{\mathfrak{U}}: H^1(\mathfrak{U}, \mathscr{F}) \to H^1(\mathfrak{U}, \mathscr{G}).$$

The collection of $\tilde{\alpha}_{\mathfrak{U}}$, where $\mathfrak{U}$ runs over all open coverings of X, then induces the homomorphism α^1.

15.11. The Connecting Homomorphism. Suppose

$$0 \to \mathscr{F} \xrightarrow{\alpha} \mathscr{G} \xrightarrow{\beta} \mathscr{H} \to 0$$

is an exact sequence of sheaves on the topological space X. A "connecting homomorphism"

$$\delta^*: H^0(X, \mathscr{H}) \to H^1(X, \mathscr{F})$$

is defined as follows. Suppose

$$h \in H^0(X, \mathscr{H}) = \mathscr{H}(X).$$

Since all the homomorphisms $\beta_x: \mathscr{G}_x \to \mathscr{H}_x$ are surjective, there exists an open covering $\mathfrak{U} = (U_i)_{i \in I}$ of X and a cochain $(g_i) \in C^0(\mathfrak{U}, \mathscr{G})$ such that

$$\beta(g_i) = h \mid U_i \quad \text{for every } i \in I. \tag{1}$$

Hence $\beta(g_j - g_i) = 0$ on $U_i \cap U_j$. By Lemma (15.8) there exists $f_{ij} \in \mathscr{F}(U_i \cap U_j)$ such that

$$\alpha(f_{ij}) = g_j - g_i. \tag{2}$$

On $U_i \cap U_j \cap U_k$ one has $\alpha(f_{ij} + f_{jk} - f_{ik}) = 0$ and thus by (15.6) $f_{ij} + f_{jk} = f_{ik}$, i.e.,

$$(f_{ij}) \in Z^1(\mathfrak{U}, \mathscr{F}).$$

Now let $\delta^* h \in H^1(X, \mathscr{F})$ be the cohomology class represented by (f_{ij}). One can easily check that this definition is independent of the various choices made.

15.12. Theorem. *Suppose X is a topological space and*

$$0 \to \mathscr{F} \xrightarrow{\alpha} \mathscr{G} \xrightarrow{\beta} \mathscr{H} \to 0$$

is a short exact sequence of sheaves on X. Then the induced sequence of cohomology groups

$$0 \longrightarrow H^0(X, \mathscr{F}) \xrightarrow{\alpha^0} H^0(X, \mathscr{G}) \xrightarrow{\beta^0} H^0(X, \mathscr{H}) \xrightarrow{\delta^*}$$
$$\xrightarrow{\delta^*} H^1(X, \mathscr{F}) \xrightarrow{\alpha^1} H^1(X, \mathscr{G}) \xrightarrow{\beta^1} H^1(X, \mathscr{H})$$

is exact.

PROOF

(a) The exactness at $H^0(X, \mathscr{F})$ and $H^0(X, \mathscr{G})$ follow from Lemma (15.8).

(b) Im $\beta^0 \subset$ Ker δ^*. Suppose $g \in H^0(X, \mathscr{G})$ and $h := \beta^0(g)$. In the construction of $\delta^* h$ described in (15.11) one can choose $g_i = g \mid U_i$. But then $f_{ij} = 0$ and thus $\delta^* h = 0$.

(c) Ker $\delta^* \subset$ Im β^0. Suppose $h \in$ Ker δ^*. Using the notation of (15.11) one can represent $\delta^* h$ by the cocycle $(f_{ij}) \in Z^1(\mathfrak{U}, \mathscr{F})$. Since $\delta^* h = 0$ there exists a cochain $(f_i) \in C^0(\mathfrak{U}, \mathscr{F})$ such that $f_{ij} = f_j - f_i$ on $U_i \cap U_j$. Set $\tilde{g}_i := g_i - \alpha(f_i)$. Then $\tilde{g}_i = \tilde{g}_j$ on $U_i \cap U_j$ because $\alpha(f_{ij}) = g_j - g_i$. Thus the $\tilde{g}_i$ are restrictions of some global element $g \in H^0(X, \mathscr{G})$. On U_i one has $\beta(g) = \beta(\tilde{g}_i) = \beta(g_i - \alpha(f_i)) = \beta(g_i) = h$, i.e., $h \in$ Im β^0.

(d) Im $\delta^* \subset$ Ker α^1. This follows from condition (2) in (15.11).

(e) Ker $\alpha^1 \subset$ Im δ^*. Suppose $\xi \in$ Ker α^1 is represented by the cocycle $(f_{ij}) \in Z^1(\mathfrak{U}, \mathscr{F})$. Since $\alpha^1(\xi) = 0$, there exists a cochain $(g_i) \in C^0(\mathfrak{U}, \mathscr{G})$ such that $\alpha(f_{ij}) = g_j - g_i$ on $U_i \cap U_j$. This implies

$$0 = \beta(\alpha(f_{ij})) = \beta(g_j) - \beta(g_i) \quad \text{on } U_i \cap U_j.$$

Hence there exists $h \in \mathscr{H}(X) = H^0(X, \mathscr{H})$ such that $h \mid U_i = \beta(g_i)$. The construction given in (15.11) now shows that $\delta^* h = \xi$.

(f) Im $\alpha^1 \subset$ Ker β^1. This follows from the fact that

$$\mathscr{F}(U_i \cap U_j) \xrightarrow{\alpha} \mathscr{G}(U_i \cap U_j) \xrightarrow{\beta} \mathscr{H}(U_i \cap U_j)$$

is exact by (15.8).

(g) Ker $\beta^1 \subset$ Im α^1. Suppose $\eta \in$ Ker β^1 is represented by the cocycle $(g_{ij}) \in Z^1(\mathfrak{U}, \mathscr{G})$, where $\mathfrak{U} = (U_i)_{i \in I}$. Then there is a cochain $(h_i) \in C^0(\mathfrak{U}, \mathscr{H})$ such that $\beta(g_{ij}) = h_j - h_i$. For every $x \in X$ choose $\tau x \in I$ such that $x \in U_{\tau x}$. Since $\beta_x : \mathscr{G}_x \to \mathscr{H}_x$ is surjective, there is an open neighborhood $V_x \subset U_{\tau x}$ of x and an element $g_x \in \mathscr{G}(V_x)$ such that $\beta(g_x) = h_{\tau x} \mid V_x$. Let $\mathfrak{V} = (V_x)_{x \in X}$ and $\tilde{g}_{xy} = g_{\tau x, \tau y} \mid V_x \cap V_y$. Then $(\tilde{g}_{xy}) \in Z^1(\mathfrak{V}, \mathscr{G})$ is a cocycle which also represents the cohomology class η. Let $\psi_{xy} := \tilde{g}_{xy} - g_y + g_x$. The cocycle (ψ_{xy}) is cohomologous to $(\tilde{g}_{xy})$ and $\beta(\psi_{xy}) = 0$. Thus there exists $f_{xy} \in \mathscr{F}(V_x \cap V_y)$ such that $\alpha(f_{xy}) = \psi_{xy}$. Since

$$\alpha : \mathscr{F}(V_x \cap V_y \cap V_z) \to \mathscr{G}(V_x \cap V_y \cap V_z)$$

is injective by (15.6), $(f_{xy}) \in Z^1(\mathfrak{V}, \mathscr{F})$. Thus the cohomology class $\xi \in H^1(X, \mathscr{F})$ of (f_{xy}) satisfies $\alpha^1(\xi) = \eta$. This completes the proof. □

15.13. Theorem. *Suppose* $0 \to \mathscr{F} \xrightarrow{\alpha} \mathscr{G} \xrightarrow{\beta} \mathscr{H} \to 0$ *is an exact sequence of sheaves on the topological space* X *such that* $H^1(X, \mathscr{G}) = 0$. *Then*

$$H^1(X, \mathscr{F}) \cong \mathscr{H}(X)/\beta\mathscr{G}(X).$$

PROOF. Since $H^1(X, \mathscr{G}) = 0$, by Theorem (15.12) one has the exact sequence

$$\mathscr{G}(X) \xrightarrow{\beta} \mathscr{H}(X) \xrightarrow{\delta*} H^1(X, \mathscr{F}) \to 0.$$

The result is now obvious. □

For many applications it is important to be able to describe the isomorphism.

$$\Phi: H^1(X, \mathscr{F}) \xrightarrow{\sim} \mathscr{H}(X)/\beta\mathscr{G}(X)$$

explicitly. By Lemma (15.8) we can always assume that $\mathscr{F} = \operatorname{Ker} \beta$ and $\alpha: \mathscr{F} \to \mathscr{G}$ is the inclusion map.

Suppose $\xi \in H^1(X, \mathscr{F})$ is a cohomology class which is represented by the cocycle $(f_{ij}) \in Z^1(\mathfrak{U}, \mathscr{F}) \subset Z^1(\mathfrak{U}, \mathscr{G})$. Since $H^1(\mathfrak{U}, \mathscr{G}) = 0$, there exists a cochain $(g_i) \in C^0(\mathfrak{U}, \mathscr{G})$ such that $f_{ij} = g_j - g_i$ on $U_i \cap U_j$. Since $\beta(f_{ij}) = 0$, $\beta(g_j)$ and $\beta(g_i)$ agree on $U_i \cap U_j$. Thus there exists a global element $h \in \mathscr{H}(X)$ with $h | U_i = \beta(g_i)$. Then $\Phi(\xi)$ is the coset of h modulo $\beta\mathscr{G}(X)$. The fact that the mapping Φ described above is the inverse of the isomorphism $\mathscr{H}(X)/\beta\mathscr{G}(X) \xrightarrow{\sim} H^1(X, \mathscr{F})$ induced by the exact cohomology sequence follows from part (e) in the proof of (15.12).

15.14. Dolbeault's Theorem. *Let X be a Riemann surface. Then there are isomorphisms*

(a) $H^1(X, \mathscr{O}) \cong \mathscr{E}^{0,1}(X)/d''\mathscr{E}(X)$,
(b) $H^1(X, \Omega) \cong \mathscr{E}^{(2)}(X)/d\mathscr{E}^{1,0}(X)$.

Since $H^1(X, \mathscr{E}) = H^1(X, \mathscr{E}^{1,0}) = 0$, one may apply Theorem (15.13) to the exact sequences given in (15.9.a) and (15.9.d) respectively.

Remark. Theorem (13.4) is a special case of Dolbeault's Theorem.

15.15. The deRham Groups. On every Riemann surface X every exact 1-form is closed but every closed form is not necessarily exact. Consequently one is interested in the quotient group

$$\operatorname{Rh}^1(X) := \frac{\operatorname{Ker}(\mathscr{E}^{(1)}(X) \xrightarrow{d} \mathscr{E}^{(2)}(X))}{\operatorname{Im}(\mathscr{E}(X) \xrightarrow{d} \mathscr{E}^{(1)}(X))}$$

of closed 1-forms modulo exact 1-forms. Two closed differential forms which determine the same element in $\operatorname{Rh}^1(X)$, i.e., whose difference is exact, are said to be cohomologous. $\operatorname{Rh}^1(X)$ is called the 1st deRham group of X. Note that $\operatorname{Rh}^1(X) = 0$ precisely if every closed 1-form $\omega \in \mathscr{E}^{(1)}(X)$ has a primitive. If X is simply connected, then $\operatorname{Rh}^1(X) = 0$ by (10.7).

deRham's Theorem. *Let X be a Riemann surface. Then*

$$H^1(X, \mathbb{C}) \cong \mathrm{Rh}^1(X).$$

This follows from (15.13) applied to the exact sequence in (15.9.b). Theorem (12.7.a) is a special case of deRham's Theorem.

Remark. The theorems of deRham and Dolbeault are proved here only for Riemann surfaces. But they are also valid in a more general form on differentiable (resp. complex) manifolds of arbitrary dimension. More details can be found in any book on several complex variables, e.g., [30], [31], [32], [33], [34], [35]. In §§6, 12 and 15 we have considered only the most basic ideas about sheaves and sheaf cohomology. A systematic introduction can be found in [41].

EXERCISES (§15)

15.1. Let X be a Riemann surface and $\mathscr{H}$ be the sheaf of harmonic functions on X. Verify that the sequence

$$0 \longrightarrow \mathscr{H} \longrightarrow \mathscr{E} \xrightarrow{d'd''} \mathscr{E}^{(2)} \longrightarrow 0$$

is exact.

15.2. Show that on any Riemann surface the sequence

$$0 \longrightarrow \mathbb{C}^* \longrightarrow \mathcal{O}^* \xrightarrow{d\log} \Omega \longrightarrow 0$$

is exact, where $(d\log)f := f^{-1}\,df$.

15.3. On a Riemann surface X let $\mathscr{Q} \subset \mathscr{M}^{(1)}$ be the sheaf of meromorphic 1-forms which have residue 0 at every point. Show that the sequence

$$0 \longrightarrow \mathbb{C} \longrightarrow \mathscr{M} \xrightarrow{d} \mathscr{Q} \longrightarrow 0$$

is exact.

15.4. Let $X = \mathbb{C}/\Gamma$ be a torus. Prove that

$$H^1(X, \mathbb{C}) \cong \mathrm{Rh}^1(X) \cong \mathbb{C}^2$$

and that the classes of dz and $d\bar{z}$ form a basis of $\mathrm{Rh}^1(X)$.
[*Hint*: Let $\omega \in \mathscr{E}^{(1)}(X)$ be a closed 1-form. Show that for suitable $c_1, c_2 \in \mathbb{C}$ all the periods of $\omega - c_1\,dz - c_2\,d\bar{z}$ vanish.]

§16. The Riemann–Roch Theorem

The Riemann–Roch Theorem is central in the theory of compact Riemann surfaces. Roughly speaking it tells us how many linearly independent meromorphic functions there are having certain restrictions on their poles.

16.1. Divisors. Let X be a Riemann surface. A divisor on X is a mapping

$$D: X \to \mathbb{Z}$$

such that for any compact subset $K \subset X$ there are only finitely many points $x \in K$ such that $D(x) \neq 0$. With respect to addition the set of all divisors on X is an abelian group which we denote by $\mathrm{Div}(X)$. As well there is a partial ordering on $\mathrm{Div}(X)$. For $D, D' \in \mathrm{Div}(X)$, set $D \leq D'$ if $D(x) \leq D'(x)$ for every $x \in X$.

16.2. Divisors of Meromorphic Functions and 1-forms. Suppose X is a Riemann surface and Y is an open subset of X. For a meromorphic function $f \in \mathcal{M}(Y)$ and $a \in Y$ define

$$\operatorname{ord}_a(f) := \begin{cases} 0, & \text{if } f \text{ is holomorphic and non-zero at } a, \\ k, & \text{if } f \text{ has a zero of order } k \text{ at } a. \\ -k, & \text{if } f \text{ has a pole of order } k \text{ at } a, \\ \infty, & \text{if } f \text{ is identically zero in a} \\ & \quad \text{neighborhood of } a. \end{cases}$$

Thus for any meromorphic function $f \in \mathcal{M}(X) \backslash \{0\}$, the mapping $x \mapsto \operatorname{ord}_x(f)$ is a divisor on X. It is called the divisor of f and will be denoted by (f).

The function f is said to be a *multiple* of the divisor D if $(f) \geq D$. Then f is holomorphic precisely if $(f) \geq 0$.

For a meromorphic 1-form $\omega \in \mathcal{M}^{(1)}(Y)$ one can define its order at a point $a \in Y$ as follows. Choose a coordinate neighborhood (U, z) of a. Then on $U \cap Y$ one may write $\omega = f\,dz$, where f is a meromorphic function. Set $\operatorname{ord}_a(\omega) = \operatorname{ord}_a(f)$. It is easy to check that this is independent of the choice of chart. For 1-forms $\omega \in \mathcal{M}^{(1)}(X) \backslash \{0\}$ the mapping $x \mapsto \operatorname{ord}_x(\omega)$ is again a divisor on X, denoted by (ω).

For $f, g \in \mathcal{M}(X) \backslash \{0\}$ and $\omega \in \mathcal{M}^{(1)}(X) \backslash \{0\}$ one has the following relations:

$$(fg) = (f) + (g), \qquad (1/f) = -(f), \qquad (f\omega) = (f) + (\omega).$$

A divisor $D \in \mathrm{Div}(X)$ is called a *principal divisor* if there exists a function $f \in \mathcal{M}(X) \backslash \{0\}$ such that $D = (f)$. Two divisors $D, D' \in \mathrm{Div}(X)$ are said to be *equivalent* if their difference $D - D'$ is a principal divisor.

By a *canonical divisor* one means the divisor (ω) of a meromorphic 1-form $\omega \in \mathcal{M}^{(1)}(X) \backslash \{0\}$. Any two canonical divisors are equivalent. For, if ω_1, $\omega_2 \in \mathcal{M}^{(1)}(X) \backslash \{0\}$ then there exists a function $f \in \mathcal{M}(X) \backslash \{0\}$ such that $\omega_1 = f\omega_2$ and thus $(\omega_1) - (\omega_2) = (f)$.

16.3. The Degree of a Divisor. Suppose now that X is a *compact* Riemann surface. Then for every $D \in \mathrm{Div}(X)$ there are only finitely many $x \in X$ such that $D(x) \neq 0$. Hence one can define a mapping

$$\deg: \mathrm{Div}(X) \to \mathbb{Z}$$

called the degree, by letting

$$\deg D := \sum_{x \in X} D(x).$$

The mapping deg is a group homomorphism. Note that $\deg(f) = 0$ for any principal divisor (f) on a compact Riemann surface since a meromorphic function has as many zeros as poles. Hence equivalent divisors have the same degree.

16.4. The Sheaves $\mathcal{O}_D$. Suppose D is a divisor on the Riemann surface X. For any open set $U \subset X$ define $\mathcal{O}_D(U)$ to be the set of all those meromorphic functions on U which are multiples of the divisor $-D$, i.e.,

$$\mathcal{O}_D(U) := \{f \in \mathcal{M}(U) : \operatorname{ord}_x(f) \geq -D(x) \quad \text{for every } x \in U\}.$$

Together with the natural restriction mappings $\mathcal{O}_D$ is a sheaf. In the special case of the zero divisor $D = 0$ one has $\mathcal{O}_0 = \mathcal{O}$. If $D, D' \in \operatorname{Div}(X)$ are equivalent divisors, then $\mathcal{O}_D$ and $\mathcal{O}_{D'}$ are isomorphic. An isomorphism can be defined as follows. Pick $\psi \in \mathcal{M}(X)\backslash\{0\}$ such that $D - D' = (\psi)$. Then the sheaf homomorphism induced by multiplication by ψ, i.e.,

$$\mathcal{O}_D \to \mathcal{O}_{D'}, \qquad f \mapsto \psi f,$$

is an isomorphism.

16.5. Theorem. *Suppose X is a compact Riemann surface and $D \in \operatorname{Div}(X)$ is a divisor with $\deg D < 0$. Then $H^0(X, \mathcal{O}_D) = 0$.*

PROOF. Suppose, to the contrary, that there exists an $f \in H^0(X, \mathcal{O}_D)$ with $f \neq 0$. Then $(f) \geq -D$ and thus

$$\deg(f) \geq -\deg D > 0.$$

However this contradicts the fact that $\deg(f) = 0$. □

16.6. The Skyscraper Sheaf $\mathbb{C}_P$. Suppose P is a point of a Riemann surface X. Define a sheaf $\mathbb{C}_P$ on X by

$$\mathbb{C}_P(U) := \begin{cases} \mathbb{C} & \text{if } P \in U, \\ 0 & \text{if } P \notin U, \end{cases}$$

where the restriction maps are the obvious homomorphisms. Then

(i) $H^0(X, \mathbb{C}_P) \cong \mathbb{C}$,
(ii) $H^1(X, \mathbb{C}_P) = 0$.

Now assertion (i) is trivial. In order to prove (ii), consider a cohomology class $\xi \in H^1(X, \mathbb{C}_P)$ which is represented by a cocycle in $Z^1(\mathfrak{U}, \mathbb{C}_P)$. The covering $\mathfrak{U}$ has a refinement $\mathfrak{V} = (V_\alpha)_{\alpha \in A}$ such that the point P is contained in only one V_α. But then $Z^1(\mathfrak{V}, \mathbb{C}_P) = 0$ and hence $\xi = 0$.

16.7. Now suppose D is an arbitrary divisor on X. For $P \in X$ denote by the same letter P the divisor which takes the value 1 at P and is zero otherwise. Then $D \le D + P$ and there is a natural inclusion map $\mathcal{O}_D \to \mathcal{O}_{D+P}$. Let (V, z) be a local coordinate on X about P such that $z(P) = 0$. Define a sheaf homomorphism

$$\beta\colon \mathcal{O}_{D+P} \to \mathbb{C}_P$$

as follows. Suppose $U \subset X$ is an open set. If $P \notin U$, then β_U is the zero homomorphism. If $P \in U$ and $f \in \mathcal{O}_{D+P}(U)$, then the function f admits a Laurent series expansion about P, with respect to the local coordinate z,

$$f = \sum_{n=-k-1}^{\infty} c_n z^n,$$

where $k = D(P)$. Set

$$\beta_U(f) := c_{-k-1} \in \mathbb{C} = \mathbb{C}_P(U).$$

Obviously β is a sheaf epimorphism and

$$0 \to \mathcal{O}_D \to \mathcal{O}_{D+P} \xrightarrow{\beta} \mathbb{C}_P \to 0$$

is a short exact sequence. By Theorem (15.12) this induces an exact sequence

$$\begin{aligned} 0 &\to H^0(X, \mathcal{O}_D) \to H^0(X, \mathcal{O}_{D+P}) \to \mathbb{C} \\ &\to H^1(X, \mathcal{O}_D) \to H^1(X, \mathcal{O}_{D+P}) \to 0. \end{aligned} \tag{*}$$

16.8. Corollary. *Let $D \le D'$ be divisors on a compact Riemann surface X. Then the inclusion map $\mathcal{O}_D \to \mathcal{O}_{D'}$ induces an epimorphism*

$$H^1(X, \mathcal{O}_D) \to H^1(X, \mathcal{O}_{D'}) \to 0.$$

PROOF. If $D' = D + P$, where P is the divisor given by a single point, then the assertion follows from (16.7). In general $D' = D + P_1 + \cdots + P_m$ with $P_j \in X$ and the assertion follows by induction. □

16.9. The Riemann–Roch Theorem. *Suppose D is a divisor on a compact Riemann surface X of genus g. Then $H^0(X, \mathcal{O}_D)$ and $H^1(X, \mathcal{O}_D)$ are finite dimensional vector spaces and*

$$\dim H^0(X, \mathcal{O}_D) - \dim H^1(X, \mathcal{O}_D) = 1 - g + \deg D.$$

PROOF

(a) First the result holds for the divisor $D = 0$. For, $H^0(X, \mathcal{O}) = \mathcal{O}(X)$ consists of only constant functions and thus $\dim H^0(X, \mathcal{O}) = 1$. As well $\dim H^1(X, \mathcal{O}) = g$ by definition.

(b) Keeping the same notation as in 16.7, suppose D is a divisor, $P \in X$ and $D' = D + P$. Suppose that the result holds for one of the divisors D, D'.

The exact cohomology sequence (*) in (16.7) can be split into two short exact sequences. For, let

$$V := \mathrm{Im}(H^0(X, \mathcal{O}_{D'}) \to \mathbb{C})$$

$$W := \mathbb{C}/V.$$

Then $\dim V + \dim W = 1 = \deg D' - \deg D$ and the sequences

$$0 \to H^0(X, \mathcal{O}_D) \to H^0(X, \mathcal{O}_{D'}) \to V \to 0,$$

$$0 \to W \to H^1(X, \mathcal{O}_D) \to H^1(X, \mathcal{O}_{D'}) \to 0$$

are exact. Thus all the vector spaces occurring are finite dimensional and one has the following equations relating the various dimensions

$$\dim H^0(X, \mathcal{O}_{D'}) = \dim H^0(X, \mathcal{O}_D) + \dim V$$

$$\dim H^1(X, \mathcal{O}_D) = \dim H^1(X, \mathcal{O}_{D'}) + \dim W.$$

Adding one gets

$$\dim H^0(X, \mathcal{O}_{D'}) - \dim H^1(X, \mathcal{O}_{D'}) - \deg D'$$
$$= \dim H^0(X, \mathcal{O}_D) - \dim H^1(X, \mathcal{O}_D) - \deg D.$$

This implies that if the Riemann–Roch formula holds for one of the two divisors, then it also holds for the other. Thus by (a) the Theorem holds for every divisor $D' \geq 0$.

(c) An arbitrary divisor D on X may be written

$$D = P_1 + \cdots + P_m - P_{m+1} - \cdots - P_n,$$

where the $P_j \in X$ are points. Starting with the zero divisor and using (b) one now proves by induction that the Riemann–Roch Theorem holds for the divisor D. □

16.10. The Index of Speciality. One calls

$$i(D) := \dim H^1(X, \mathcal{O}_D)$$

the index of speciality of the divisor D. Thus the Riemann–Roch Theorem may be written in the form

$$\dim H^0(X, \mathcal{O}_D) = 1 - g + \deg D + i(D).$$

In (17.16) we will show that $i(D) = 0$ whenever $\deg D > 2g - 2$. In any case $i(D) \geq 0$ and thus $\dim H^0(X, \mathcal{O}_D)$ is bounded from below. From Theorem (16.5) it follows that

$$i(D) = g - 1 - \deg D \quad \text{if } \deg D < 0.$$

16.11. Theorem. *Suppose X is a compact Riemann surface of genus g and a is a point of X. Then there is a non-constant meromorphic function f on X which has a pole of order $\leq g + 1$ at a and is otherwise holomorphic.*

PROOF. Let $D: X \to \mathbb{Z}$ be the divisor with $D(a) = g + 1$ and $D(x) = 0$ for $x \neq a$. By the Riemann–Roch Theorem

$$\dim H^0(X, \mathcal{O}_D) \geq 1 - g + \deg D = 2.$$

Thus there exists a non-constant function $f \in H^0(X, \mathcal{O}_D)$ and clearly this function fulfills the requirements of the theorem. □

16.12. Corollary. *Suppose X is a Riemann surface of genus g. Then there exists a holomorphic covering mapping $f: X \to \mathbb{P}^1$ with at most $g + 1$ sheets.*

PROOF. The function f found in Theorem (16.11) is by Theorem (4.24) such a covering mapping since the value ∞ is assumed with multiplicity $\leq g + 1$. □

16.13. Corollary. *Every Riemann surface of genus zero is isomorphic to the Riemann sphere.*

This follows from the fact that a one-sheeted covering map is a biholomorphism.

EXERCISES (§16)

16.1. Let D be a divisor on the Riemann sphere $\mathbb{P}^1$. Prove

(a) $\dim H^0(\mathbb{P}^1, \mathcal{O}_D) = \max(0, 1 + \deg D)$
(b) $\dim H^1(\mathbb{P}^1, \mathcal{O}_D) = \max(0, -1 - \deg D)$.

16.2. Let $X = \mathbb{C}/\Gamma$ be a torus, $x_0 \in X$ a point and P the divisor

$$P(x) = \begin{cases} 1 & \text{if } x = x_0, \\ 0 & \text{if } x \neq x_0. \end{cases}$$

Show

$$\dim H^0(X, \mathcal{O}_{nP}) = \begin{cases} 0 & \text{for } n < 0, \\ 1 & \text{for } n = 0, \\ n & \text{for } n \geq 1. \end{cases}$$

[*Hint*: Use the Weierstrass $\wp$-function (Ex. 2.1).]

16.3. Let X be a compact Riemann surface, D a divisor on X and $\mathfrak{U} = (U_i)$ an open covering of X such that every U_i is isomorphic to a disk. Show that $\mathfrak{U}$ is a Leray covering for the sheaf $\mathcal{O}_D$, cf. (12.8).

16.4. (a) On a Riemann surface X let $\mathfrak{D}$ be the sheaf of divisors, i.e., for $U \subset X$ open $\mathfrak{D}(U)$ consists of all maps

$$D: U \to \mathbb{Z}$$

such that for every compact set $K \subset U$ there are only finitely many $x \in K$ with $D(x) \neq 0$. Show that $\mathfrak{D}$ together with the natural restriction morphisms is actually a sheaf and that

$$H^1(X, \mathfrak{D}) = 0.$$

[*Hint*: Imitate the proof of Theorem (12.6), using a (discontinuous) integer-valued partition of unity.]

(b) Let $\beta: \mathcal{M}^* \to \mathfrak{D}$ be the map which assigns to every meromorphic function $f \in \mathcal{M}^*(U)$ its divisor $(f) \in \mathfrak{D}(U)$ and let $\alpha: \mathcal{O}^* \to \mathcal{M}^*$ be the natural inclusion map. Show that

$$0 \to \mathcal{O}^* \xrightarrow{\alpha} \mathcal{M}^* \xrightarrow{\beta} \mathfrak{D} \to 0$$

is an exact sequence of sheaves and thus that there is an exact sequence of groups

$$0 \to H^0(X, \mathcal{O}^*) \to H^0(X, \mathcal{M}^*) \to \operatorname{Div}(X) \to H^1(X, \mathcal{O}^*) \to H^1(X, \mathcal{M}^*) \to 0.$$

§17. The Serre Duality Theorem

The Serre Duality Theorem allows a simpler interpretation of the cohomology groups $H^1(X, \mathcal{O}_D)$ in terms of differential forms. In fact, $\dim H^1(X, \mathcal{O}_D)$ is equal to the maximum number of linearly independent meromorphic 1-forms which are multiples of the divisor D. One consequence is the Riemann–Hurwitz formula, which allows one to calculate the genus of a covering from the number of sheets it has and its branching order. Another consequence is a vanishing theorem which asserts that $H^1(X, \mathcal{O}_D) = 0$, if $\deg D > 2g - 2$. This vanishing theorem itself has interesting applications and we will use it to prove an embedding theorem for compact Riemann surfaces into $\mathbb{P}^N$.

17.1. Definition of a Linear Form Res: $H^1(X, \Omega) \to \mathbb{C}$. Suppose X is a compact Riemann surface. By (15.14) the exact sequence

$$0 \to \Omega \to \mathcal{E}^{1,0} \xrightarrow{d} \mathcal{E}^{(2)} \to 0$$

induces an isomorphism $H^1(X, \Omega) \cong \mathcal{E}^{(2)}(X)/d\mathcal{E}^{1,0}(X)$. Suppose $\xi \in H^1(X, \Omega)$ and $\omega \in \mathcal{E}^{(2)}(X)$ is a representative of ξ via this isomorphism. Set

$$\operatorname{Res}(\xi) := \frac{1}{2\pi i} \iint_X \omega.$$

Because of Theorem (10.20) this definition is independent of the choice of the representative ω.

17.2. Mittag–Leffler Distributions of Differential Forms. Suppose X is a Riemann surface, $\mathscr{M}^{(1)}$ is the sheaf of meromorphic 1-forms on X and $\mathfrak{U} = (U_i)_{i \in I}$ is an open covering of X. A cochain $\mu = (\omega_i) \in C^0(\mathfrak{U}, \mathscr{M}^{(1)})$ is called a *Mittag-Leffler distribution* if the differences $\omega_j - \omega_i$ are holomorphic on $U_i \cap U_j$, i.e., $\delta\mu \in Z^1(\mathfrak{U}, \Omega)$. Denote by $[\delta\mu] \in H^1(X, \Omega)$ the cohomology class of $\delta\mu$.

Let a be a point of X. The *residue* of the Mittag–Leffler distribution $\mu = (\omega_i)$ at the point a is defined as follows. Choose $i \in I$ such that $a \in U_i$ and set

$$\operatorname{Res}_a(\mu) := \operatorname{Res}_a(\omega_i).$$

If $a \in U_i \cap U_j$, the difference $\omega_i - \omega_j$ is holomorphic and ω_i and ω_j have the same residue at a. Thus the definition is independent of the choice of $i \in I$.

Now assume that the Riemann surface X is *compact*. Then $\operatorname{Res}_a(\mu) \neq 0$ for only finitely many points a. Thus one can define

$$\operatorname{Res}(\mu) := \sum_{a \in X} \operatorname{Res}_a(\mu).$$

We will now show that this residue is related to the mapping Res defined in (17.1).

17.3. Theorem. *Assume the notation is the same as above. Then*

$$\operatorname{Res}(\mu) = \operatorname{Res}([\delta\mu]).$$

PROOF. In order to compute $\operatorname{Res}([\delta\mu])$ we have to construct the isomorphism $H^1(X, \Omega) \cong \mathscr{E}^{(2)}(X)/d\mathscr{E}^{1,0}(X)$ explicitly, cf. (15.13).

Since $\delta\mu = (\omega_j - \omega_i) \in Z^1(\mathfrak{U}, \Omega) \subset Z^1(\mathfrak{U}, \mathscr{E}^{1,0})$ and $H^1(X, \mathscr{E}^{1,0}) = 0$, there exists a cochain $(\sigma_i) \in C^0(\mathfrak{U}, \mathscr{E}^{1,0})$ such that

$$\omega_j - \omega_i = \sigma_j - \sigma_i \quad \text{on } U_i \cap U_j.$$

Then $d(\omega_j - \omega_i) = d''(\omega_j - \omega_i) = 0$ implies $d\sigma_i = d\sigma_j$ on $U_i \cap U_j$. Thus there exists a global 2-form $\tau \in \mathscr{E}^{(2)}(X)$ such that $\tau \mid U_i = d\sigma_i$. This differential form represents the cohomology class $[\delta\mu]$ and thus

$$\operatorname{Res}([\delta\mu]) = \frac{1}{2\pi i} \iint_X \tau.$$

Suppose $a_1, \ldots, a_n \in X$ are the finitely many poles of μ and let $X' = X \setminus \{a_1, \ldots, a_n\}$. On $X' \cap U_i \cap U_j$ one has $\sigma_i - \omega_i = \sigma_j - \omega_j$. Thus there exists a differential form $\sigma \in \mathscr{E}^{1,0}(X')$ such that $\sigma = \sigma_i - \omega_i$ on $X' \cap U_i$. Hence $\tau = d\sigma$ on X'.

For every a_k there is an $i(k) \in I$ such that $a_k \in U_{i(k)}$. Choose a coordinate neighborhood (V_k, z_k) such that $V_k \subset U_{i(k)}$ and $z_k(a_k) = 0$. We may assume that the V_k are pairwise disjoint and that each $z_k(V_k) \subset \mathbb{C}$ is a disk. For every k

choose a function $f_k \in \mathscr{E}(X)$ such that $\operatorname{Supp}(f_k) \subset V_k$ and such that there is an open neighborhood $V'_k \subset V_k$ of a_k with $f_k \mid V'_k = 1$. Set

$$g := 1 - (f_1 + \cdots + f_n).$$

Since $g \mid V'_k = 0$, $g\sigma$ may be continued across the points a_k by defining it to be zero there and thus may be considered as an element of $\mathscr{E}^{1,0}(X)$. By (10.20) one has

$$\iint_X d(g\sigma) = 0.$$

On $V'_k \backslash \{a_k\}$ one has $d(f_k\sigma) = d\sigma = d(\sigma_{i(k)} - \omega_{i(k)}) = d\sigma_{i(k)}$. Thus $d(f_k\sigma)$ may be continued differentiably across a_k. Since $f_k\sigma$ vanishes on $X' \backslash \operatorname{Supp}(f_k)$, $d(f_k\sigma)$ may be considered as an element of $\mathscr{E}^{(2)}(X)$. Then $\tau = d(g\sigma) + \sum d(f_k\sigma)$ implies

$$\iint_X \tau = \sum_{k=1}^{n} \iint_X d(f_k\sigma) = \sum_{k=1}^{n} \iint_{V_k} d(f_k\sigma_{i(k)} - f_k\omega_{i(k)}).$$

Using (10.20) again, one has

$$\iint_{V_k} d(f_k\sigma_{i(k)}) = 0$$

and as in (10.21) one can show

$$\iint_{V_k} d(f_k\omega_{i(k)}) = -2\pi i \operatorname{Res}_{a_k}(\omega_{i(k)}).$$

Combining everything, one gets

$$\frac{1}{2\pi i}\iint_X \tau = \sum_{k=1}^{n} \operatorname{Res}_{a_k}(\omega_{i(k)}) = \operatorname{Res}(\mu).$$ □

17.4. The Sheaves Ω_D. Let X be a compact Riemann surface. For any divisor $D \in \operatorname{Div}(X)$ we denote by Ω_D the sheaf of meromorphic 1-forms which are multiples of $-D$. Thus for any open set $U \subset X$ the set $\Omega_D(U)$ consists of all differential forms $\omega \in \mathscr{M}^{(1)}(U)$ such that $\operatorname{ord}_x(\omega) \geq -D(x)$ for every $x \in U$. In particular $\Omega_0 = \Omega$ is the sheaf of all holomorphic 1-forms.

Suppose $\omega \in \mathscr{M}^{(1)}(X)$ is a non-trivial meromorphic 1-form on X, e.g., $\omega = df$, where $f \in \mathscr{M}(X)$ is a non-constant meromorphic function. Let K be the divisor of ω. Then for an arbitrary divisor $D \in \operatorname{Div}(X)$ multiplication by ω induces a sheaf isomorphism

$$\mathcal{O}_{D+K} \xrightarrow{\sim} \Omega_D, \qquad f \mapsto f\omega.$$

Lemma. *There is a constant* $k_0 \in \mathbb{Z}$ *such that*

$$\dim H^0(X, \Omega_D) \geq \deg D + k_0$$

for every $D \in \operatorname{Div}(X)$.

PROOF. Suppose ω and K are as above and g is the genus of X. Set $k_0 := 1 - g + \deg K$. Then by Riemann–Roch

$$\begin{aligned} \dim H^0(X, \Omega_D) &= \dim H^0(X, \mathcal{O}_{D+K}) \\ &= \dim H^1(X, \mathcal{O}_{D+K}) + 1 - g + \deg(D + K) \\ &\geq \deg D + k_0. \end{aligned}$$

□

17.5. Definition of a Dual Pairing. Suppose X is a compact Riemann surface and $D \in \operatorname{Div}(X)$ is a divisor. The product

$$\Omega_{-D} \times \mathcal{O}_D \to \Omega, \qquad (\omega, f) \mapsto \omega f,$$

induces a mapping

$$H^0(X, \Omega_{-D}) \times H^1(X, \mathcal{O}_D) \to H^1(X, \Omega).$$

The composition of this mapping with Res: $H^1(X, \Omega) \to \mathbb{C}$ produces a bilinear mapping

$$\langle\ ,\ \rangle \colon H^0(X, \Omega_{-D}) \times H^1(X, \mathcal{O}_D) \to \mathbb{C},$$
$$\langle \omega, \xi \rangle := \operatorname{Res}(\omega\xi).$$

Hence this mapping induces a linear mapping

$$\iota_D \colon H^0(X, \Omega_{-D}) \to H^1(X, \mathcal{O}_D)^*$$

of $H^0(X, \Omega_{-D})$ into the dual of $H^1(X, \mathcal{O}_D)$. The Serre Duality Theorem asserts that $\langle\ ,\ \rangle$ is a dual pairing, i.e., ι_D is an isomorphism. This will be proved in (17.6) and (17.9).

17.6. Theorem. *The mapping* ι_D *is injective.*

PROOF. We have to show that for any non-zero $\omega \in H^0(X, \Omega_{-D})$ there exists $\xi \in H^1(X, \mathcal{O}_D)$ such that $\langle \omega, \xi \rangle \neq 0$. Let $a \in X$ be a point such that $D(a) = 0$ and (U_0, z) be a coordinate neighborhood of a with $z(a) = 0$ and $D \mid U_0 = 0$. On U_0 one can write ω as $\omega = f\,dz$ where $f \in \mathcal{O}(U_0)$. We may assume U_0 is so small that f has no zeros in $U_0 \setminus \{a\}$. Set $U_1 = X \setminus \{a\}$ and $\mathfrak{U} = (U_0, U_1)$. Let $\eta = (f_0, f_1) \in C^0(\mathfrak{U}, \mathcal{M})$, where $f_0 = (zf)^{-1}$ and $f_1 = 0$. Then

$$\omega\eta = \left(\frac{dz}{z}, 0\right) \in C^0(\mathfrak{U}, \mathcal{M}^{(1)})$$

is a Mittag-Leffler distribution with $\operatorname{Res}(\omega\eta) = 1$. One has $\delta\eta \in Z^1(\mathfrak{U}, \mathcal{O}_D)$. Let $\xi = [\delta\eta] \in H^1(X, \mathcal{O}_D)$ be the cohomology class of $\delta\eta$. Since $\omega\xi = \omega \cdot [\delta\eta] = [\delta(\omega\eta)]$, it follows from Theorem (17.3) that

$$\langle \omega, \xi \rangle = \operatorname{Res}(\omega\xi) = \operatorname{Res}([\delta(\omega\eta)]) = \operatorname{Res}(\omega\eta) = 1. \qquad \square$$

17.7. Suppose $D, D' \in \operatorname{Div}(X)$ are two divisors on the compact Riemann surface X with $D' \leq D$. Then by (16.8) the inclusion $0 \to \mathcal{O}_{D'} \to \mathcal{O}_D$ induces an epimorphism

$$H^1(X, \mathcal{O}_{D'}) \to H^1(X, \mathcal{O}_D) \to 0.$$

This then induces a monomorphism of the duals

$$0 \to H^1(X, \mathcal{O}_D)^* \xrightarrow{i^D_{D'}} H^1(X, \mathcal{O}_{D'})^*.$$

One can easily check that the diagram

$$\begin{array}{ccc} 0 \to H^1(X, \mathcal{O}_D)^* & \xrightarrow{i^D_{D'}} & H^1(X, \mathcal{O}_{D'})^* \\ \uparrow \iota_D & & \uparrow \iota_{D'} \\ 0 \to H^0(X, \Omega_{-D}) & \longrightarrow & H^0(X, \Omega_{-D'}) \end{array}$$

commutes, where the vertical arrows are the maps defined in (17.5).

Lemma. *Using the same notation as above suppose* $\lambda \in H^1(X, \mathcal{O}_D)^*$ *and* $\omega \in H^0(X, \Omega_{-D'})$ *satisfy*

$$i^D_{D'}(\lambda) = \iota_{D'}(\omega).$$

Then ω *is also contained in* $H^0(X, \Omega_{-D})$ *and* $\lambda = \iota_D(\omega)$.

PROOF. Suppose, to the contrary, that ω is not an element of $H^0(X, \Omega_{-D})$. Then there is a point $a \in X$ such that $\operatorname{ord}_a(\omega) < D(a)$. Let (U_0, z) be a coordinate neighborhood of a with $z(a) = 0$. On U_0 one may write ω as $\omega = f\,dz$, where $f \in \mathcal{M}(U_0)$. We may suppose U_0 is sufficiently small so that

(i) $D \mid U_0 \backslash \{a\} = 0, \qquad D' \mid U_0 \backslash \{a\} = 0.$
(ii) f has no zeros or poles in $U_0 \backslash \{a\}$.

Set $U_1 = X \backslash \{a\}$ and $\mathfrak{U} = (U_0, U_1)$. Let $\eta = (f_0, f_1) \in C^0(\mathfrak{U}, \mathcal{M})$, where $f_0 = (zf)^{-1}$ and $f_1 = 0$. Because $\operatorname{ord}_a(\omega) < D(a)$, one even has $\eta \in C^0(\mathfrak{U}, \mathcal{O}_D)$. Thus

$$\delta\eta \in Z^1(\mathfrak{U}, \mathcal{O}) = Z^1(\mathfrak{U}, \mathcal{O}_D) = Z^1(\mathfrak{U}, \mathcal{O}_{D'}).$$

Denote the cohomology class of $\delta\eta$ in $H^1(X, \mathcal{O}_{D'})$ by ξ' and in $H^1(X, \mathcal{O}_D)$ by ξ. Note that $\xi = 0$. By assumption

$$\langle \omega, \xi' \rangle = i^D_{D'}(\lambda)(\xi') = \lambda(\xi) = 0.$$

On the other hand, since $\omega\eta = ((dz/z), 0)$, one has

$$\langle \omega, \xi' \rangle = \operatorname{Res}(\omega\eta) = 1, \quad \text{a contradiction!}$$

Thus the assumption is false and $\omega \in H^0(X, \Omega_{-D})$.

Since $i_{D'}^D(\lambda) = \iota_{D'}(\omega) = i_{D'}^D(\iota_D(\omega))$, the equality $\lambda = \iota_D(\omega)$ follows from the fact that $i_{D'}^D$ is one-to-one. □

17.8. Suppose D and B are two divisors on the compact Riemann surface X. Given a meromorphic function $\psi \in H^0(X, \mathcal{O}_B)$ the sheaf morphism

$$\mathcal{O}_{D-B} \xrightarrow{\psi} \mathcal{O}_D, \qquad f \mapsto \psi \cdot f,$$

induces a linear mapping $H^1(X, \mathcal{O}_{D-B}) \to H^1(X, \mathcal{O}_D)$ and thus a linear mapping

$$H^1(X, \mathcal{O}_D)^* \to H^1(X, \mathcal{O}_{D-B})^*,$$

which we also denote by ψ. By definition

$$(\psi\lambda)(\xi) = \lambda(\psi\xi) \quad \text{for } \lambda \in H^1(X, \mathcal{O}_D)^*, \qquad \xi \in H^1(X, \mathcal{O}_{D-B}).$$

The diagram

$$\begin{array}{ccc} H^1(X, \mathcal{O}_D)^* & \xrightarrow{\psi} & H^1(X, \mathcal{O}_{D-B})^* \\ \uparrow{\scriptstyle \iota_D} & & \uparrow{\scriptstyle \iota_{D-B}} \\ H^0(X, \Omega_{-D}) & \xrightarrow{\psi} & H^0(X, \Omega_{-D+B}) \end{array}$$

commutes, where the arrow in the second row is also defined as multiplication by ψ. This follows since $\langle \psi\omega, \xi \rangle = \langle \omega, \psi\xi \rangle$.

Lemma. *If $\psi \in H^0(X, \mathcal{O}_B)$ is not the zero element, then the mapping*

$$\psi \colon H^1(X, \mathcal{O}_D)^* \to H^1(X, \mathcal{O}_{D-B})^*$$

is injective.

PROOF. Let $A := (\psi) \geq -B$ be the divisor of ψ. The mapping $\mathcal{O}_{D-B} \xrightarrow{\psi\cdot} \mathcal{O}_D$ factors through $\mathcal{O}_{D+A}$, i.e., one has

$$\mathcal{O}_{D-B} \to \mathcal{O}_{D+A} \xrightarrow{\psi\cdot} \mathcal{O}_D,$$

where $\mathcal{O}_{D+A} \xrightarrow{\psi\cdot} \mathcal{O}_D$ is an isomorphism. Since the mapping $H^1(X, \mathcal{O}_{D-B}) \to H^1(X, \mathcal{O}_{D+A})$ induced by the inclusion $\mathcal{O}_{D-B} \to \mathcal{O}_{D+A}$ is an epimorphism (16.8), it follows that

$$H^1(X, \mathcal{O}_{D-B}) \xrightarrow{\psi\cdot} H^1(X, \mathcal{O}_D)$$

is also an epimorphism. The result follows from this. □

17.9. The Duality Theorem of Serre. *For any divisor D on a compact Riemann surface X the mapping*

$$\iota_D\colon H^0(X, \Omega_{-D}) \to H^1(X, \mathcal{O}_D)^*$$

defined in (17.5) *is an isomorphism.*

PROOF. Because of (17.6) only the surjectivity of ι_D remains to be proved. Suppose $\lambda \in H^1(X, \mathcal{O}_D)^*$ with $\lambda \neq 0$. We want to show that λ lies in the image of ι_D.

Suppose P is a divisor with $\deg P = 1$. For any natural number n let

$$D_n := D - nP.$$

Denote by $\Lambda \subset H^1(X, \mathcal{O}_{D_n})^*$ the vector subspace of all linear forms of the form $\psi\lambda$, where $\psi \in H^0(X, \mathcal{O}_{nP})$. By Lemma (17.8) Λ is isomorphic to $H^0(X, \mathcal{O}_{nP})$. It thus follows from the Riemann–Roch Theorem that

$$\dim \Lambda \geq 1 - g + n,$$

where g denotes the genus of X. By Lemma (17.4) the vector subspace $\operatorname{Im}(\iota_{D_n}) \subset H^1(X, \mathcal{O}_{D_n})^*$ satisfies

$$\dim \operatorname{Im}(\iota_{D_n}) = \dim H^0(X, \Omega_{-D_n}) \geq n + k_0 - \deg D.$$

For $n > \deg D$ one has $\deg D_n < 0$ and thus $H^0(X, \mathcal{O}_{D_n}) = 0$. The Riemann–Roch Theorem implies

$$\dim H^1(X, \mathcal{O}_{D_n})^* = g - 1 - \deg D_n = n + (g - 1 - \deg D).$$

If one chooses n sufficiently large, then

$$\dim \Lambda + \dim \operatorname{Im}(\iota_{D_n}) > \dim H^1(X, \mathcal{O}_{D_n})^*.$$

This implies $\Lambda \cap \operatorname{Im}(\iota_{D_n}) \neq 0$. Thus there exists $\psi \in H^0(X, \mathcal{O}_{nP})$, $\psi \neq 0$, and $\omega \in H^0(X, \Omega_{-D_n})$ with $\psi\lambda = \iota_{D_n}(\omega)$. Let $A := (\psi)$ be the divisor of ψ, i.e., $1/\psi \in H^0(X, \mathcal{O}_A)$, and let $D' := D_n - A$. Then

$$\iota_{D'}^{D}(\lambda) = \frac{1}{\psi}(\psi\lambda) = \frac{1}{\psi}\iota_{D_n}(\omega) = \iota_{D'}\left(\frac{1}{\psi}\omega\right).$$

From Lemma (17.7) one gets $\omega_0 := (1/\psi)\omega \in H^0(X, \Omega_{-D})$ and $\lambda = \iota_D(\omega_0)$. □

17.10. Remark. Frequently one only uses the Serre Duality Theorem to obtain equality of the dimensions

$$\dim H^1(X, \mathcal{O}_D) = \dim H^0(X, \Omega_{-D}).$$

In particular for $D = 0$ one has

$$g = \dim H^1(X, \mathcal{O}) = \dim H^0(X, \Omega).$$

Thus the genus of a compact Riemann surface X is equal to the maximum number of linearly independent holomorphic 1-forms on X.

One can now formulate the Riemann–Roch Theorem as follows:

$$\dim H^0(X, \mathcal{O}_{-D}) - \dim H^0(X, \Omega_D) = 1 - g - \deg D,$$

or in words: On a compact Riemann surface of genus g the maximum number of linearly independent meromorphic functions which are multiples of a divisor D minus the maximum number of linearly independent meromorphic 1-forms which are multiples of $-D$ is equal to $1 - g - \deg D$.

17.11. Theorem. *Suppose D is a divisor on the compact Riemann surface X. Then*

$$H^0(X, \mathcal{O}_{-D}) \cong H^1(X, \Omega_D)^*.$$

PROOF. Let $\omega_0 \neq 0$ be a meromorphic 1-form on X and let K be its divisor. By (17.4) one has $\Omega_D \cong \mathcal{O}_{D+K}$ and $\mathcal{O}_{-D} \cong \Omega_{-D-K}$. Hence the result follows from the Serre Duality Theorem. □

Consequence. In particular, for $D = 0$ one has $\dim H^1(X, \Omega) = \dim H^0(X, \mathcal{O}) = 1$. This implies that the mapping

$$\text{Res}: H^1(X, \Omega) \to \mathbb{C}$$

is an isomorphism, for it is clear that it is not identically zero.

17.12. Theorem. *The divisor of a non-vanishing meromorphic 1-form ω on a compact Riemann surface of genus g satisfies*

$$\deg(\omega) = 2g - 2.$$

PROOF. Let $K = (\omega)$. By Riemann–Roch

$$\dim H^0(X, \mathcal{O}_K) - \dim H^1(X, \mathcal{O}_K) = 1 - g + \deg K.$$

By (17.4) one has $\Omega \cong \mathcal{O}_K$. Thus

$$1 - g + \deg K = \dim H^0(X, \Omega) - \dim H^1(X, \Omega) = g - 1$$

and so $\deg K = 2(g - 1)$. □

17.13. Corollary. *For any lattice $\Gamma \subset \mathbb{C}$ the torus $\mathbb{C}/\Gamma$ has genus one.*

PROOF. The 1-form dz on $\mathbb{C}$ induces a 1-form ω on $\mathbb{C}/\Gamma$ having no zeros or poles (see 10.14). Thus $\deg(\omega) = 2g - 2 = 0$ and hence $g = 1$. □

17.14. The Riemann–Hurwitz Formula. Suppose X and Y are compact Riemann surfaces and $f: X \to Y$ is a non-constant holomorphic mapping.

For $x \in X$ let $v(x, f)$ be the multiplicity with which f takes the value $f(x)$ at the point x, cf. (2.2) and (4.23). The number

$$b(f, x) := v(f, x) - 1$$

is called the *branching order* of f at the point x. Note that $b(f, x) = 0$ precisely if f is unbranched at x. Since X is compact, there are only finitely many points $x \in X$ such that $b(f, x) \neq 0$. Thus

$$b(f) := \sum_{x \in X} b(f, x),$$

the *total branching order* of f, is well-defined.

Theorem. *Suppose f: $X \to Y$ is an n-sheeted holomorphic covering mapping between compact Riemann surfaces X and Y with total branching order $b = b(f)$. Let g be the genus of X and g' be the genus of Y. Then*

$$g = \frac{b}{2} + n(g' - 1) + 1.$$

This is known as the "Riemann–Hurwitz formula."

PROOF. Suppose ω is a non-vanishing meromorphic 1-form on Y. Then $\deg(\omega) = 2g' - 2$ and $\deg(f^*\omega) = 2g - 2$.

Suppose $x \in X$ and $f(x) = y$. By Theorem (2.1) there is a coordinate neighborhood (U, z) of x (resp. (U', w) of y) with $z(x) = 0$ (resp. $w(y) = 0$) such that with respect to these coordinates one can write f as $w = z^k$, where $k = v(f, x)$. On U' let $\omega = \psi(w)\, dw$. Then on U one has

$$f^*\omega = \psi(z^k)\, dz^k = kz^{k-1}\psi(z^k)\, dz.$$

This implies

$$\mathrm{ord}_x(f^*\omega) = b(f, x) + v(f, x)\mathrm{ord}_y(\omega).$$

Since

$$\sum_{x \in f^{-1}(y)} v(f, x) = n,$$

for any $y \in Y$ one has

$$\sum_{x \in f^{-1}(y)} \mathrm{ord}_x(f^*\omega) = \sum_{x \in f^{-1}(y)} b(f, x) + n\, \mathrm{ord}_y(\omega).$$

Thus

$$\begin{aligned}\deg(f^*\omega) &= \sum_{x \in X} \mathrm{ord}_x(f^*\omega) = \sum_{y \in Y} \sum_{x \in f^{-1}(y)} \mathrm{ord}_x(f^*\omega) \\ &= \sum_{x \in X} b(f, x) + n \sum_{y \in Y} \mathrm{ord}_y(\omega) = b(f) + n \deg(\omega).\end{aligned}$$

This implies $2g - 2 = b + n(2g' - 2)$ and the result follows. □

17.15. Coverings of the Riemann Sphere. For the special case of an n-sheeted covering $\pi: X \to \mathbb{P}^1$ of the Riemann sphere with total branching order b one gets the genus g of X from the Riemann–Hurwitz formula, i.e.,

$$g = \frac{b}{2} - n + 1.$$

If one has a double covering of $\mathbb{P}^1$, then b is equal the number of branch points and $g = (b/2) - 1$. A compact Riemann surface of genus >1 which admits a double covering of $\mathbb{P}^1$ is called *hyperelliptic*.

For example, let $\pi: X \to \mathbb{P}^1$ be the Riemann surface of $\sqrt{P(z)}$, where

$$P(z) = (z - a_1) \cdot \cdots \cdot (z - a_k)$$

is a polynomial of degree k which has distinct roots a_j (cf. 8.10). Since b must be even, we see that X is branched over ∞ precisely if k is odd. This was proved earlier. The genus of X is $g = [(k-1)/2]$, where $[x]$ denotes the largest integer $\leq x$. One can give an explicit basis $\omega_1, \ldots, \omega_g$ for the vector space of holomorphic 1-forms on X as follows

$$\omega_j := \frac{z^{j-1}\, dz}{\sqrt{P(z)}}, \qquad 1 \leq j \leq g = [(k-1)/2],$$

where z is simply another notation for the meromorphic function $\pi: X \to \mathbb{P}^1$. Using local coordinates at the critical points one can easily show that the ω_j are holomorphic on all of X. Clearly $\omega_1, \ldots, \omega_g$ are linearly independent.

17.16. Theorem. *Suppose X is a compact Riemann surface of genus g and D is a divisor on X. Then*

$$H^1(X, \mathcal{O}_D) = 0 \quad \textit{whenever } \deg D > 2g - 2.$$

Proof. Suppose ω is a non-vanishing meromorphic 1-form on X and K is its divisor. Then by (17.4) there is an isomorphism $\Omega_{-D} \cong \mathcal{O}_{K-D}$. Thus $H^1(X, \mathcal{O}_D)^* \cong H^0(X, \Omega_{-D}) \cong H^0(X, \mathcal{O}_{K-D})$. If $\deg D > 2g - 2$, then $\deg(K - D) < 0$. Thus $H^0(X, \mathcal{O}_{K-D}) = 0$ by Theorem (16.5).

17.17. Corollary. *Suppose X is a compact Riemann surface and $\mathcal{M}$ is the sheaf of meromorphic functions on X. Then*

$$H^1(X, \mathcal{M}) = 0.$$

Proof. Let $\xi \in H^1(X, \mathcal{M})$ be a cohomology class which is represented by a cocycle $(f_{ij}) \in Z^1(\mathfrak{U}, \mathcal{M})$. Passing to a refinement of $\mathfrak{U}$, if necessary, one may assume without loss of generality that the total number of poles of all the f_{ij} is finite. Hence there is a divisor D with $\deg D > 2g - 2$ such that $(f_{ij}) \in Z^1(\mathfrak{U}, \mathcal{O}_D)$. By (17.16) the cocycle (f_{ij}) is cohomologous to zero relative to the sheaf $\mathcal{O}_D$ and thus also relative to the sheaf $\mathcal{M}$. □

Remark. The sheaf $\mathcal{M}^{(1)}$ of meromorphic 1-forms on X is isomorphic to $\mathcal{M}$. An isomorphism $\mathcal{M} \xrightarrow{\sim} \mathcal{M}^{(1)}$ is given by $f \mapsto f\omega$, where $\omega \neq 0$ is a fixed element of $\mathcal{M}^{(1)}$. Thus $H^1(X, \mathcal{M}^{(1)}) = 0$ as well.

This can be used to give a definition, without the use of integrals, of the residue mapping $\text{Res}: H^1(X, \Omega) \to \mathbb{C}$ introduced in (17.1). For, suppose $\xi \in H^1(X, \Omega)$ is represented by the cocycle $(\omega_{ij}) \in Z^1(\mathfrak{U}, \Omega)$. Since $H^1(X, \mathcal{M}^{(1)}) = 0$, this cocycle splits relative to the sheaf $\mathcal{M}^{(1)}$. Thus there is a Mittag-Leffler distribution $\mu \in C^0(\mathfrak{U}, \mathcal{M}^{(1)})$ with $[\delta\mu] = \xi$. Then

$$\text{Res}(\xi) = \text{Res}(\mu)$$

by Theorem (17.3).

17.18. We are now going to give some other applications of Theorem (17.16), but we first consider the following notion. Let D be a divisor on a Riemann surface X. We say that the sheaf $\mathcal{O}_D$ is *globally generated*, if for every $x \in X$ there exists an $f \in H^0(X, \mathcal{O}_D)$ such that

$$\mathcal{O}_{D,x} = \mathcal{O}_x f,$$

i.e., every germ $\varphi \in \mathcal{O}_{D,x}$ may be written $\varphi = \psi f$ with $\psi \in \mathcal{O}_x$. The condition $\mathcal{O}_{D,x} = \mathcal{O}_x f$ is equivalent to

$$\text{ord}_x(f) = -D(x).$$

17.19. Theorem. *Let X be a compact Riemann surface of genus g and D be a divisor on X with* $\deg D \geq 2g$. *Then $\mathcal{O}_D$ is globally generated.*

PROOF. Suppose $x \in X$ is a fixed point and let D' be the divisor defined by

$$D'(y) = \begin{cases} D(y) & \text{for } y \neq x, \\ D(y) - 1 & \text{for } y = x. \end{cases}$$

Since $\deg D > \deg D' > 2g - 2$, by Theorem (17.16) we have

$$H^1(X, \mathcal{O}_D) = H^1(X, \mathcal{O}_{D'}) = 0.$$

The Riemann–Roch Theorem now implies

$$\dim H^0(X, \mathcal{O}_D) > \dim H^0(X, \mathcal{O}_{D'}),$$

and hence there exists an element $f \in H^0(X, \mathcal{O}_D) \backslash H^0(X, \mathcal{O}_{D'})$. This element satisfies the condition $\text{ord}_x(f) = -D(x)$. □

17.20. Embedding into Projective Space. Denote by $\mathbb{P}^N$ the N-dimensional projective space which is defined as $\mathbb{P}^N = (\mathbb{C}^{N+1} \backslash 0)/\sim$, where $\sim$ is the following equivalence relation:

$$(z_0, \ldots, z_N) \sim (z'_0, \ldots, z'_N) \Leftrightarrow \exists \lambda \in \mathbb{C}^*: z_\nu = \lambda z'_\nu \quad \text{for } \nu = 0, \ldots, N.$$

Denote by $(z_0 : \cdots : z_N) \in \mathbb{P}^N$ the equivalence class of $(z_0, \ldots, z_N) \in \mathbb{C}^{N+1} \backslash 0$. Equipped with the quotient topology, $\mathbb{P}^N$ is a compact Hausdorff space. For $j = 0, \ldots, N$ let

$$U_j := \{(z_0 : \cdots : z_N) \in \mathbb{P}^N : z_j \neq 0\}.$$

The family $(U_0, \ldots, U_n)$ forms an open covering of $\mathbb{P}^N$. Let

$$\varphi_j : U_j \to \mathbb{C}^N$$

be defined by

$$\varphi_j(z_0 : \cdots : z_N) := \left(\frac{z_0}{z_j}, \ldots, \frac{z_{j-1}}{z_j}, \frac{z_{j+1}}{z_j}, \ldots, \frac{z_N}{z_j}\right).$$

It is easy to see that φ_j is well-defined and maps U_j homeomorphically onto $\mathbb{C}^N$.

Now suppose X is a compact Riemann surface and

$$F : X \to \mathbb{P}^N$$

is a continuous map. Then $W_j := F^{-1}(U_j)$ is an open subset of X for $j = 0, \ldots, N$, and we can consider the maps

$$F_j := \varphi_j \circ F : W_j \to \mathbb{C}^N.$$

Then every F_j is an N-dimensional vector $F_j = (F_{j1}, \ldots, F_{jN})$ of functions $F_{j\nu} : W_j \to \mathbb{C}$. The map $F : X \to \mathbb{P}^N$ is said to be *holomorphic* if all of the functions $F_{j\nu}$ are holomorphic. F is called an *immersion* if it is holomorphic and for every point $x \in X$ there exists at least one $F_{j\nu}$ such that $x \in W_j$ and $dF_{j\nu}(x) \neq 0$. A holomorphic map $F : X \to \mathbb{P}^N$ is called an *embedding* if it is an injective immersion.

17.21. Examples of holomorphic mappings $F : X \to \mathbb{P}^N$ can be obtained in the following way. Let $f_0, \ldots, f_N \in \mathcal{M}(X)$ be meromorphic functions on X which do not vanish identically. Define

$$F = (f_0 : f_1 : \cdots : f_N) : X \to \mathbb{P}^N$$

as follows. For $x \in X$ let (V, z) be a coordinate neighborhood with $z(x) = 0$ and let

$$k := \min_j \operatorname{ord}_x(f_j).$$

On V we can write $f_j = z^k g_j$, where g_j is holomorphic in a neighborhood of x and for at least one j we have $g_j(x) \neq 0$. Set

$$F(x) := (g_0(x) : \cdots : g_N(x)).$$

Of course this definition is independent of the local coordinate chosen. If $g_j(x) \neq 0$, then $F(x) \in U_j$ and hence $x \in W_j$ and the map $F_j: W_j \to \mathbb{C}^N$, as defined in (17.20), has the following form in a neighborhood of x:

$$F_j = \left(\frac{g_0}{g_j}, \ldots, \frac{g_{j-1}}{g_j}, \frac{g_{j+1}}{g_j}, \ldots, \frac{g_N}{g_j}\right).$$

This shows that F is holomorphic.

17.22. Theorem. *On a compact Riemann surface X of genus g let D be a divisor of degree $\geq 2g + 1$. Let $f_0, \ldots, f_N$ be a basis of $H^0(X, \mathcal{O}_D)$. Then*

$$F = (f_0 : \cdots : f_N): X \to \mathbb{P}^N$$

is an embedding.

PROOF

(a) First let us show F is injective. Suppose $x_1 \neq x_2$ are two points of X. Let D' be the divisor defined by

$$D'(x) := \begin{cases} D(x) & \text{for } x \neq x_2, \\ D(x) - 1 & \text{for } x = x_2. \end{cases}$$

Since $\deg D' = \deg D - 1 \geq 2g$, the sheaf $\mathcal{O}_{D'}$ is globally generated by Theorem (17.19), hence there exists an $f \in H^0(X, \mathcal{O}_{D'})$ such that

$$\operatorname{ord}_{x_1}(f) = -D(x_1). \qquad (*)$$

By the definition of D' we have

$$\operatorname{ord}_{x_2}(f) \geq -D(x_2) + 1. \qquad (**)$$

Of course f also belongs to $H^0(X, \mathcal{O}_D)$, so $f = \sum \lambda_j f_j$ for certain coefficients $\lambda_j \in \mathbb{C}$. Let (V_1, z_1) and (V_2, z_2) be coordinate neighborhoods of x_1 and x_2 resp. such that $z_\mu(x_\mu) = 0$, $\mu = 1, 2$. Since $\mathcal{O}_D$ is globally generated, we have

$$k_\mu := \min_j \operatorname{ord}_{x_\mu}(f_j) = -D(x_\mu).$$

Write $f_j = z_\mu^{k_\mu} g_{\mu j}$ and $f = z_\mu^{k_\mu} g_\mu$ in a neighborhood of x_μ. Then

$$F(x_\mu) = (g_{\mu 0}(x_\mu) : \cdots : g_{\mu N}(x_\mu))$$

and

$$\sum_{j=0}^{N} \lambda_j g_{\mu j}(x_\mu) = g_\mu(x_\mu).$$

But from (*) and (**) it follows that $g_1(x_1) \neq 0$ and $g_2(x_2) = 0$. This shows $F(x_1) \neq F(x_2)$.

(b) We now prove that F is an immersion. Let $x_0 \in X$ be a given point and consider the divisor D' defined by

$$D'(x) := \begin{cases} D(x) & \text{for } x \neq x_0, \\ D(x) - 1 & \text{for } x = x_0. \end{cases}$$

Then D' is globally generated and hence there exists an $f \in H^0(X, \mathcal{O}_{D'})$ such that

$$\operatorname{ord}_{x_0}(f) = -D(x_0) + 1.$$

As above $f = \sum \lambda_j f_j$ for certain $\lambda_j \in \mathbb{C}$. Let (V, z) be a coordinate neighborhood of x_0 such that $z(x_0) = 0$ and set

$$f_j = z^k g_j, \qquad f = z^k g,$$

where $k = \min \operatorname{ord}_{x_0}(f_j) = -D(x_0)$. Let ν be an index such that $g_\nu(x_0) \neq 0$. We may assume $\nu = 0$. The map $F_0 = \varphi_0 \circ F\colon W_0 \to \mathbb{C}^N$ considered in (17.20), is now given in a neighborhood of x_0 by

$$F_0 = (F_{01}, \ldots, F_{0N}) = \left(\frac{g_1}{g_0}, \ldots, \frac{g_N}{g_0}\right)$$

and we get

$$\sum_{j=1}^{N} \lambda_j F_{0j} = \sum_{j=1}^{N} \lambda_j \left(\frac{g_j}{g_0}\right) = \frac{g}{g_0} - \lambda_0.$$

Hence

$$\sum \lambda_j \, dF_{0j} = d\left(\frac{g}{g_0}\right).$$

Since $g_0(x_0) \neq 0$ and g has a zero of first order at x_0, we have $d(g/g_0)(x_0) \neq 0$. Hence $dF_{0j}(x_0) \neq 0$ for at least one index j. This shows that F is an immersion. □

Remark. It can be shown that if $\deg D \geq 2g + 1$, then there exist elements $\varphi_0, \ldots, \varphi_3 \in H^0(X, \mathcal{O}_D)$ such that $(\varphi_0 : \cdots : \varphi_3)\colon X \to \mathbb{P}^3$ is an embedding. Thus every compact Riemann surface admits an embedding into $\mathbb{P}^3$.

EXERCISES (§17)

17.1. Let $X \to \mathbb{P}^1$ be the Riemann surface of the algebraic function $\sqrt[n]{1 - z^n}$, i.e., the algebraic function defined by the polynomial

$$P(T) = T^n + z^n - 1 \in \mathcal{M}(\mathbb{P}^1)[T],$$

where $z \in \mathcal{M}(\mathbb{P}^1)$ is the canonical coordinate function. Show that the genus of X is

$$g = \frac{(n-1)(n-2)}{2}.$$

17.2. Let X be a compact Riemann surface. Let $\mathcal{Q}(X) \subset \mathcal{M}^{(1)}(X)$ be the space of all meromorphic 1-forms on X whose residues vanish at every point. Using Ex. 15.3 show

$$H^1(X, \mathbb{C}) \cong \mathcal{Q}(X)/d\mathcal{M}(X).$$

17.3. Let $X = \mathbb{C}/\Gamma$ be a torus. Show that the classes of dz and $\wp_\Gamma\, dz$ form a basis of $\mathscr{Q}(X) \bmod d\mathscr{M}(X)$.

17.4. Let D be a divisor on the compact Riemann surface X of genus g. Show

$$\dim H^0(X, \mathcal{O}_D) = 0 \qquad \text{for } \deg D \le -1$$

$$0 \le \dim H^0(X, \mathcal{O}_D) \le 1 + \deg D \qquad \text{for } -1 \le \deg D \le g - 1$$

$$1 - g + \deg D \le \dim H^0(X, \mathcal{O}_D) \le g \quad \text{for } g - 1 \le \deg D \le 2g - 1$$

$$\dim H^0(X, \mathcal{O}_D) = 1 - g + \deg D \qquad \text{for } \deg D \ge 2g - 1.$$

17.5. Let K be a canonical divisor on a compact Riemann surface X of genus > 0, and let $D \ge K$ be a divisor with $\deg D = \deg K + 1$. Show that the sheaf $\mathcal{O}_K$ is globally generated, but $\mathcal{O}_D$ is not.

17.6. Let $\Gamma \subset \mathbb{C}$ be a lattice and let $\wp$ be the Weierstrass $\wp$-function with respect to Γ. Interpret $\wp$ and its derivative $\wp'$ as meromorphic functions on $\mathbb{C}/\Gamma$. Show that

$$(1\colon \wp\colon \wp')\colon \mathbb{C}/\Gamma \to \mathbb{P}^2$$

is an embedding.

17.7. Let X be a compact Riemann surface of genus two. Suppose ω_1 and ω_2 form a basis of $H^0(X, \Omega)$ and define $f \in \mathscr{M}(X)$ by $\omega_1 = f\omega_2$. Show that $f\colon X \to \mathbb{P}^1$ is a 2-sheeted (branched) covering map.

§18. Functions and Differential Forms with Prescribed Principal Parts

As is well known, the classical theorem of Mittag–Leffler asserts that in the complex plane there always exists a meromorphic function having suitably prescribed principal parts. Our present goal is to look at the analogous problem on compact Riemann surfaces. Here the problem does not always have a solution. But from the Serre Duality Theorem one can derive necessary and sufficient conditions for a solution to exist.

18.1. Mittag–Leffler Distributions of Meromorphic Functions. Suppose X is a Riemann surface and $\mathfrak{U} = (U_i)_{i \in I}$ is an open covering of X. A cochain $\mu = (f_i) \in C^0(\mathfrak{U}, \mathscr{M})$ is called a *Mittag–Leffler distribution* if the differences $f_j - f_i$ are holomorphic on $U_i \cap U_j$, i.e., $\delta\mu \in Z^1(\mathfrak{U}, \mathcal{O})$. Thus the functions f_i and f_j have the same principal parts on their common domain of definition. By a *solution* of μ is meant a global meromorphic function $f \in \mathscr{M}(X)$ which has the same principal parts as μ, i.e. $f \mid U_i - f_i \in \mathcal{O}(U_i)$ for every $i \in I$. Denote by $[\delta\mu] \in H^1(X, \mathcal{O})$ the cohomology class represented by the cocycle $\delta\mu$.

Theorem. *A Mittag–Leffler distribution μ has a solution if and only if $[\delta\mu] = 0$.*

PROOF

(a) Suppose $f \in \mathcal{M}(X)$ is a solution of $\mu = (f_i)$. Set $g_i := f_i - f \in \mathcal{O}(U_i)$. Then on $U_i \cap U_j$ one has

$$f_j - f_i = g_j - g_i.$$

This means that the cocycle $\delta\mu = (f_j - f_i)$ is contained in $B^1(\mathfrak{U}, \mathcal{O})$, i.e., $[\delta\mu] = 0$.

(b) Suppose $[\delta\mu] = 0$ and thus $\delta\mu \in B^1(\mathfrak{U}, \mathcal{O})$. Then there exists a cochain $(g_i) \in C^0(\mathfrak{U}, \mathcal{O})$ such that

$$f_j - f_i = g_j - g_i \quad \text{on } U_i \cap U_j.$$

This implies $f_i - g_i = f_j - g_j$ on $U_i \cap U_j$. Thus the $f_i - g_i$ piece together to form a global meromorphic function $f \in \mathcal{M}(X)$. Since $f | U_i - f_i = -g_i \in \mathcal{O}(U_i)$, f is a solution of μ. □

Remark. By (17.17) on every compact Riemann surface $H^1(X, \mathcal{M}) = 0$. This implies that given any cohomology class $\xi \in H^1(X, \mathcal{O})$ there exists a Mittag–Leffler distribution $\mu \in C^0(\mathfrak{U}, \mathcal{M})$ such that $\xi = [\delta\mu]$, for a suitably chosen covering $\mathfrak{U}$. Thus on every compact Riemann surface of genus ≥ 1 there are Mittag–Leffler problems which have no solution. But on the Riemann sphere $H^1(\mathbb{P}^1, \mathcal{O}) = 0$ and every Mittag–Leffler distribution has a solution. This is also easy to see directly.

18.2. Now suppose X is a compact Riemann surface and $\mu \in C^0(\mathfrak{U}, \mathcal{M})$ is a Mittag–Leffler distribution of meromorphic functions on X. Then for every holomorphic 1-form $\omega \in \Omega(X)$ the product $\omega\mu \in C^0(\mathfrak{U}, \mathcal{M}^{(1)})$ is a Mittag–Leffler distribution of 1-forms and thus by (17.2) the residue $\text{Res}(\omega\mu)$ is defined. This allows us to formulate the criterion alluded to above which tells us when μ has a solution.

Theorem. *Suppose $\mu \in C^0(\mathfrak{U}, \mathcal{M})$ is a Mittag–Leffler distribution of meromorphic functions on the compact Riemann surface X. Then μ has a solution if and only if*

$$\text{Res}(\omega\mu) = 0 \quad \textit{for every } \omega \in \Omega(X).$$

PROOF. Now $[\delta\mu] \in H^1(X, \mathcal{O})$ vanishes if and only if $\lambda([\delta\mu]) = 0$ for every $\lambda \in H^1(X, \mathcal{O})^*$. By the Serre Duality Theorem this is the case exactly if

$$\langle \omega, [\delta\mu] \rangle = 0 \quad \text{for every } \omega \in \Omega(X).$$

By Theorem (17.3) one has $\langle \omega, [\delta\mu] \rangle = \text{Res}(\omega[\delta\mu]) = \text{Res}(\omega\mu)$. Thus the result follows from Theorem (18.1). □

Remarks

(a) If $\omega_1, \ldots, \omega_g$ is a basis of $\Omega(X)$, then $\operatorname{Res}(\omega\mu) = 0$ for every $\omega \in \Omega(X)$ if and only if

$$\operatorname{Res}(\omega_k \mu) = 0 \quad \text{for } k = 1, \ldots, g.$$

Thus μ has a solution if and only if g linear equations hold, where g is the genus of X.

(b) If μ has a solution and $f_1, f_2 \in \mathcal{M}(X)$ are two solutions, then $f_1 - f_2$ is holomorphic on X and thus constant. Hence the solution is unique up to an additive constant.

18.3. Application to Doubly Periodic Functions. Suppose $\gamma_1, \gamma_2 \in \mathbb{C}$ are linearly independent over $\mathbb{R}$ and let

$$P := \{t_1\gamma_1 + t_2\gamma_2 : 0 \le t_1 < 1, 0 \le t_2 < 1\}.$$

Suppose that at the points $a_1, \ldots, a_n \in P$ principal parts

$$\sum_{\nu=-r_j}^{-1} c_\nu^{(j)}(z - a_j)^\nu, \quad \text{for } j = 1, \ldots, n,$$

are prescribed. Then there exists a meromorphic function $f \in \mathcal{M}(\mathbb{C})$ doubly periodic with respect to $\Gamma = \mathbb{Z}\gamma_1 + \mathbb{Z}\gamma_2$ and having poles with the prescribed principal parts at the points $a_1, \ldots, a_n$ if and only if

$$\sum_{j=1}^{n} c_{-1}^{(j)} = 0.$$

PROOF. Any function doubly periodic with respect to Γ may be considered as a function on the torus $X = \mathbb{C}/\Gamma$. The prescribed principal parts then give rise to a Mittag-Leffler distribution μ on X. The differential form ω on X induced by the 1-form dz on $\mathbb{C}$ (cf. 10.14) is a basis of $\Omega(X)$, since $\dim \Omega(X) = 1$. Now

$$\operatorname{Res}(\omega\mu) = \sum_{j=1}^{n} c_{-1}^{(j)}$$

and the result follows from the above theorem. □

In particular, this implies that there are no doubly periodic meromorphic functions having precisely one pole of order one in any period parallelogram. For, such a function would have a non-zero residue (cf. 5.7.c).

We now consider whether on a Riemann surface of genus $g > 1$ there exist meromorphic functions which have one pole of order $\le g$ but are otherwise holomorphic. To do this we need some preliminaries.

18.4. The Wronskian Determinant. Suppose $f_1, \ldots, f_g$ are holomorphic functions on a domain $U \subset \mathbb{C}$. By the Wronskian determinant of $f_1, \ldots, f_g$ one means the determinant of the matrix of derivatives $f_k^{(m)}$, where $0 \le m \le g-1$, $1 \le k \le g$, i.e.,

$$W(f_1, \ldots, f_g) := \det \begin{pmatrix} f_1 & f_2 & \cdots & f_g \\ f'_1 & f'_2 & \cdots & f'_g \\ \vdots & \vdots & & \vdots \\ f_1^{(g-1)} & f_2^{(g-1)} & \cdots & f_g^{(g-1)} \end{pmatrix}.$$

If the functions $f_1, \ldots, f_g$ are linearly independent over $\mathbb{C}$, then the Wronskian determinant is not identically zero. This can be proved by induction on g. For, suppose that we have already shown that $W(f_1, \ldots, f_{g-1}) \not\equiv 0$. Consider the differential equation

$$W(f_1, \ldots, f_{g-1}, w) = \det \begin{pmatrix} f_1 & f_2 & \cdots & f_{g-1} & w \\ f'_1 & f'_2 & \cdots & f'_{g-1} & w' \\ \vdots & \vdots & & \vdots & \vdots \\ f_1^{(g-1)} & f_2^{(g-1)} & \cdots & f_{g-1}^{(g-1)} & w^{(g-1)} \end{pmatrix}$$

$$= 0$$

for some unknown function w. If one expands by cofactors about the last column, then one gets

$$a_0 w^{(g-1)} + a_1 w^{(g-2)} + \cdots + a_{g-1} w = 0, \qquad (*)$$

where $a_0 = W(f_1, \ldots, f_{g-1})$. Clearly $f_1, \ldots, f_{g-1}$ are solutions of this differential equation. If $W(f_1, \ldots, f_g)$ vanishes identically, then f_g is another solution of (*). Hence f_g is a linear combination of $f_1, \ldots, f_{g-1}$ over $U' := \{z \in U : a_0(z) \neq 0\}$ and by the Identity Theorem over U as well. But this is a contradiction.

Now suppose X is a compact Riemann surface of genus $g \ge 1$ and $\omega_1, \ldots, \omega_g$ is a basis of $\Omega(X)$. For any coordinate neighborhood (U, z) we can define a holomorphic function $W_z(\omega_1, \ldots, \omega_g)$ on U as follows. The 1-forms ω_k may be written $\omega_k = f_k\, dz$ on U. Set

$$W_z(\omega_1, \ldots, \omega_g) := W(f_1, \ldots, f_g),$$

where the derivatives of the functions f_k on the right-hand side are taken with respect to z. How the Wronskian determinant of $\omega_1, \ldots, \omega_g$ behaves under a change of coordinates is answered by the next theorem.

18.5. Theorem. *Suppose (U, z) and $(\tilde{U}, \tilde{z})$ are two coordinate neighborhoods on X. Then on $U \cap \tilde{U}$ one has*

$$W_z(\omega_1, \ldots, \omega_g) = \left(\frac{d\tilde{z}}{dz}\right)^N W_{\tilde{z}}(\omega_1, \ldots, \omega_g), \quad \textit{where } N = \frac{g(g+1)}{2}.$$

PROOF. Set $\psi := (d\tilde{z}/dz) \in \mathcal{O}^*(U \cap \tilde{U})$. Define the functions f_k and $\tilde{f}_k$ on $U \cap \tilde{U}$ by

$$\omega_k = f_k\, dz = \tilde{f}_k\, d\tilde{z}.$$

Then $f_k = \psi \tilde{f}_k$. By induction on m one can now show that

$$\frac{d^m f_k}{dz^m} = \psi^{m+1} \frac{d^m \tilde{f}_k}{d\tilde{z}^m} + \sum_{\mu=0}^{m-1} \varphi_{m\mu} \frac{d^\mu \tilde{f}_k}{d\tilde{z}^\mu}$$

where the $\varphi_{m\mu}$ are holomorphic functions on $U \cap \tilde{U}$ which are independent of k. From this one gets

$$\det\left(\frac{d^m f_k}{dz^m}\right)_{m=0,\ldots,g-1,\, k=1,\ldots,g} = \det\left(\psi^{m+1} \frac{d^m \tilde{f}_k}{d\tilde{z}^m}\right)_{m=0,\ldots,g-1,\, k=1,\ldots,g}$$

Since $1 + 2 + \cdots + g = g(g+1)/2$, the result follows. □

If $\tilde{\omega}_1, \ldots, \tilde{\omega}_g$ is another basis of $\Omega(X)$, then there exist constants $c_{jk} \in \mathbb{C}$ with $\det(c_{jk}) =: c \neq 0$ such that $\omega_j = \sum_k c_{jk} \tilde{\omega}_k$. Then

$$W_z(\omega_1, \ldots, \omega_g) = c W_z(\tilde{\omega}_1, \ldots, \tilde{\omega}_g).$$

Hence the following definition is meaningful, i.e., it does not depend on the choice of basis of $\Omega(X)$ nor on the choice of local coordinate.

18.6. Definition. Suppose X is a compact Riemann surface of genus $g \geq 1$. A point $p \in X$ is called a *Weierstrass point*, if for a basis $\omega_1, \ldots, \omega_g$ of $\Omega(X)$ and a coordinate neighborhood (U, z) of p, the Wronskian determinant $W_z(\omega_1, \ldots, \omega_g)$ has a zero at p. The order of this zero is called the weight of the Weierstrass point. By definition a Riemann surface of genus 0, i.e., $\mathbb{P}^1$, does not have any Weierstrass points.

18.7. Theorem. *Suppose X is a compact Riemann surface of genus g and p is a point of X. Then there exists a non-constant meromorphic function $f \in \mathcal{M}(X)$ which has a pole of order $\leq g$ at p and is holomorphic on $X \setminus \{p\}$ if and only if p is a Weierstrass point.*

PROOF. We will use the criterion of Theorem (18.2). Suppose $\omega_1, \ldots, \omega_g$ is a basis of $\Omega(X)$ and (U, z) is a coordinate neighborhood of p with $z(p) = 0$. The ω_k may be expanded in series

$$\omega_k = \sum_{\nu=0}^{\infty} a_{k\nu} z^\nu\, dz, \qquad k = 1, \ldots, g,$$

about p. The function f which we are looking for has a principal part at p of the form

$$h = \sum_{\nu=0}^{g-1} \frac{c_\nu}{z^{1+\nu}}, \qquad (c_0, \ldots, c_{g-1}) \neq (0, \ldots, 0)$$

and thus is a solution of the Mittag-Leffler distribution

$$\mu = (h, 0) \in C^0(\mathfrak{U}, \mathcal{M}), \qquad \mathfrak{U} = (U, X\setminus\{p\}).$$

Now

$$\operatorname{Res}(\omega_k \mu) = \operatorname{Res}_p(\omega_k h) = \sum_{\nu=0}^{g-1} a_{k\nu} c_\nu .$$

Thus the equations $\operatorname{Res}_p(\omega_k h) = 0$ have a non-trivial solution $(c_0, \ldots, c_{g-1})$ if and only if $\det(a_{k\nu}) = 0$. But this is equivalent to

$$W_z(\omega_1, \ldots, \omega_g)(p) = 0. \qquad \square$$

18.8. Theorem. *On a compact Riemann surface X of genus g the number of Weierstrass points, counted according to their weights, is $(g-1)g(g+1)$.*

PROOF. Suppose (U_i, z_i), $i \in I$, is a covering of X by coordinate neighborhoods. On $U_i \cap U_j$ the function $\psi_{ij} := (dz_j/dz_i)$ is holomorphic and has no zeros. With respect to a fixed basis $\omega_1, \ldots, \omega_g$ of $\Omega(X)$ let

$$W_i := W_{z_i}(\omega_1, \ldots, \omega_g) \in \mathcal{O}(U_i).$$

By Theorem (18.5) one has

$$W_i = \psi_{ij}^N W_j \quad \text{on } U_i \cap U_j, \quad \text{where } N = g(g+1)/2. \tag{1}$$

Setting $D(x) := \operatorname{ord}_x(W_i)$ for $x \in U_i$, defines the divisor D on X corresponding to the Weierstrass points together with their respective weights. Thus $\deg D$ is the total of the weights of the Weierstrass points and the proof is complete once we show $\deg D = (g-1)g(g+1)$.

Let D_1 be the divisor of ω_1. Then $\deg D_1 = 2g - 2$ by Theorem (17.12). If we set $\omega_1 = f_{1i}\, dz_i$ on U_i, then $D_1(x) = \operatorname{ord}_x(f_{1i})$ for every $x \in U_i$. Moreover

$$f_{1i} = \psi_{ij} f_{1j} \quad \text{on } U_i \cap U_j. \tag{2}$$

From (1) and (2) it follows that

$$W_i f_{1i}^{-N} = W_j f_{1j}^{-N} \quad \text{on } U_i \cap U_j.$$

Thus there exists a global meromorphic function $f \in \mathcal{M}(X)$ with $f \mid U_i = W_i f_{1i}^{-N}$. For the divisor of f one has

$$(f) = D - ND_1.$$

Since $\deg(f) = 0$, it follows that

$$\deg D = N \deg D_1 = \frac{g(g+1)}{2}(2g-2) = (g-1)g(g+1). \qquad \square$$

18.9. Corollary. *Every compact Riemann surface X of genus $g \geq 2$ admits a holomorphic covering mapping $f\colon X \to \mathbb{P}^1$ having at most g sheets. In particular every compact Riemann surface of genus 2 is hyperelliptic.*

Remark. In fact, every compact Riemann surface of genus $g \geq 2$ admits a covering of $\mathbb{P}^1$ with $[(g+3)/2]$ or fewer sheets, see [56].

18.10. Differential Forms with Prescribed Principal Parts. Suppose X is a Riemann surface, $\mathfrak{U} = (U_i)_{i \in I}$ is an open covering of X and $\mu = (\omega_i) \in C^0(\mathfrak{U}, \mathcal{M}^{(1)})$ is a Mittag-Leffler distribution of meromorphic 1-forms on X, cf. (17.2). By a *solution* of μ we mean a global meromorphic 1-form $\omega \in \mathcal{M}^{(1)}(X)$ which has the same principal parts as μ, i.e., $\omega | U_i - \omega_i \in \Omega(U_i)$ for every $i \in I$. As in (18.1) one can prove that μ has a solution if and only if the cohomology class $[\delta\mu] \in H^1(X, \Omega)$ vanishes.

18.11. Theorem. *On a compact Riemann surface X a Mittag-Leffler distribution $\mu \in C^0(\mathfrak{U}, \mathcal{M}^{(1)})$ of meromorphic 1-forms has a solution if and only if* $\operatorname{Res}(\mu) = 0$.

PROOF. By Theorem (17.3) one has $\operatorname{Res}(\mu) = \operatorname{Res}([\delta\mu])$. By the consequence in (17.11) the mapping $\operatorname{Res}: H^1(X, \Omega) \to \mathbb{C}$ is an isomorphism. Thus $[\delta\mu] = 0$ is equivalent to $\operatorname{Res}(\mu) = 0$ and the result follows. □

18.12. Corollary. *Suppose X is a compact Riemann surface.*

(a) *For any point $p \in X$ and any natural number $n \geq 2$ there exists a meromorphic 1-form on X which has a pole of order n at p and is otherwise holomorphic ("an elementary differential of the second kind.").*

(b) *For any two distinct points $p_1, p_2 \in X$ there exists a meromorphic 1-form on X which has poles of first order at p_1 and p_2 with residues $+1$ and -1, respectively, and is otherwise holomorphic ("an elementary differential of the third kind").*

EXERCISES (§18)

18.1. Let $U := \{z \in \mathbb{C} : |z| < r\}$, $r > 0$, and let $f: U \to \mathbb{C}$ be a holomorphic function with $f(0) \neq 0$.

(a) Define $f_j(z) := z^{j-1} f(z)$ for $j = 1, \ldots, g$. Prove that the Wronskian determinant $W(f_1, \ldots, f_g)$ does not vanish at the origin.

(b) Define $\varphi_j(z) := z^{2j-2} f(z)$ for $j = 1, \ldots, g$. Prove that the Wronskian determinant $W(\varphi_1, \ldots, \varphi_g)$ has a zero of order $(g(g-1)/2)$ at the origin.

18.2. Let $\pi: X \to \mathbb{P}^1$ be a hyperelliptic Riemann surface of genus $g \geq 2$.

(a) Show that all the ramification points $p_1, \ldots, p_{2g+2} \in X$ of π are Weierstrass points of X.

(b) Prove that there are no other Weierstrass points and that every Weierstrass point p_j has weight $(g(g-1)/2)$.
[*Hint*: Use Ex. 18.1.]

18.3. Let X be a compact Riemann surface of genus $g \geq 1$ and suppose $\omega_1, \ldots, \omega_g$ is a basis of $\Omega(X)$. Let $D \geq 0$ be a non-negative divisor on X. Denote by M_D the set of all Mittag–Leffler distributions $\mu \geq -D$ on X, i.e. the set of all Mittag–Leffler distributions lying in $C^0(\mathfrak{U}, \mathcal{O}_D)$ for some open covering $\mathfrak{U}$ of X. Define a linear map

$$R \colon M_D \to \mathbb{C}^g$$

by

$$R(\mu) := (\operatorname{Res}(\mu\omega_1), \ldots, \operatorname{Res}(\mu\omega_g)).$$

Prove

$$\dim H^1(X, \mathcal{O}_D) = g - \dim R(M_D).$$

§19. Harmonic Differential Forms

With the help of the results obtained so far it is now easy to derive the most important results about harmonic differential forms on compact Riemann surfaces X. In particular every closed differential form on X may be uniquely written as the sum of a harmonic and an exact differential form. This implies that the 1st deRham group of X is isomorphic to the vector space of harmonic differential forms on X. Using this one can show that the genus is a topological invariant.

19.1. Complex Conjugation. For any 1-form $\omega \in \mathcal{E}^{(1)}(X)$ on a Riemann surface X, the complex conjugation of functions induces a conjugate complex differential 1-form $\bar{\omega} \in \mathcal{E}^{(1)}(X)$. For, locally $\omega = \sum f_j\, dg_j$, where the functions f_j and g_j are differentiable. Thus $\bar{\omega} = \sum \bar{f}_j\, d\bar{g}_j$. A 1-form $\omega \in \mathcal{E}^{(1)}(X)$ is said to be real if $\omega = \bar{\omega}$. In general the real part of a differential form ω is defined by

$$\operatorname{Re}(\omega) = \tfrac{1}{2}(\omega + \bar{\omega}).$$

Clearly ω is real if and only if $\omega = \operatorname{Re}(\omega)$. If c is a curve on X, then

$$\overline{\int_c \omega} = \int_c \bar{\omega}, \quad \text{and thus } \operatorname{Re}\left(\int_c \omega\right) = \int_c \operatorname{Re}(\omega).$$

(If ω is not closed, then we assume that c is piecewise continuously differentiable.) If $\omega \in \Omega(X)$ is a holomorphic 1-form, then $\bar{\omega}$ is called *anti-holomorphic*. We denote the vector space of all anti-holomorphic 1-forms on X by $\bar{\Omega}(X)$.

19.2. The $*$-operator. Any 1-form $\omega \in \mathcal{E}^{(1)}(X)$ may be uniquely decomposed as

$$\omega = \omega_1 + \omega_2 \quad \text{where } \omega_1 \in \mathcal{E}^{1,0}(X), \quad \omega_2 \in \mathcal{E}^{0,1}(X).$$

Set

$$*\omega := i(\bar{\omega}_1 - \bar{\omega}_2).$$

The mapping $*: \mathscr{E}^{(1)}(X) \to \mathscr{E}^{(1)}(X)$ is an $\mathbb{R}$-linear isomorphism which maps $\mathscr{E}^{1,0}(X)$ onto $\mathscr{E}^{0,1}(X)$ and vice versa.

For $\omega \in \mathscr{E}^{(1)}(X)$, $\omega_1 \in \mathscr{E}^{1,0}(X)$, $\omega_2 \in \mathscr{E}^{0,1}(X)$ and $f \in \mathscr{E}(X)$ one has the following:

(a) $**\omega = -\omega, \qquad \overline{*\omega} = *\bar{\omega}$,
(b) $d*(\omega_1 + \omega_2) = id'\bar{\omega}_1 - id''\bar{\omega}_2$,
(c) $*d'f = id''\bar{f}, \qquad *d''f = -id'\bar{f}$,
(d) $d*df = 2id'd''\bar{f}$.

19.3. Harmonic Differential Forms. A 1-form $\omega \in \mathscr{E}^{(1)}(X)$ on a Riemann surface X is called *harmonic* if

$$d\omega = d*\omega = 0.$$

Theorem. *Suppose $\omega \in \mathscr{E}^{(1)}(X)$. Then the following conditions are equivalent:*

(i) *ω is harmonic,*
(ii) *$d'\omega = d''\omega = 0$,*
(iii) *$\omega = \omega_1 + \omega_2$ where $\omega_1 \in \Omega(X)$ and $\omega_2 \in \bar{\Omega}(X)$,*
(iv) *given any point $a \in X$ there exists an open neighborhood U of a and a harmonic function f on U such that $\omega = df$.*

PROOF. The equivalence of (i), (ii) and (iii) follows from (19.2).

(i) ⇒ (iv). Since in particular a harmonic differential form is closed, locally $\omega = df$, where f is a differentiable function. Since $0 = d*\omega = d*df = 2id'd''\bar{f}$, it follows that f is harmonic.

(iv) ⇒ (i). If $\omega = df$ and f is harmonic, then $d\omega = ddf = 0$ and $d*\omega = d*df = 0$. □

Notation. The vector space of all harmonic 1-forms on the Riemann surface X will be denoted by $\text{Harm}^1(X)$. Thus

$$\text{Harm}^1(X) = \Omega(X) \oplus \bar{\Omega}(X).$$

Thus if X is a compact Riemann surface of genus g, then

$$\dim \text{Harm}^1(X) = 2g.$$

19.4. Theorem. *Every real harmonic 1-form $\sigma \in \text{Harm}^1(X)$ is the real part of precisely one holomorphic 1-form $\omega \in \Omega(X)$.*

PROOF. Suppose $\sigma = \omega_1 + \bar{\omega}_2$ with $\omega_1, \omega_2 \in \Omega(X)$. Because $\sigma = \omega_1 + \bar{\omega}_2 = \bar{\sigma} = \bar{\omega}_1 + \omega_2$, it follows that $\omega_1 = \omega_2$. Thus $\sigma = \text{Re}(2\omega_1)$.

To prove the uniqueness, suppose $\omega \in \Omega(X)$ and $\mathrm{Re}(\omega) = 0$. Since locally $\omega = df$, where f is a holomorphic function, it follows that f has constant real part. Then f itself is constant and thus $\omega = 0$. □

19.5. Scalar Products in $\mathscr{E}^{(1)}(X)$. We now assume that X is a *compact* Riemann surface. For $\omega_1, \omega_2 \in \mathscr{E}^{(1)}(X)$ let

$$\langle \omega_1, \omega_2 \rangle := \iint_X \omega_1 \wedge *\omega_2 .$$

Clearly the mapping $(\omega_1, \omega_2) \mapsto \langle \omega_1, \omega_2 \rangle$ is linear in the first and semi-linear in the second argument and

$$\langle \omega_2, \omega_1 \rangle = \overline{\langle \omega_1, \omega_2 \rangle}.$$

We now claim that $\langle\ ,\ \rangle$ is positive definite. For, suppose $\omega \in \mathscr{E}^{(1)}(X)$. With respect to a local chart (U, z), where $z = x + iy$, suppose

$$\omega = f\,dz + g\,d\bar{z}.$$

Then

$$*\omega = i(\bar{f}\,d\bar{z} - \bar{g}\,dz)$$

and

$$\omega \wedge *\omega = i(|f|^2 + |g|^2)\,dz \wedge d\bar{z} = 2(|f|^2 + |g|^2)\,dx \wedge dy.$$

This shows that $\langle \omega, \omega \rangle \geq 0$ and $\langle \omega, \omega \rangle = 0$ only if $\omega = 0$. Hence with this scalar product $\mathscr{E}^{(1)}(X)$ becomes a unitary vector space. However it is not a Hilbert space, since it is not complete.

19.6. Lemma. *Suppose X is a compact Riemann surface.*

(a) *$d'\mathscr{E}(X)$, $d''\mathscr{E}(X)$, $\Omega(X)$ and $\bar{\Omega}(X)$ are pairwise orthogonal vector subspaces of $\mathscr{E}^{(1)}(X)$.*

(b) *$d\mathscr{E}(X)$ and $*d\mathscr{E}(X)$ are orthogonal vector subspaces of $\mathscr{E}^{(1)}(X)$ and*

$$d\mathscr{E}(X) \oplus *d\mathscr{E}(X) = d'\mathscr{E}(X) \oplus d''\mathscr{E}(X).$$

PROOF

(a) Since $\mathscr{E}^{1,0}(X)$ and $\mathscr{E}^{0,1}(X)$ are trivially orthogonal, it suffices to show that $d'\mathscr{E}(X) \perp \Omega(X)$ and $d''\mathscr{E}(X) \perp \bar{\Omega}(X)$.

Suppose $f \in \mathscr{E}(X)$ and $\omega \in \Omega(X)$. Then

$$\omega \wedge *d'f = i\omega \wedge d''\bar{f} = i\omega \wedge d\bar{f} = -i\,d(\bar{f}\omega).$$

Thus

$$\langle \omega, d'f \rangle = -i \iint_X d(\bar{f}\omega) = 0$$

by Theorem (10.20). Similarly one can show $\langle \bar{\omega}, d''f \rangle = 0$.

(b) Suppose $f, g \in \mathscr{E}(X)$. Then

$$df \wedge *(*dg) = -df \wedge dg = -d(f\,dg).$$

Thus

$$\langle df, *dg \rangle = -\iint_X d(f\,dg) = 0.$$

The equality $d\mathscr{E}(X) \oplus *d\mathscr{E}(X) = d'\mathscr{E}(X) \oplus d''\mathscr{E}(X)$ follows from (19.2.c). □

19.7. Corollary. *On a compact Riemann surface X every exact differential form $\sigma \in \mathrm{Harm}^1(X)$ vanishes and every harmonic function $f \in \mathscr{E}(X)$ is constant.*

This follows since $d\mathscr{E}(X)$ is orthogonal to $\mathrm{Harm}^1(X) = \Omega(X) \oplus \bar{\Omega}(X)$.

19.8. Corollary. *Suppose X is a compact Riemann surface and $\sigma \in \mathrm{Harm}^1(X)$, $\omega \in \Omega(X)$. If for every closed curve γ on X one has*

$$\int_\gamma \sigma = 0, \qquad \text{resp. } \mathrm{Re}\left(\int_\gamma \omega\right) = 0,$$

then $\sigma = 0$, resp. $\omega = 0$.

PROOF. Since σ (resp. $\mathrm{Re}(\omega)$) is exact by Theorem (10.15), the result follows from (19.7) and (19.4). □

19.9. Theorem. *On any compact Riemann surface X there is an orthogonal decomposition*

$$\mathscr{E}^{0,1}(X) = d''\mathscr{E}(X) \oplus \bar{\Omega}(X).$$

PROOF. Let g be the genus of X. Since $H^1(X, \mathscr{O}) \cong \mathscr{E}^{0,1}(X)/d''\mathscr{E}(X)$ by Dolbeault's Theorem (15.14), one has

$$\dim \mathscr{E}^{0,1}(X)/d''\mathscr{E}(X) = g.$$

On the other hand, $\dim \bar{\Omega}(X) = g$ by (17.10). The result now follows from Lemma (19.6.a). □

19.10. Corollary. *Suppose X is a compact Riemann surface and $\sigma \in \mathscr{E}^{0,1}(X)$. The equation $d''f = \sigma$ has a solution $f \in \mathscr{E}(X)$ if and only if*

$$\iint_X \sigma \wedge \omega = 0 \quad \text{for every } \omega \in \Omega(X).$$

The given condition is equivalent to $\sigma \perp \bar{\Omega}(X)$.

19.11. Theorem. *On any compact Riemann surface X there is an orthogonal decomposition*

$$\mathscr{E}^{(1)}(X) = *d\mathscr{E}(X) \oplus d\mathscr{E}(X) \oplus \mathrm{Harm}^1(X).$$

PROOF. Taking complex conjugates in (19.9) one gets $\mathscr{E}^{1,0}(X) = d'\mathscr{E}(X) \oplus \Omega(X)$. Thus

$$\mathscr{E}^{(1)}(X) = d'\mathscr{E}(X) \oplus d''\mathscr{E}(X) \oplus \Omega(X) \oplus \bar{\Omega}(X).$$

Hence the result follows from (19.6). □

19.12. Theorem. *Suppose X is a compact Riemann surface. Then*

$$\mathrm{Ker}(\mathscr{E}^{(1)}(X) \xrightarrow{d} \mathscr{E}^{(2)}(X)) = d\mathscr{E}(X) \oplus \mathrm{Harm}^1(X).$$

PROOF. Since $\mathscr{Z}(X) := \mathrm{Ker}(\mathscr{E}^{(1)}(X) \xrightarrow{d} \mathscr{E}^{(2)}(X)) \supset d\mathscr{E}(X) \oplus \mathrm{Harm}^1(X)$, it suffices by Theorem (19.11) to show that

$$\mathscr{Z}(X) \perp *d\mathscr{E}(X).$$

Suppose $\omega \in \mathscr{Z}(X)$ and $f \in \mathscr{E}(X)$. Then

$$\omega \wedge *(*df) = -\omega \wedge df = d(f\omega).$$

Hence

$$\langle \omega, *df \rangle = \iint_X d(f\omega) = 0.$$ □

19.13. Corollary. *Suppose X is a compact Riemann surface. Then a differential form $\sigma \in \mathscr{E}^{(1)}(X)$ is exact if and only if for every closed 1-form $\omega \in \mathscr{E}^{(1)}(X)$ one has*

$$\iint_X \sigma \wedge \omega = 0.$$

PROOF. The given condition is equivalent to $\langle \omega, *\sigma \rangle = 0$ for every closed 1-form ω. But by (19.11) this means $*\sigma \in *d\mathscr{E}(X)$, i.e., $\sigma \in d\mathscr{E}(X)$. □

19.14. Theorem (deRham–Hodge). *Suppose X is a compact Riemann surface. Then*

$$H^1(X, \mathbb{C}) \cong \mathrm{Rh}^1(X) \cong \mathrm{Harm}^1(X).$$

Because of (19.12) this follows directly from deRham's Theorem (15.15).

Remark. Since the sheaf $\mathbb{C}$ of locally constant complex-valued functions on X depends only on the topological structure of X, it follows that

$$b_1(X) := \dim H^1(X, \mathbb{C}),$$

the *first Betti number* of X, is a topological invariant. From (19.14) one has

$$b_1(X) = 2g,$$

where g is the genus of X. Thus the genus is a topological invariant.

There is a topological classification of connected orientable compact two-dimensional manifolds (Riemann surfaces are orientable), which depends only on the first Betti number. Every such surface X with $b_1(X) = 2g$ is homeomorphic to a sphere with g handles (cf. Seifert–Threlfall [46] or [42]).

It should also be noted that for every genus ≥ 1, there are Riemann surfaces which are homeomorphic but which are not holomorphically equivalent. The holomorphic equivalence classes of Riemann surfaces of genus g depend on one complex parameter when $g = 1$ and on $3g - 3$ complex parameters when $g \geq 2$.

This Teichmüller theory will not be dealt with here; for this see [50].

Exercises (§19)

19.1. Let X be a compact Riemann surface. Prove

(a) $d\mathscr{E}^{0,1}(X) = d'd''\mathscr{E}(X) \subset \mathscr{E}^{(2)}(X)$.

(b) Let $\mathscr{H}$ be the sheaf of harmonic functions on X. Then

$$H^1(X, \mathscr{H}) \cong \mathscr{E}^{(2)}(X)/d'd''\mathscr{E}(X) \cong \mathbb{C}.$$

(c) Let $\omega \in \mathscr{E}^{(2)}(X)$. Prove that there exists a function $f \in \mathscr{E}(X)$ such that

$$d'd''f = \omega$$

if and only if

$$\iint_X \omega = 0.$$

19.2. Let $X = \mathbb{C}/\Gamma$ be a torus. For a function $f \in \mathscr{E}(X)$ define its mean value $M(f)$ by

$$M(f) := \left(\iint_X f\, dz \wedge d\bar{z}\right)\left(\iint_X dz \wedge d\bar{z}\right)^{-1}$$

For $\omega = f\, dz + g\, d\bar{z} \in \mathscr{E}^{(1)}(X)$ let $M(\omega) := M(f)\, dz + M(g)\, d\bar{z}$. Show

(a) If $\omega \in \mathscr{Z}(X) := \operatorname{Ker}(\mathscr{E}^{(1)}(X) \xrightarrow{d} \mathscr{E}^{(2)}(X))$, then ω and $M(\omega)$ are cohomologous.

(b) The mapping

$$M\colon \mathscr{Z}(X) \to \operatorname{Harm}^1(X)$$

induces an isomorphism

$$\operatorname{Rh}^1(X) \xrightarrow{\sim} \operatorname{Harm}^1(X).$$

§20. Abel's Theorem

In this section we investigate when there exist meromorphic functions with prescribed zeros and poles on a compact Riemann surface X. Clearly it is necessary that the total order of the zeros equal the total order of the poles. However on Riemann surfaces of genus $g \geq 1$ this condition is not sufficient. Abel's Theorem gives a necessary and sufficient condition for the existence of such functions.

20.1 Functions with Prescribed Divisors. Suppose X is a Riemann surface and D is a divisor on X. A meromorphic function $f \in \mathcal{M}(X)$ is said to be a *solution* of D if $(f) = D$. Thus the function f has precisely the zeros and poles prescribed by the divisor D. If X is compact, then it is possible for D to have a solution only if $\deg D = 0$.

We also need the notion of a weak solution of D. Let

$$X_D := \{x \in X : D(x) \geq 0\}.$$

By a *weak solution* of D we mean a function $f \in \mathcal{E}(X_D)$ with the following property. For every point $a \in X$ there exists a coordinate neighborhood (U, z) with $z(a) = 0$ and a function $\psi \in \mathcal{E}(U)$ with $\psi(a) \neq 0$, such that

$$f = \psi z^k \quad \text{on } U \cap X_D, \quad \text{where } k = D(a). \tag{*}$$

Clearly a weak solution f is a proper, i.e., meromorphic, solution precisely if f is holomorphic on X_D. Two weak solutions f and g of D differ by a factor $\varphi \in \mathcal{E}(X)$ which never vanishes.

If f_1 (resp. f_2) is a weak solution of D_1 (resp. D_2), then $f := f_1 f_2$ is a weak solution of the divisor $D := D_1 + D_2$. At those points $a \in X$ where

$$D(a) \geq 0, \quad \text{but } D_1(a) < 0 \quad \text{or } D_2(a) < 0,$$

the product $f_1 f_2$ is not defined, but using continuity it may be extended to such points. Similarly f_1/f_2 is a weak solution of the divisor $D_1 - D_2$.

20.2. Logarithmic Differentiation. Suppose f is a weak solution of the divisor D. Then the logarithmic derivative df/f is a smooth 1-form on the complement of

$$\operatorname{Supp}(D) = \{x \in X : D(x) \neq 0\}.$$

If $a \in \operatorname{Supp}(D)$ and $k = D(a)$, then using (*) one has the representation

$$\frac{df}{f} = k\frac{dz}{z} + \frac{d\psi}{\psi}.$$

Now $d\psi/\psi$ is differentiable in a neighborhood of a and thus has no singularities. As in (13.1) this implies that for any 1-form $\sigma \in \mathscr{E}^{(1)}(X)$ with compact support the integral

$$\iint_X \frac{df}{f} \wedge \sigma$$

exists. For later use we also note that the 1-form $d''f/f$ is differentiable on all of X, for the local representation $f = \psi z^k$ implies $d''f/f = d''\psi/\psi$.

20.3. Lemma. *Suppose $a_1, \ldots, a_n$ are distinct points on the Riemann surface X and $k_1, \ldots, k_n \in \mathbb{Z}$. Suppose $D \in \mathrm{Div}(X)$ is the divisor with $D(a_j) = k_j$ for $j = 1, \ldots, n$ and $D(x) = 0$ otherwise. Let f be a weak solution of D. Then for any $g \in \mathscr{E}(X)$ with compact support*

$$\frac{1}{2\pi i} \iint_X \frac{df}{f} \wedge dg = \sum_{j=1}^{n} k_j g(a_j).$$

PROOF. Choose disjoint coordinate neighborhoods (U_j, z_j) of the a_j with $z(a_j) = 0$ such that on U_j one may write f as

$$f = \psi_j z_j^{k_j} \quad \text{with } \psi_j \in \mathscr{E}(U_j), \qquad \psi_j(x) \neq 0 \quad \text{for every } x \in U_j.$$

We may assume that $z_j(U_j) \subset \mathbb{C}$ is the unit disk for $j = 1, \ldots, n$.

Suppose $0 < r_1 < r_2 < 1$. There exist functions $\varphi_j \in \mathscr{E}(X)$ with

$$\mathrm{Supp}(\varphi_j) \subset \{|z_j| < r_2\} \quad \text{and } \varphi_j|\{|z_j| \le r_1\} = 1.$$

Let $g_j := \varphi_j g$ for $j = 1, \ldots, n$ and $g_0 := g - (g_1 + \cdots + g_n)$. Since $\mathrm{Supp}(g_0)$ is compact in $X' := X\backslash\{a_1, \ldots, a_n\}$, it follows from (10.20) that

$$\iint_X \frac{df}{f} \wedge dg_0 = -\iint_{X'} d\left(g_0 \frac{df}{f}\right) = 0.$$

Thus

$$\iint_X \frac{df}{f} \wedge dg = \sum_{j=1}^{n} \iint_{U_j} \frac{df}{f} \wedge dg_j = \sum_{j=1}^{n} k_j \iint_{U_j} \frac{dz_j}{z_j} \wedge dg_j.$$

Now Stokes' Theorem implies

$$\iint_{U_j} \frac{dz_j}{z_j} \wedge dg_j = -\lim_{\varepsilon \to 0} \iint_{\varepsilon \le |z_j| \le r_2} d\left(g_j \frac{dz_j}{z_j}\right)$$

$$= \lim_{\varepsilon \to 0} \int_{|z_j| = \varepsilon} g_j \frac{dz_j}{z_j} = 2\pi i g_j(a_j) = 2\pi i g(a_j). \qquad \square$$

20.4. Chains, Cycles and Homology. By a 1-chain on a Riemann surface X we mean a formal finite linear combination with integer coefficients,

$$c = \sum_{j=1}^{k} n_j c_j, \qquad n_j \in \mathbb{Z},$$

where the c_j: $[0, 1] \to X$ are curves. The integral over c of a closed differential form $\omega \in \mathscr{E}^{(1)}(X)$ is defined by

$$\int_c \omega := \sum_{j=1}^{k} n_j \int_{c_j} \omega.$$

The set of all 1-chains on X, which in a natural way is an abelian group, will be denoted by $C_1(X)$. A boundary operator

$$\partial\colon C_1(X) \to \operatorname{Div}(X)$$

is defined as follows. Suppose c: $[0, 1] \to X$ is a curve. Set $\partial c = 0$ if $c(0) = c(1)$. Otherwise let ∂c be the divisor with value $+1$ at $c(1)$ and -1 at $c(0)$ and zero at all other points. For an arbitrary 1-chain $c = \sum n_j c_j$ let $\partial c := \sum n_j \, \partial c_j$. Clearly

$$\deg(\partial c) = 0 \quad \text{for every } c \in C_1(X).$$

Conversely on a compact Riemann surface given any divisor D with $\deg D = 0$ there exists a 1-chain c such that $\partial c = D$. For, a divisor D of degree zero may be written as a sum $D = D_1 + \cdots + D_k$, where each D_j takes the value $+1$ at some point b_j, -1 at some other point a_j and is zero otherwise. Let c_j be a curve from a_j to b_j and $c := c_1 + \cdots + c_k$. Then $\partial c = D$.

The kernel of the mapping ∂,

$$Z_1(X) := \operatorname{Ker}\left(C_1(X) \xrightarrow{\partial} \operatorname{Div}(X)\right),$$

is called the group of 1-*cycles* on X. In particular every closed curve is a 1-cycle.

Two cycles c, $c' \in Z_1(X)$ are said to be *homologous* if for every closed differential form $\omega \in \mathscr{E}^{(1)}(X)$ one has

$$\int_c \omega = \int_{c'} \omega.$$

The set of all homology classes of 1-cycles forms an additive group $H_1(X)$, the 1*st homology group* of X. For $\gamma \in H_1(X)$ and a closed differential form $\omega \in \mathscr{E}^{(1)}(X)$, the integral $\int_\gamma \omega$ is well-defined.

Two closed curves which are homotopic are also homologous. Hence there is a group homomorphism $\pi_1(X) \to H_1(X)$. One can easily check that this mapping is surjective. However it is not in general injective, since the fundamental group is not always abelian.

20.5. Lemma. *Suppose X is a Riemann surface, $c\colon [0, 1] \to X$ is a curve and U is a relatively compact open neighborhood of $c([0, 1])$. Then there exists a weak solution f of the divisor ∂c with $f \mid X \setminus U = 1$, such that for every closed differential form $\omega \in \mathscr{E}^{(1)}(X)$ one has*

$$\int_c \omega = \frac{1}{2\pi i} \iint_X \frac{df}{f} \wedge \omega.$$

Remark. Since $df/f = 0$ on $X \setminus U$, the integral over X exists.

PROOF

(a) We first consider the case where (U, z) is a coordinate neighborhood on X such that $z(U) \subset \mathbb{C}$ is the unit disk and the curve c lies entirely in U. For simplicity identify U with the unit disk.

Let $a := c(0)$ and $b := c(1)$. There exists $r < 1$ such that $c([0, 1]) \subset \{|z| < r\}$. The function $\log((z - b)/(z - a))$ has a well-defined branch in $\{r < |z| < 1\}$. Choose a function $\psi \in \mathscr{E}(U)$ with $\psi \mid \{|z| \le r\} = 1$ and $\psi \mid \{|z| \ge r'\} = 0$, where $r < r' < 1$ and define $f_0 \in \mathscr{E}(U \setminus \{a\})$ by

$$f_0 = \begin{cases} \exp\left(\psi \log \dfrac{z - b}{z - a}\right) & \text{if } r < |z| < 1, \\[2ex] \dfrac{z - b}{z - a} & \text{if } |z| \le r. \end{cases}$$

Since $f_0 \mid \{r' < |z| < 1\} = 1$, one can continuously extend f_0 to a function $f \in \mathscr{E}(X \setminus \{a\})$, by defining it to be 1 on $X \setminus U$. By construction f is a weak solution of the divisor ∂c. Now suppose $\omega \in \mathscr{E}^{(1)}(X)$ is a closed differential form. Since ω has a primitive on U, there exists a function $g \in \mathscr{E}(X)$ with compact support, such that $\omega = dg$ on $\{|z| \le r'\}$. Thus from Lemma (20.3)

$$\frac{1}{2\pi i} \iint_X \frac{df}{f} \wedge \omega = \frac{1}{2\pi i} \iint_X \frac{df}{f} \wedge dg = g(b) - g(a) = \int_c \omega.$$

(b) In the general case there exists a partition

$$0 = t_0 < t_1 < \cdots < t_n = 1$$

of the interval $[0, 1]$ and coordinate neighborhoods (U_j, z_j), $j = 1, \ldots, n$, on X with the following properties:

(i) $c([t_{j-1}, t_j]) \subset U_j \subset U$,
(ii) $z_j(U_j) \subset \mathbb{C}$ is the unit disk.

Letting c_j denote the curve $c \mid [t_{j-1}, t_j]$ and using (a) one can construct a weak solution f_j of the divisor ∂c_j such that $f_j \mid X \setminus U_j = 1$ and

$$\int_{c_j} \omega = \frac{1}{2\pi i} \iint_X \frac{df_j}{f_j} \wedge \omega$$

for every closed differential form $\omega \in \mathscr{E}^{(1)}(X)$. The product $f := f_1 \cdots f_n$ then satisfies the conditions of the lemma. □

20.6. Corollary. *Suppose X is a compact Riemann surface. Then given any closed curve α on X there exists a unique harmonic differential form $\sigma_\alpha \in \mathrm{Harm}^1(X)$ such that*

$$\int_\alpha \omega = \iint_X \sigma_\alpha \wedge \omega$$

for every closed differential form $\omega \in \mathscr{E}^{(1)}(X)$.

PROOF. Suppose f is a weak solution of the divisor $\partial\alpha = 0$ which satisfies the conditions of Lemma (20.5). Since f does not vanish, df/f is differentiable and closed on all of X. By Theorem (19.12) there exists a differential form $\sigma_\alpha \in \mathrm{Harm}^1(X)$ and a function $g \in \mathscr{E}(X)$ such that

$$\frac{1}{2\pi i}\frac{df}{f} = \sigma_\alpha + dg.$$

If $\omega \in \mathscr{E}^{(1)}(X)$ is closed, then $dg \wedge \omega = d(g\omega)$ and thus by Theorem (10.20)

$$\int_\alpha \omega = \frac{1}{2\pi i}\iint_X \frac{df}{f} \wedge \omega = \iint_X \sigma_\alpha \wedge \omega.$$

To prove the uniqueness, suppose $\sigma' \in \mathrm{Harm}^1(X)$ is a second solution of the problem. Then for the difference $\tau := \sigma_\alpha - \sigma'$ one has

$$\iint_X \tau \wedge \omega = 0 \quad \text{for every closed } \omega \in \mathscr{E}^{(1)}(X).$$

In particular one can choose $\omega = *\tau$ and thus $\langle \tau, \tau \rangle = 0$, i.e., $\tau = \sigma_\alpha - \sigma' = 0$. □

20.7. Abel's Theorem. *Suppose D is a divisor on a compact Riemann surface X with $\deg D = 0$. Then D has a solution if and only if there exists a 1-chain $c \in C_1(X)$ with $\partial c = D$ such that*

$$\int_c \omega = 0 \quad \text{for every } \omega \in \Omega(X). \tag{*}$$

Remark. Clearly the condition $\int_c \omega = 0$ only has to be checked for a basis of $\Omega(X)$. If $\gamma \in C_1(X)$ is an arbitrary 1-chain with $\partial\gamma = D$, then the condition may be formulated as follows. There exists a cycle $\alpha \in Z_1(X)$, namely $\alpha = \gamma - c$, such that

$$\int_\gamma \omega_j = \int_\alpha \omega_j, \qquad j = 1, \ldots, g,$$

where $\omega_1, \ldots, \omega_g$ is a basis of $\Omega(X)$.

PROOF

(a) First we show that the condition is sufficient. Suppose $c \in C_1(X)$ is a 1-chain with $\partial c = D$ and which satisfies (*). By Lemma (20.5) there is a weak solution f of the divisor D such that

$$\int_c \omega = \frac{1}{2\pi i} \iint_X \frac{df}{f} \wedge \omega \quad \text{for every } \omega \in \mathscr{E}^{(1)}(X) \text{ with } d\omega = 0.$$

For every $\omega \in \Omega(X)$ one has by (*)

$$0 = \int_c \omega = \frac{1}{2\pi i} \iint_X \frac{df}{f} \wedge \omega = \frac{1}{2\pi i} \iint_X \frac{d''f}{f} \wedge \omega.$$

As noted in (20.2) one has $\sigma := d''f/f \in \mathscr{E}^{0,1}(X)$. By (19.10) there is a function $g \in \mathscr{E}(X)$ with $d''g = d''f/f$. Set

$$F := e^{-g}f.$$

Like f the function F is a weak solution of D and

$$d''F = (d''e^{-g})f + e^{-g}d''f = -e^{-g}fd''g + e^{-g}d''f = 0.$$

Thus F is even a meromorphic solution of D.

(b) We now prove the necessity of the condition. We may assume that $D \neq 0$. Let f be a meromorphic function on X with $(f) = D$. The function f defines an n-sheeted covering $f\colon X \to \mathbb{P}^1$ for some $n \geq 1$. Suppose $a_1, \ldots, a_r \in X$ are the branch points of f and let $Y := \mathbb{P}^1 \backslash \{f(a_1), \ldots, f(a_r)\}$. For every differential form $\omega \in \Omega(X)$ we construct a holomorphic differential form $\sigma = \text{Trace}(\omega)$ on $\mathbb{P}^1$ in the following way. Every point $y \in Y$ has an open neighborhood V such that $f^{-1}(V)$ is the disjoint union of open sets $U_1, \ldots, U_n \subset X$ and all the mappings $f \mid U_\nu \to V$ are biholomorphic. Let $\varphi\colon V \to U_\nu$ be the inverse of $f \mid U_\nu \to V$. Now let

$$\text{Trace}(\omega) \mid V := \varphi_1^*\omega + \cdots + \varphi_n^*\omega.$$

If one carries out the same construction on an open neighborhood V' of another point of Y, then on the intersection one gets the same differential form. As in (8.2) one sees that one can holomorphically continue $\text{Trace}(\omega)$ to all of $\mathbb{P}^1$. Since $\Omega(\mathbb{P}^1) = 0$, $\text{Trace}(\omega) = 0$.

Now let γ be a curve on $\mathbb{P}^1$ from ∞ to 0 which with the possible exception of its end points lies entirely in Y. The preimage of γ under f consists of n curves $c_1, \ldots, c_n$ which join the poles of f with the zeros of f. Then letting $c := c_1 + \cdots + c_n$ one has $\partial c = D$ and for every $\omega \in \Omega(X)$

$$\int_c \omega = \int_\gamma \text{Trace}(\omega) = 0. \qquad \square$$

20.8. Application to Doubly-Periodic Functions. Suppose $\gamma_1, \gamma_2 \in \mathbb{C}$ are linearly independent over $\mathbb{R}$ and

$$P := \{t_1\gamma_1 + t_2\gamma_2 : 0 \le t_1 < 1, 0 \le t_2 < 1\}.$$

Suppose zeros $a_1, \ldots, a_n \in P$ and poles $b_1, \ldots, b_n \in P$ are prescribed, where each point appears as often as its multiplicity demands. Then there exists a meromorphic function which is doubly-periodic with respect to $\Gamma = \mathbb{Z}\gamma_1 + \mathbb{Z}\gamma_2$ and has zeros $a_1, \ldots, a_n$ and poles $b_1, \ldots, b_n$ if and only if

$$\sum_{k=1}^{n} (a_k - b_k) \in \Gamma.$$

PROOF. Let D be the divisor on $\mathbb{C}/\Gamma$ determined by the prescribed zeros and poles. Choose curves c_k from b_k to a_k in $\mathbb{C}$, e.g. straight line segments. Let $\pi: \mathbb{C} \to \mathbb{C}/\Gamma$ be the canonical projection and

$$c := \pi \circ c_1 + \cdots + \pi \circ c_n \in C_1(\mathbb{C}/\Gamma).$$

Then $\partial c = D$. Let $\omega \in \Omega(\mathbb{C}/\Gamma)$ be the differential form on the torus induced by the differential form dz on $\mathbb{C}$. Then

$$\int_c \omega = \sum_{k=1}^{n} \int_{c_k} dz = \sum_{k=1}^{n} (a_k - b_k).$$

Hence the result follows from the Remark right after the statement of Abel's Theorem. □

EXERCISES (§20)

20.1. Let X be a compact Riemann surface, α and β be closed curves in X and σ_α and σ_β be the harmonic 1-forms associated to α and β according to Corollary (20.6). Show that

$$\iint_X \sigma_\alpha \wedge \sigma_\beta$$

is an integer. (This integer is the "intersection number" of α and β.) [*Hint*: Show that for $f \in \mathcal{E}(X)$ and α a closed curve

$$\frac{1}{2\pi i} \int_\alpha \frac{df}{f} \text{ is an integer.}]$$

20.2. Let $\Gamma = \mathbb{Z}\gamma_1 + \mathbb{Z}\gamma_2 \subset \mathbb{C}$ be a lattice, $X = \mathbb{C}/\Gamma$ and

$$\alpha_j : [0, 1] \to X$$

be the closed curves defined by

$$\alpha_j(t) := \pi(t\gamma_j),$$

where $\pi: \mathbb{C} \to \mathbb{C}/\Gamma$ is the canonical projection. Find the harmonic forms σ_{α_j}.

20.3. Let X be a compact Riemann surface of genus g. Show that there exist closed curves $\alpha_1, \ldots, \alpha_{2g}$ on X such that

$$\mathrm{Harm}^1(X) = \sum_{j=1}^{2g} \mathbb{C}\sigma_{\alpha_j}.$$

20.4. Let $\Gamma \subset \mathbb{C}$ be a lattice. A *theta function* with respect to Γ is a holomorphic function

$$F\colon \mathbb{C} \to \mathbb{C}$$

satisfying

$$F(z + \gamma) = e^{L_\gamma(z)}F(z) \quad \text{for every } z \in \mathbb{C} \text{ and } \gamma \in \Gamma,$$

where $L_\gamma(z) = a_\gamma z + b_\gamma$ is an affine linear function depending on γ.

(a) The Weierstrass σ-function $\sigma\colon \mathbb{C} \to \mathbb{C}$ is defined by

$$\sigma(z) := z \prod_{\gamma \in \Gamma \setminus 0} \left(1 - \frac{z}{\gamma}\right) \exp\left(\frac{z}{\gamma} + \frac{z^2}{2\gamma^2}\right).$$

Show that σ is a theta function with zeros of first order precisely at the points of Γ.

(b) Show that every doubly periodic meromorphic function with respect to Γ is the quotient of two theta functions.

§21. The Jacobi Inversion Problem

Abel's Theorem tells us when a divisor of degree zero on a compact Riemann surface has a solution which is a meromorphic function, i.e., when a divisor is a principal divisor. In this section we will be concerned with a more detailed study of the quotient group of divisors of degree zero modulo the subgroup of principal divisors. It turns out that this group is isomorphic to a complex g-dimensional torus, where g is the genus of the Riemann surface.

21.1. Lattices. Suppose V is an N-dimensional vector space over $\mathbb{R}$. An additive subgroup $\Gamma \subset V$ is called a *lattice* if there exist N vectors $\gamma_1, \ldots, \gamma_N \in V$, which are linearly independent over $\mathbb{R}$, such that

$$\Gamma = \mathbb{Z}\gamma_1 + \cdots + \mathbb{Z}\gamma_N.$$

Theorem. *A subgroup $\Gamma \subset V$ is a lattice precisely if both of the following conditions hold:*

(i) *Γ is discrete,* i.e., *there exists a neighborhood U of zero such that $\Gamma \cap U = (0)$.*

(ii) *Γ is contained in no proper vector subspace of V.*

Remark. Every real N-dimensional vector space V has a unique topology such that every isomorphism $V \xrightarrow{\sim} \mathbb{R}^N$ is a homeomorphism.

PROOF. Clearly a lattice $\Gamma \subset V$ satisfies conditions (i) and (ii). Now conversely suppose $\Gamma \subset V$ is a subgroup satisfying conditions (i) and (ii). By induction on $N = \dim_{\mathbb{R}} V$ we will show that there exist linearly independent vectors $\gamma_1, \ldots, \gamma_N \in V$ such that

$$\Gamma = \mathbb{Z}\gamma_1 + \cdots + \mathbb{Z}\gamma_N.$$

This is trivial for $N = 0$.

Now consider the induction step $N - 1 \to N$. Since Γ is not contained in any proper vector subspace of V, there exist N linearly independent vectors $x_1, \ldots, x_N \in \Gamma$. Let V_1 be the vector subspace of V spanned by $x_1, \ldots, x_{N-1}$ and let $\Gamma_1 := \Gamma \cap V_1$. The induction hypothesis may be applied to Γ_1. Thus there exist linearly independent vectors $\gamma_1, \ldots, \gamma_{N-1} \in \Gamma_1 \subset \Gamma$ such that

$$\Gamma_1 = \mathbb{Z}\gamma_1 + \cdots + \mathbb{Z}\gamma_{N-1}.$$

Every vector $x \in \Gamma$ may be written uniquely in the form

$$x = c_1(x)\gamma_1 + \cdots + c_{N-1}(x)\gamma_{N-1} + c(x)x_N,$$

where $c_j(x)$ and $c(x)$ are real numbers. Since the parallelotope

$$P := \{\lambda_1\gamma_1 + \cdots + \lambda_{N-1}\gamma_{N-1} + \lambda x_N : \lambda_j, \lambda \in [0, 1]\}$$

is compact, $\Gamma \cap P$ is finite. Hence there exists a vector $\gamma_N \in (\Gamma \cap P)\backslash V_1$ such that

$$c(\gamma_N) = \min\{c(x) : x \in (\Gamma \cap P)\backslash V_1\} \in]0, 1].$$

Now we claim that $\Gamma = \Gamma_1 + \mathbb{Z}\gamma_N$. For, suppose $x \in \Gamma$ is arbitrary. Then there exist $n_j \in \mathbb{Z}$ such that

$$x' := x - \sum_{j=1}^{N} n_j\gamma_j = \sum_{j=1}^{N-1} \lambda_j\gamma_j + \lambda x_N,$$

where

$$0 \le \lambda_j < 1 \quad \text{for } j = 1, \ldots, N-1$$

and

$$0 \le \lambda < c(\gamma_N).$$

Since $x' \in \Gamma \cap P$, it follows from the definition of γ_N that $\lambda = 0$. Thus $x' \in \Gamma \cap V_1 = \Gamma_1$. Hence all λ_j are integers and thus are zero. This implies $x' = 0$, i.e.,

$$x = \sum_{j=1}^{N} n_j\gamma_j \in \mathbb{Z}\gamma_1 + \cdots + \mathbb{Z}\gamma_N. \qquad \square$$

21.2. Period Lattices. Now suppose X is a compact Riemann surface of genus $g \ge 1$ and $\omega_1, \ldots, \omega_g$ is a basis of the vector space $\Omega(X)$ of holomorphic 1-forms on X. Define a subgroup

$$\mathrm{Per}(\omega_1, \ldots, \omega_g) \subset \mathbb{C}^g$$

as follows. $\mathrm{Per}(\omega_1, \ldots, \omega_g)$ consists of all vectors

$$\left(\int_\alpha \omega_1, \int_\alpha \omega_2, \ldots, \int_\alpha \omega_g\right) \in \mathbb{C}^g,$$

where α runs through the fundamental group $\pi_1(X)$ (cf. 10.11).

Remark. It is also true (cf. 20.4) that

$$\mathrm{Per}(\omega_1, \ldots, \omega_g) = \left\{\left(\int_\alpha \omega_1, \ldots, \int_\alpha \omega_g\right) : \alpha \in H_1(X)\right\}.$$

We will show that $\mathrm{Per}(\omega_1, \ldots, \omega_g)$ is a lattice in $\mathbb{C}^g$, where $\mathbb{C}^g$ is considered as a real $2g$-dimensional vector space. This lattice is called the period lattice of X relative to the basis $(\omega_1, \ldots, \omega_g)$.

For the proof we need a lemma.

21.3. Lemma. *Suppose X is a compact Riemann surface of genus g. Then there are g distinct points $a_1, \ldots, a_g \in X$ with the following property: Every holomorphic 1-form $\omega \in \Omega(X)$ which vanishes at all the points $a_1, \ldots, a_g$ is identically zero.*

PROOF. For $a \in X$, let

$$H_a := \{\omega \in \Omega(X) \colon \omega(a) = 0\}.$$

Every H_a is either equal to $\Omega(X)$ or else has codimension one in $\Omega(X)$. Since the intersection of all the H_a is zero and $\Omega(X)$ has dimension g, there exist g points $a_1, \ldots, a_g \in X$ such that

$$H_{a_1} \cap \cdots \cap H_{a_g} = 0.$$

These points satisfy the conditions of the Lemma. □

21.4. Theorem. *Suppose X is a compact Riemann surface of genus $g \geq 1$ and $\omega_1, \ldots, \omega_g$ is a basis of $\Omega(X)$. Then $\Gamma := \mathrm{Per}(\omega_1, \ldots, \omega_g)$ is a lattice in $\mathbb{C}^g$.*

PROOF

(a) Choose points $a_1, \ldots, a_g$ as in the Lemma and disjoint simply connected coordinate neighborhoods (U_j, z_j) of a_j with $z_j(a_j) = 0$, for $j = 1, \ldots, g$. With respect to these coordinates let

$$\omega_i = \varphi_{ij}\, dz_j \quad \text{on } U_j.$$

By Lemma (21.3) the matrix

$$A := (\varphi_{ij}(a_j))_{1 \leq i, j \leq g}$$

has rank g. Now define a mapping

$$F \colon U_1 \times \cdots \times U_g \to \mathbb{C}^g$$

as follows: For $x = (x_1, \ldots, x_g) \in U_1 \times \cdots \times U_g$ let

$$F(x_1, \ldots, x_g) = (F_1(x), \ldots, F_g(x))$$

where

$$F_i(x) := \sum_{j=1}^{g} \int_{a_j}^{x_j} \omega_i, \qquad i = 1, \ldots, g.$$

Here the integral $\int_{a_j}^{x_j} \omega_j$ is along any curve from a_j to x_j which lies in U_j; since U_j is simply connected, the integral is independent of the curve chosen.

The map F is complex differentiable with respect to the coordinates $z_1, \ldots, z_g$ and has Jacobian matrix

$$J_F(x) = \left(\frac{\partial F_i}{\partial z_j}(x)\right) = (\varphi_{ij}(x_j)).$$

Thus at the point $a = (a_1, \ldots, a_g)$ the matrix $J_F(a) = A$ and is invertible. Hence

$$W := F(U_1 \times \cdots \times U_g) \subset \mathbb{C}^g$$

is a neighborhood of $F(a) = 0$.

(b) Now we will show that $\Gamma \cap W = 0$. For, suppose to the contrary that there exists a point $t \in \Gamma \cap (W \backslash 0)$. Then there exists

$$x = (x_1, \ldots, x_g) \in U_1 \times \cdots \times U_g, \qquad x \neq a,$$

with $F(x) \in \Gamma$. Renumbering, if necessary, we may assume

$$x_j \neq a_j \quad \text{for } 1 \leq j \leq k \quad \text{and } x_j = a_j \quad \text{for } j > k,$$

where $1 \leq k \leq g$. By Abel's Theorem there exists a meromorphic function f on X which has a pole of first order at a_j, $1 \leq j \leq k$, a zero of first order at x_j, $1 \leq j \leq k$ and is holomorphic otherwise. Let $c_j z_j^{-1}$ be the principal part of f at a_j. Of course $c_j \neq 0$ for $1 \leq j \leq k$. By the Residue Theorem (10.21)

$$0 = \operatorname{Res}(f\omega_i) = \sum_{j=1}^{k} c_j \varphi_{ij}(a_j) \quad \text{for } i = 1, \ldots, g.$$

But this is not possible since the matrix $(\varphi_{ij}(a_j))$ has rank g. Thus the assumption is false and we have shown that Γ is a discrete subgroup of $\mathbb{C}^g$.

(c) Now we will show that Γ is not contained in any proper real vector subspace of $\mathbb{C}^g$. Otherwise, there would exist a non-trivial real linear form on $\mathbb{C}^g$, which vanished identically on Γ. Since every real linear form is the real part of a complex linear form, one thus gets a vector $(c_1, \ldots, c_g) \in \mathbb{C}^g \backslash 0$ such that

$$\operatorname{Re}\left(\sum_{j=1}^{g} c_j \int_\alpha \omega_j\right) = 0 \quad \text{for every } \alpha \in \pi_1(X).$$

But from Corollary (19.8) it then follows that

$$\omega := c_1\omega_1 + \cdots + c_g\omega_g = 0, \quad \text{a contradiction!}$$

This proves that Γ is a lattice in $\mathbb{C}^g$. □

21.5. Remark. Theorem (21.4) tells us that there are $2g$ closed curves $\alpha_1, \ldots, a_{2g}$ on X such that the vectors

$$\gamma_\nu := \left(\int_{\alpha_\nu} \omega_1, \ldots, \int_{\alpha_\nu} \omega_g\right) \in \mathbb{C}^g, \qquad \nu = 1, \ldots, 2g,$$

are linearly independent over the reals and

$$\operatorname{Per}(\omega_1, \ldots, \omega_g) = \mathbb{Z}\gamma_1 + \cdots + \mathbb{Z}\gamma_{2g}.$$

One can easily see from this that the homology classes of $\alpha_1, \ldots, \alpha_{2g}$ in $H_1(X)$ are linearly independent over $\mathbb{Z}$ and generate $H_1(X)$. Thus $H_1(X) \cong \mathbb{Z}^{2g}$.

21.6. The Jacobi Variety and the Picard Group. Suppose X is a compact Riemann surface of genus g and $\omega_1, \ldots, \omega_g$ is a basis of $\Omega(X)$. Then

$$\operatorname{Jac}(X) := \mathbb{C}^g/\operatorname{Per}(\omega_1, \ldots, \omega_g)$$

is called the *Jacobi variety* of X. Here we are considering $\operatorname{Jac}(X)$ only as an abelian group. It also has the structure of a compact complex manifold (a complex g-dimensional torus), similar to the tori defined in (1.5.d). This structure will not be dealt with here. Note that the definition depends on the choice of basis $\omega_1, \ldots, \omega_g$, but the choice of a different basis leads to an isomorphic $\operatorname{Jac}(X)$.

Let $\operatorname{Div}_0(X) \subset \operatorname{Div}(X)$ denote the subgroup of divisors of degree zero and $\operatorname{Div}_P(X) \subset \operatorname{Div}_0(X)$ the subgroup of principal divisors. The quotient

$$\operatorname{Pic}(X) := \operatorname{Div}(X)/\operatorname{Div}_P(X)$$

is called the *Picard group* of X. We will also consider the subgroup

$$\operatorname{Pic}_0(X) := \operatorname{Div}_0(X)/\operatorname{Div}_P(X)$$

of $\operatorname{Pic}(X)$. Since $\operatorname{Div}(X)/\operatorname{Div}_0(X) = \mathbb{Z}$, we have an exact sequence

$$0 \to \operatorname{Pic}_0(X) \to \operatorname{Pic}(X) \to \mathbb{Z} \to 0.$$

Define a map

$$\Phi\colon \operatorname{Div}_0(X) \to \operatorname{Jac}(X)$$

as follows. Suppose $D \in \operatorname{Div}_0(X)$ and $c \in C_1(X)$ is a chain with $\partial c = D$. The vector

$$\left(\int_c \omega_1, \ldots, \int_c \omega_g\right) \in \mathbb{C}^g$$

is determined uniquely by D up to equivalence modulo $\operatorname{Per}(\omega_1, \ldots, \omega_g)$. By definition $\Phi(D)$ is its equivalence class. Clearly Φ is a group homomorphism. Now Abel's Theorem says that the kernel of the mapping Φ is equal to $\operatorname{Div}_P(X)$. Hence by passing to the quotient we get an injective mapping

$$j\colon \operatorname{Pic}_0(X) \to \operatorname{Jac}(X).$$

The Jacobi inversion problem asks if this map is surjective. Actually this is the case!

21.7. Theorem. *For every compact Riemann surface X the mapping*

$$j\colon \operatorname{Pic}_0(X) \to \operatorname{Jac}(X)$$

is an isomorphism.

PROOF. Let $p \in \operatorname{Jac}(X)$ be an arbitrary point which is represented by the vector $\xi \in \mathbb{C}^g$. For N a sufficiently large natural number, the vector $(1/N)\xi$ lies in the image of the mapping F considered in part (a) of the proof of Theorem (21.4). This means that there exist points $a_j, x_j \in X$ and curves γ_j from a_j to x_j, for $j = 1, \ldots, g$, such that if $c := \gamma_1 + \cdots + \gamma_g$, then

$$\left(\int_c \omega_1, \ldots, \int_c \omega_g\right) = \frac{1}{N}\xi.$$

Thus for the divisor $D := \partial c$ one has

$$\Phi(D) = \frac{1}{N}\xi \bmod \operatorname{Per}(\omega_1, \ldots, \omega_g).$$

Now if θ is the point of $\operatorname{Pic}_0(X)$ represented by the divisor ND, then $j(\theta) = p$. This proves that j is surjective and thus an isomorphism. □

21.8. Suppose X is a compact Riemann surface of genus g and $a_1, \ldots, a_g \in X$ are arbitrarily chosen points. Define a mapping

$$\psi\colon X^g \to \operatorname{Pic}_0(X)$$

in the following way. For $(x_1, \ldots, x_g) \in X^g$ let

$$\psi(x_1, \ldots, x_g) := \sum_{j=1}^{g} (D_{x_j} - D_{a_j}) \bmod \operatorname{Div}_P(X);$$

where D_x, for $x \in X$, is the divisor which has the value $+1$ at x but is otherwise zero. Let

$$J\colon X^g \to \operatorname{Jac}(X)$$

be the composition of the mappings $\psi: X^g \to \mathrm{Pic}_0(X)$ and $j: \mathrm{Pic}_0(X) \to \mathrm{Jac}(X)$. Recalling the various definitions, one sees that

$$J(x_1, \ldots, x_g) = \left(\sum_{j=1}^{g} \int_{a_j}^{x_j} \omega_i \right)_{1 \le i \le g} \mod \mathrm{Per}(\omega_1, \ldots, \omega_g).$$

One has a sharper version of (21.7).

21.9. Theorem. *With the same notation as above, the mapping*

$$J: X^g \to \mathrm{Jac}(X)$$

is surjective.

Proof. It suffices to show that $\psi: X^g \to \mathrm{Pic}_0(X)$ is surjective. But this is the same as saying that every divisor $D \in \mathrm{Div}_0(X)$ is equivalent modulo $\mathrm{Div}_P(X)$ to a divisor of the form

$$\sum_{j=1}^{g} (D_{x_j} - D_{a_j}), \qquad (x_1, \ldots, x_g) \in X^g.$$

One sees this as follows. Let

$$D' := D + D_{a_1} + \cdots + D_{a_g}.$$

Then $\deg D' = g$ and by the Riemann-Roch Theorem (16.9) one has $\dim H^0(X, \mathcal{O}_{D'}) \ge 1$. Thus there exists a meromorphic function $f \ne 0$ on X with $(f) \ge -D'$, i.e.

$$D'' := (f) + D' \ge 0.$$

Since $\deg D'' = g$, there are points $x_1, \ldots, x_g \in X$ such that

$$D'' = D_{x_1} + \cdots + D_{x_g}.$$

Thus

$$\sum_{j=1}^{g} (D_{x_j} - D_{a_j}) = D + (f). \qquad \square$$

Remark. It follows directly from the definition of the mapping $J: X^g \to \mathrm{Jac}(X)$ that $J(x_1, \ldots, x_g)$ remains invariant under any permutation of $x_1, \ldots, x_g$. Hence J induces a mapping $S^gX \to \mathrm{Jac}(X)$ of the g-fold symmetric product of X into the Jacobi variety. One can define on S^gX, as well as on $\mathrm{Jac}(X)$, the structure of a compact complex g-dimensional manifold. Then the mapping $S^gX \to \mathrm{Jac}(X)$ is holomorphic. Note that it is not bijective, but one can show that it is bimeromorphic, i.e. induces an isomorphism between the fields of meromorphic functions of $\mathrm{Jac}(X)$ and S^gX. For details, see [16].

21.10. Theorem. *For every compact Riemann surface X of genus 1 the mapping $J: X \to \mathrm{Jac}(X)$ is an isomorphism.*

Remark. Together with Corollary (17.13) this shows that the compact Riemann surfaces of genus 1 are precisely the tori $\mathbb{C}/\Gamma$.

PROOF. The mapping J may be described in the following way. Suppose $\omega \in \Omega(X)\backslash 0$, $\Gamma := \mathrm{Per}(\omega)$ and $a \in X$. Then for $x \in X$, one has

$$J(x) = \int_a^x \omega \bmod \Gamma \in \mathbb{C}/\Gamma = \mathrm{Jac}(X).$$

Clearly J is a holomorphic mapping. By (21.9) J is surjective. By the way this also follows directly from Theorem (2.7). The mapping J is also injective, for otherwise by Abel's Theorem there would exist a meromorphic function f on X having a single pole of order one. This is impossible. For, in that case X would be isomorphic to $\mathbb{P}^1$. □

Remark. Suppose $P(z)$ is a polynomial of degree 3 or 4 without repeated roots and let X be the Riemann surface of the algebraic function $\sqrt{P(z)}$. Then X has genus one (cf. 17.15) and

$$\omega = \frac{dz}{\sqrt{P(z)}}$$

is a basis of $\Omega(X)$. Let $\Gamma \subset \mathbb{C}$ be the period matrix of ω. The mapping $J: X \to \mathrm{Jac}(X) = \mathbb{C}/\Gamma$ is then given by the "elliptic integral of the first kind"

$$J(x) = \int_a^x \frac{dz}{\sqrt{P(z)}} \bmod \Gamma \in \mathbb{C}/\Gamma.$$

Let $F: \mathbb{C}/\Gamma \to X$ be the inverse of J and let $\pi: \mathbb{C} \to \mathbb{C}/\Gamma$ and $p: X \to \mathbb{P}^1$ be the canonical projections. Then

$$f := p \circ F \circ \pi: \mathbb{C} \to \mathbb{P}^1$$

is a doubly-periodic meromorphic function. It was the great discovery of Abel and Jacobi that the study of elliptic integrals could be replaced by the study of doubly-periodic functions. The generalization of this question to hyperelliptic integrals then lead to the Jacobi inversion problem. An account of the history of this problem can be found in [63].

EXERCISES (§21)

21.1. Let X be a compact Riemann surface and $Y \subset X$ be an open subset such that $X\backslash Y$ has non-empty interior. Let D be a divisor on X. Show that there exists a function $f \in \mathcal{M}^*(X)$ such that

$$\mathrm{ord}_x(f) = D(x) \quad \text{for every } x \in Y.$$

[*Hint*: Find a divisor D' with support in $X\backslash Y$ such that $D + D'$ is a principal divisor.]

21.2. (a) Show that the polynomial

$$F(z) := 4z^3 - g_2 z - g_3, \qquad g_2, g_3 \in \mathbb{C},$$

has 3 distinct roots if and only if

$$g_2^3 - 27g_3^2 \neq 0.$$

(b) Let $\Gamma \subset \mathbb{C}$ be a lattice and

$$\wp(z) = \frac{1}{z^2} + \sum_{\omega \in \Gamma \backslash 0} \left(\frac{1}{(z-\omega)^2} - \frac{1}{\omega^2} \right)$$

be the associated Weierstrass $\wp$-function. Show that $\wp$ satisfies the differential equation

$$\wp'^2 = 4\wp^3 - g_2 \wp - g_3,$$

where

$$g_2 = 60 \sum_{\omega \in \Gamma \backslash 0} \frac{1}{\omega^4} \quad \text{and} \quad g_3 = 140 \sum_{\omega \in \Gamma \backslash 0} \frac{1}{\omega^6}$$

and that the torus $\mathbb{C}/\Gamma$ is isomorphic to the Riemann surface $X \to \mathbb{P}^1$ of the algebraic function $\sqrt{4z^3 - g_2 z - g_3}$.

(c) Given $g_2, g_3 \in \mathbb{C}$ with $g_2^3 \neq 27g_3^2$, show that there is a lattice $\Gamma \subset \mathbb{C}$ such that

$$g_2 = 60 \sum_{\omega \in \Gamma \backslash 0} \frac{1}{\omega^4} \quad \text{and} \quad g_3 = 140 \sum_{\omega \in \Gamma \backslash 0} \frac{1}{\omega^6}.$$

[*Hint*: Use part (b) and Theorem (21.10).]

CHAPTER 3

Non-compact Riemann Surfaces

In many respects, function theory on non-compact Riemann surfaces is similar to function theory on domains in the complex plane. Thus for non-compact Riemann surfaces one has analogues of the Mittag–Leffler Theorem and the Weierstrass Theorem as well as the Riemann Mapping Theorem.

In this chapter we will first consider the Dirichlet Boundary Value Problem for harmonic functions on Riemann surfaces. This will then serve as a tool in proving that every Riemann surface has a countable topology. Also it will be needed later in the proof of the Riemann Mapping Theorem. With the help of Weyl's Lemma we will prove Runge's Approximation Theorem. And then from Runge's Approximation Theorem we easily derive the Theorems of Mittag–Leffler and Weierstrass. Also in this chapter we complete the discussion, begun in §§10 and 11, concerning the existence of holomorphic functions with prescribed summands of automorphy. We also look at the Riemann–Hilbert problem.

§22. The Dirichlet Boundary Value Problem

The existence theorems for holomorphic and meromorphic functions on Riemann surfaces which we have so far considered are all essentially dependent on Dolbeault's Lemma (13.2) and the Finiteness Theorem (14.9). We now prove another existence theorem on Riemann surfaces which is entirely independent of these previous results, namely the solution of the Dirichlet Problem for harmonic functions using Perron's method.

22.1. Suppose Y is an open subset of a Riemann surface X. Then a differentiable function $u \in \mathscr{E}(Y)$ is called harmonic if $d'd''u = 0$, cf. (9.14). With respect to a local coordinate $z = x + iy$ this is equivalent to

$$\Delta u = \left(\frac{\partial^2}{\partial x^2} + \frac{\partial^2}{\partial y^2}\right)u = 0.$$

Every real-valued harmonic function u on a simply connected domain $G \subset X$ is the real part of a holomorphic function $f \in \mathcal{O}(G)$. For, from Theorem (19.4) it follows that the harmonic differential form du may be written $du = \operatorname{Re}(dg)$ for some $g \in \mathcal{O}(G)$. This implies $u = \operatorname{Re}(g) + \text{const.}$

This observation allows one to derive quite easily the Maximum Principle for harmonic functions. If a harmonic function $u\colon Y \to \mathbb{R}$ on the domain Y attains its maximum at a point $x_0 \in Y$, then u is a constant. For, suppose $u = \operatorname{Re}(f)$ where f is a function holomorphic on some neighborhood of x_0. Since $|e^f| = e^u$, the holomorphic function e^f attains its maximum modulus at x_0. Now the Maximum Principle for holomorphic functions implies u is constant on a neighborhood of x_0. Thus u is also constant on all of Y, since Y is connected.

22.2. By the Dirichlet Problem on a Riemann surface X we mean the following:

Suppose Y is an open subset of X and $f\colon \partial Y \to \mathbb{R}$ is a continuous function. Find a continuous function $u\colon \bar{Y} \to \mathbb{R}$ which is harmonic in Y and satisfies $u \mid \partial Y = f$. Suppose $\bar{Y}$ is compact and $\partial Y \neq \varnothing$, i.e., $Y \neq X$. If a solution exists, then it is unique. For, the difference $u_1 - u_2$ of two solution u_i has boundary values zero. Because of the Maximum Principle for harmonic functions one then has $0 \leq u_1 - u_2 \leq 0$ on Y. Thus $u_1 = u_2$.

For the disk

$$D(R) := \{z \in \mathbb{C}\colon |z| < R\}, \quad \text{where } R > 0,$$

the Dirichlet Problem can be easily solved using the *Poisson Integral.*

22.3. Theorem. *Suppose $f\colon \partial D(R) \to \mathbb{R}$ is continuous and let*

$$u(z) := \frac{1}{2\pi}\int_0^{2\pi} \frac{R^2 - |z|^2}{|Re^{i\theta} - z|^2} f(Re^{i\theta})\, d\theta \quad \textit{for } |z| < R \tag{*}$$

and $u(z) := f(z)$ for $|z| = R$. Then u is continuous on $\overline{D(R)}$ and harmonic on $D(R)$.

PROOF. For $z \neq \zeta$ let

$$P(z, \zeta) := \frac{|\zeta|^2 - |z|^2}{|\zeta - z|^2}, \qquad F(z, \zeta) := \frac{\zeta + z}{\zeta - z}.$$

Then $P(z, \zeta) = \operatorname{Re} F(z, \zeta)$. Thus (*) may be written as

$$u(z) = \frac{1}{2\pi}\int_0^{2\pi} P(z, Re^{i\theta})f(Re^{i\theta})\, d\theta$$
$$= \operatorname{Re}\left(\frac{1}{2\pi}\int_0^{2\pi} F(z, Re^{i\theta})f(Re^{i\theta})\, d\theta\right)$$
$$= \operatorname{Re}\left(\frac{1}{2\pi i}\int_{|\zeta|=R} F(z, \zeta)f(\zeta)\frac{d\zeta}{\zeta}\right).$$

Since $F(z, \zeta)$ is holomorphic as a function of z, it follows that u is the real part of a holomorphic function on $D(R)$ and thus is harmonic.

The remaining point is to verify the continuity at the boundary. Using the Residue Theorem, one has

$$\frac{1}{2\pi}\int_0^{2\pi} P(z, Re^{i\theta})\, d\theta = \operatorname{Re}\left(\frac{1}{2\pi i}\int_{|\zeta|=R} \frac{\zeta + z}{\zeta - z}\cdot\frac{d\zeta}{\zeta}\right) = 1.$$

Hence for $\zeta_0 \in \partial D(R)$, $z \in D(R)$ and letting $\zeta = Re^{i\theta}$ one has

$$u(z) - f(\zeta_0) = \frac{1}{2\pi}\int_0^{2\pi} P(z, \zeta)(f(\zeta) - f(\zeta_0))\, d\theta.$$

Suppose $\varepsilon > 0$ is given. Since f is continuous, there exists $\delta_0 > 0$ such that $|f(\zeta) - f(\zeta_0)| \leq \varepsilon/2$ for $|\zeta - \zeta_0| \leq \delta_0$. Also there is a constant $M > 0$ such that $|f(\zeta)| \leq M$ for every $\zeta \in \partial D(R)$. Now split the interval $[0, 2\pi]$ up into two subsets. Namely let α be the subset of all those $\theta \in [0, 2\pi]$ such that $|Re^{i\theta} - \zeta_0| \leq \delta_0$ and let β be the rest. Then

$$|u(z) - f(\zeta_0)| \leq \frac{1}{2\pi}\int_\alpha P(z, \zeta)\frac{\varepsilon}{2}\, d\theta + \frac{1}{2\pi}\int_\beta P(z, \zeta)2M\, d\theta$$
$$\leq \frac{\varepsilon}{2} + \frac{M}{\pi}\int_\beta P(z, Re^{i\theta})\, d\theta.$$

If $|z - \zeta_0| =: \delta \leq \delta_0/2$, then for $\theta \in \beta$ one has

$$|Re^{i\theta} - z| \geq |Re^{i\theta} - \zeta_0| - |z - \zeta_0| \geq \delta_0/2$$

and

$$P(z, Re^{i\theta}) = \frac{(R + |z|)(R - |z|)}{|Re^{i\theta} - z|^2} \leq \frac{2R\delta}{(\delta_0/2)^2} = \frac{8R}{\delta_0^2}\delta.$$

Thus

$$|u(z) - f(\zeta_0)| < \frac{\varepsilon}{2} + \frac{16RM}{\delta_0^2}\delta \leq \varepsilon,$$

whenever $|z - \zeta_0|$ is chosen sufficiently small. □

22.4. Corollary. *Suppose $u: D(R) \to \mathbb{R}$ is a harmonic function. Then*

$$u(z) = \frac{1}{2\pi} \int_0^{2\pi} \frac{r^2 - |z|^2}{|re^{i\theta} - z|^2} u(re^{i\theta})\, d\theta$$

for every $r < R$ and $|z| < r$. In particular, u satisfies the "Mean Value Principle"

$$u(0) = \frac{1}{2\pi} \int_0^{2\pi} u(re^{i\theta})\, d\theta.$$

Because of the uniqueness of the solution of the Dirichlet Boundary Value Problem this follows from (22.3).

22.5. Corollary. *Suppose $u_n: D(R) \to \mathbb{R}$, $n \in \mathbb{N}$, is a sequence of harmonic functions which converges uniformly on compact subsets to a function $u: D(R) \to \mathbb{R}$. Then u is also harmonic.*

PROOF. By (22.4) for every $r < R$ and all $|z| < r$ one has

$$u_n(z) = \frac{1}{2\pi} \int_0^{2\pi} P(z, re^{i\theta}) u_n(re^{i\theta})\, d\theta,$$

where $P(z, \zeta)$ is the kernel defined in (22.3). Since the sequence u_n converges to u uniformly on $\partial D(r)$, this integral formula is also valid for the function u. But then u is harmonic on $D(r)$ by Theorem (22.3). □

22.6. Harnack's Theorem. *Suppose $M \in \mathbb{R}$ and*

$$u_0 \le u_1 \le u_2 \le \cdots \le M$$

is a monotone increasing, bounded sequence of harmonic functions $u_n: D(R) \to \mathbb{R}$. Then the sequence converges uniformly on every compact subset of $D(R)$ to a harmonic function $u: D(R) \to \mathbb{R}$.

PROOF. Suppose $K \subset D(R)$ is compact. Then there exist constants $\rho < r < R$ such that

$$K \subset \{z \in \mathbb{C}: |z| \le \rho\}.$$

Suppose $\varepsilon > 0$ is given and let $\varepsilon' := \varepsilon(r - \rho)/(r + \rho)$. Since the sequence $(u_n(0))$ is monotone increasing and bounded, there exists an N such that

$$u_n(0) - u_m(0) \le \varepsilon' \quad \text{for every } n \ge m \ge N.$$

Now apply the Poisson Integral Formula to the positive harmonic function $u_n - u_m$. Since for $|z| \le \rho$ one has

$$0 \le P(z, re^{i\theta}) \le \frac{r + |z|}{r - |z|} \le \frac{r + \rho}{r - \rho},$$

for all $z \in K$

$$
\begin{aligned}
u_n(z) - u_m(z) &= \frac{1}{2\pi}\int_0^{2\pi} P(z, re^{i\theta})(u_n(re^{i\theta}) - u_m(re^{i\theta}))\, d\theta \\
&\le \frac{r+\rho}{r-\rho}\frac{1}{2\pi}\int_0^{2\pi} (u_n(re^{i\theta}) - u_m(re^{i\theta}))\, d\theta \\
&= \frac{r+\rho}{r-\rho}(u_n(0) - u_m(0)) \le \varepsilon.
\end{aligned}
$$

Thus the sequence (u_n) converges uniformly on K and by (22.5) its limit function is also harmonic. □

22.7. We now return to the Dirichlet Problem on an arbitrary Riemann surface X. Note that the property that a function is harmonic remains invariant under biholomorphic mappings. Thus one can also solve the Dirichlet Problem on all domains $D \subset X$ which are relatively compact and are contained in a chart (U, z) so that $z(D) \subset \mathbb{C}$ is a disk.

We need some additional notation. For any open set $Y \subset X$ let $\operatorname{Reg}(Y)$ denote the set of all subdomains $D \Subset Y$ such that the Dirichlet problem can be solved on D for arbitrary continuous boundary values $f\colon \partial D \to \mathbb{R}$. For any continuous function $u\colon Y \to \mathbb{R}$ and $D \in \operatorname{Reg}(Y)$ let $P_D u$ denote the continuous function on Y which coincides with u on $Y \backslash D$ and solves the Dirichlet problem on $\bar{D}$ for the boundary values $u \mid \partial D$.

Let $\mathscr{C}_{\mathbb{R}}(Y)$ denote the vector space of all continuous real-valued functions on Y. Clearly for every $u, v \in \mathscr{C}_{\mathbb{R}}(Y)$, $\lambda \in \mathbb{R}$, the following hold:

(i) $P_D(u + v) = P_D u + P_D v$,
(ii) $P_D(\lambda u) = \lambda P_D u$,
(iii) $u \le v \Rightarrow P_D u \le P_D v$.

A function $u \in \mathscr{C}_{\mathbb{R}}(Y)$ is harmonic precisely if $P_D(u) = u$ for every $D \in \operatorname{Reg}(Y)$.

22.8. Definition. A continuous function $u\colon Y \to \mathbb{R}$ is called *subharmonic* if

$$P_D u \ge u \quad \text{for every } D \in \operatorname{Reg}(Y).$$

It follows directly from the definition that if $u, v\colon Y \to \mathbb{R}$ are subharmonic functions and λ is a non-negative real number, then $u + v$, λu and $\sup(u, v)$ are subharmonic on Y.

A function $u\colon Y \to \mathbb{R}$ is called *locally subharmonic* if u is subharmonic on a neighborhood of every point of Y.

22.9. Theorem (The Maximum Principle for Locally Subharmonic Functions). *Suppose Y is a domain in a Riemann surface X and $u\colon Y \to \mathbb{R}$ is a locally subharmonic function. If u attains its maximum at some point $x_0 \in Y$, then u is constant.*

PROOF. Let $u(x_0) =: c$ and

$$S := \{x \in Y: u(x) = c\}.$$

If $S \neq Y$, then there exists a point $a \in \partial S \cap Y$. Since u is continuous, $u(a) = c$. In every neighborhood of a there is a point x with $u(x) < c$. Hence there is some open neighborhood $D \in \text{Reg}(Y)$ of a such that $u \mid \partial D$ is not constantly equal to c. Moreover we may assume that u is subharmonic on some neighborhood of $\bar{D}$. Thus

$$u \leq P_D u =: v.$$

The function v is harmonic in D. Because

$$v \mid \partial D = u \mid \partial D \leq c,$$

the Maximum Principle for harmonic functions implies $v \leq c$ on $\bar{D}$. Since $c = u(a) \leq v(a)$, v attains its maximum at a and thus is constantly equal to c. But this contradicts the choice of D. Thus $S = Y$. □

22.10. Corollary. *If $u: Y \to \mathbb{R}$ is locally subharmonic, then u is subharmonic.*

PROOF. Suppose $D \in \text{Reg}(Y)$ is arbitrary. Since $P_D u$ is harmonic on D,

$$v := u - P_D u$$

is locally subharmonic on D. Since $v \mid \partial D = 0$, the Maximum Principle implies $v \leq 0$ on D. Thus $P_D u \geq u$. □

22.11. Lemma. *If $u: Y \to \mathbb{R}$ is subharmonic and $B \in$ Reg(Y), then $P_B u$ is also subharmonic.*

PROOF. Set $v := P_B u$ and suppose $D \in \text{Reg}(Y)$ is arbitrary. We have to show that $P_D v \geq v$. On $Y \backslash D$ one has $P_D v = v$ and on $Y \backslash B$, because $v \geq u$, one has

$$P_D v \geq P_D u \geq u = v.$$

Thus $v - P_D v \leq 0$ on $Y \backslash (B \cap D)$. Since $v - P_D v$ is harmonic on $B \cap D$, it follows that

$$v - P_D v \leq 0 \quad \text{on } B \cap D.$$

Hence $P_D v \geq v$ on all of Y. □

22.12. Lemma (Perron). *Suppose $M \subset \mathscr{C}_{\mathbb{R}}(Y)$ is a non-empty set of subharmonic functions on Y with the following properties:*

(i) $u, v \in M \Rightarrow \sup(u, v) \in M$.
(ii) $u \in M$, $D \in \text{Reg}(Y) \Rightarrow P_D u \in M$.
(iii) *There exists a constant $K \in \mathbb{R}$ such that*

$$u \leq K \quad \textit{for every } u \in M.$$

Then the function $u^*: Y \to \mathbb{R}$ defined by

$$u^*(x) := \sup\{u(x): u \in M\}$$

is harmonic on Y.

PROOF. Suppose $a \in Y$ and $D \in \operatorname{Reg}(Y)$ is a neighborhood of a. Choose a sequence $u_n \in M$, $n \in \mathbb{N}$, with

$$\lim u_n(a) = u^*(a).$$

Because of (i) we may assume

$$u_0 \leq u_1 \leq u_2 \leq \cdots.$$

Let $v_n := P_D u_n$. Then one also has

$$v_0 \leq v_1 \leq v_2 \leq \cdots.$$

By Harnack's Theorem the sequence (v_n) converges on D to a harmonic function $v: D \to \mathbb{R}$ and the following hold

$$v(a) = u^*(a) \quad \text{and } v \leq u^* \quad \text{on } D.$$

Now we claim that $v(x) = u^*(x)$ for every $x \in D$. To see this, suppose $w_n \in M$, $n \in \mathbb{N}$, is a sequence with

$$\lim_{n \to \infty} w_n(x) = u^*(x).$$

Because of (i) and (ii) we may assume that

$$v_n \leq w_n = P_D w_n \quad \text{and } w_n \leq w_{n+1}$$

for every $n \in \mathbb{N}$. Hence the sequence (w_n) converges on D to a harmonic function $w: D \to \mathbb{R}$ with

$$v \leq w \leq u^*.$$

Since $v(a) = w(a) = u^*(a)$, the Maximum Principle applied to the harmonic function $v - w$ on D implies $v(y) = w(y)$ for every $y \in D$. In particular,

$$v(x) = w(x) = u^*(x),$$

and thus $u^* = w$ is a harmonic function on D. □

22.13. To solve the Dirichlet Problem we will now use the technique devised by Perron. Suppose

$$f: \partial Y \to \mathbb{R}$$

is a continuous bounded function (we are not assuming that $\bar{Y}$ is compact) and

$$K := \sup\{f(x): x \in \partial Y\}.$$

Denote by $\mathfrak{P}_f$ the set of all functions $u \in \mathscr{C}(\bar{Y})$ with

(i) $u \mid Y$ subharmonic,
(ii) $u \mid \partial Y \leq f, \qquad u \leq K.$

$\mathfrak{P}_f$ is called the *Perron class* of f. By the lemma

$$u^* := \sup\{u : u \in \mathfrak{P}_f\}$$

is harmonic on Y. For this to be a solution of the boundary value problem it must satisfy

$$\lim_{\substack{y \to x \\ y \in Y}} u^*(y) = f(x)$$

for every point $x \in \partial Y$. Under certain conditions this will be the case, but not in general.

22.14. Definition. A point $x \in \partial Y$ is called *regular* if there is an open neighborhood U of x and a function $\beta \in \mathscr{C}_{\mathbb{R}}(\bar{Y} \cap U)$ with the following properties:

(i) $\beta \mid Y \cap U$ is subharmonic
(ii) $\beta(x) = 0$ and $\beta(y) < 0$ for every $y \in \bar{Y} \cap U \backslash \{x\}$.

The function β is called a *barrier* at x.

Remark. Suppose $x \in \partial Y$ is a regular boundary point of Y and Y_1 is an open subset of Y with $x \in \partial Y_1$. Then x is a regular boundary point of Y_1. This follows directly from the definition. Hence, as a consequence, if Y has a regular boundary (i.e., every boundary point is regular), then every connected component of Y also has a regular boundary.

22.15. Lemma. *Suppose $x \in \partial Y$ is a regular boundary point, V is a neighborhood of x and m and c are real constants with $m \leq c$. Then there exists a function $v \in \mathscr{C}_{\mathbb{R}}(\bar{Y})$ with the following properties:*

(i) *$v \mid Y$ is subharmonic,*
(ii) *$v(x) = c, \qquad v \mid \bar{Y} \cap V \leq c,$*
(iii) *$v \mid \bar{Y} \backslash V = m.$*

PROOF. Without loss of generality we may assume $c = 0$. Suppose U is an open neighborhood of x and $\beta \in \mathscr{C}_{\mathbb{R}}(\bar{Y} \cap U)$ is a barrier at x. By shrinking V if necessary, we may assume $V \Subset U$. Then

$$\sup\{\beta(y) : y \in \partial V \cap \bar{Y}\} < 0.$$

Thus there exists a constant $k > 0$ such that

$$k\beta \mid \partial V \cap \bar{Y} < m.$$

Define

$$v := \begin{cases} \sup(m, k\beta) & \text{on } \bar{Y} \cap V \\ m & \text{on } \bar{Y} \setminus V. \end{cases}$$

Then v is continuous on $\bar{Y}$, locally subharmonic on Y, thus subharmonic, and also satisfies conditions (ii) and (iii). □

22.16. Lemma. *Suppose Y is an open subset of a Riemann surface, $f: \partial Y \to \mathbb{R}$ is a continuous bounded function and*

$$u^* = \sup\{u : u \in \mathfrak{P}_f\},$$

where $\mathfrak{P}_f$ is the Perron class of f. Then for every regular boundary point $x \in \partial Y$

$$\lim_{\substack{y \to x \\ y \in Y}} u^*(y) = f(x).$$

PROOF. Suppose $\varepsilon > 0$ is given. Then there exists a relatively compact open neighborhood V of x with

$$f(x) - \varepsilon \leq f(y) \leq f(x) + \varepsilon \quad \text{for every } y \in \partial Y \cap V.$$

Suppose k and K are real constants such that

$$k \leq f(y) \leq K \quad \text{for every } y \in \partial Y.$$

(a) Using Lemma (22.15) choose a function $v \in \mathscr{C}_{\mathbb{R}}(\bar{Y})$ which is subharmonic on Y and satisfies

$$v(x) = f(x) - \varepsilon$$
$$v \mid \bar{Y} \cap V \leq f(x) - \varepsilon$$
$$v \mid \bar{Y} \setminus V = k - \varepsilon.$$

Then $v \mid \partial Y \leq f$. Thus $v \in \mathfrak{P}_f$ and hence $v \leq u^*$. Then

$$\liminf_{y \to x} u^*(y) \geq v(x) = f(x) - \varepsilon.$$

(b) Again using Lemma (22.15) there exists a function $w \in \mathscr{C}_{\mathbb{R}}(\bar{Y})$ which is subharmonic on Y and satisfies

$$w(x) = -f(x)$$
$$w \mid \bar{Y} \cap V \leq -f(x)$$
$$w \mid \bar{Y} \setminus V = -K.$$

For every $u \in \mathfrak{P}_f$ and $y \in \partial Y \cap V$ one has $u(y) \leq f(x) + \varepsilon$. Thus

$$u(y) + w(y) \leq \varepsilon \quad \text{for } y \in \partial Y \cap V.$$

As well

$$u(z) + w(z) \leq K - K = 0 \quad \text{for every } z \in \bar{Y} \cap \partial V.$$

Applying the Maximum Principle to the function $u + w$, which is subharmonic on $Y \cap V$, one has

$$u + w \leq \varepsilon \quad \text{on } \bar{Y} \cap V.$$

Thus

$$u \mid \bar{Y} \cap V \leq \varepsilon - w \mid \bar{Y} \cap V \quad \text{for every } u \in \mathfrak{P}_f.$$

Hence

$$\limsup_{\substack{y \to x \\ y \in Y}} u^*(y) \leq \varepsilon - w(x) = f(x) + \varepsilon.$$

From (a) and (b) one has the result. □

22.17. Theorem. *Suppose Y is an open subset of a Riemann surface X such that all the boundary points of Y are regular. Then for every continuous bounded function $f: \partial Y \to \mathbb{R}$ the Dirichlet Problem on Y can be solved.*

This follows directly from Lemma (22.16).

We now point out a simple geometric condition which ensures that a boundary point is regular. Since regularity is a local condition which is invariant under biholomorphic mappings, it suffices to formulate this condition for $Y \subset \mathbb{C}$.

22.18. Theorem. *Suppose Y is an open subset of $\mathbb{C}$ and $a \in \partial Y$. Suppose there exists a disk*

$$D = \{z \in \mathbb{C}: |z - m| < r\}, \quad \text{where } m \in \mathbb{C}, r > 0,$$

such that $a \in \partial D$ and $\bar{D} \cap Y = \emptyset$. Then a is a regular boundary point of Y.

PROOF. Set $c := (a + m)/2$, see Fig. 6. Then

$$\beta(z) := \log \frac{r}{2} - \log |z - c|$$

defines a barrier at a. Thus a is a regular boundary point. □

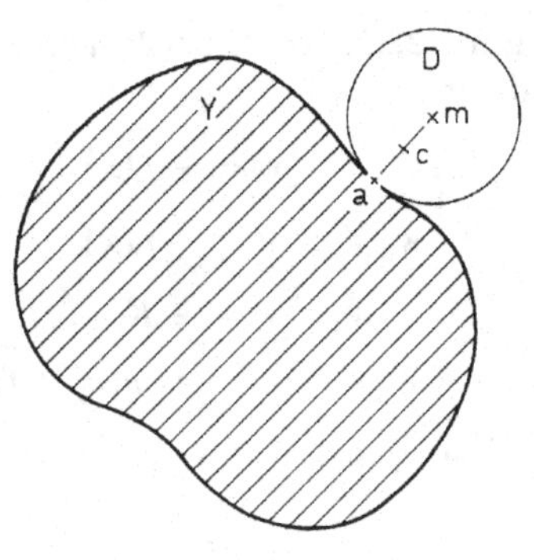

Figure 6

EXERCISES (§22)

22.1. Let $D := \{z \in \mathbb{C} : |z| < R\}$ and let $u: D \to \mathbb{R}_+$ be a nonnegative harmonic function. Prove *Harnack's inequality*

$$\frac{R - |z|}{R + |z|} u(0) \le u(z) \le \frac{R + |z|}{R - |z|} u(0) \quad \text{for every } z \in D.$$

22.2. Using Harnack's inequality prove Liouville's Theorem for harmonic functions. Let $u: \mathbb{C} \to \mathbb{R}$ be a harmonic function which is bounded from above. Then u is constant.

22.3. Suppose $Y \subset \mathbb{C}$ is open and $u: Y \to \mathbb{R}$ is a continuous function such that for every closed disk

$$D(a, r) = \{z \in \mathbb{C} : |z - a| \le r\} \subset Y$$

one has

$$u(a) \le \frac{1}{2\pi} \int_0^{2\pi} u(a + re^{i\varphi})\, d\varphi.$$

Show that u is subharmonic.

22.4. Suppose $Y \subset \mathbb{C}$ is open, $a \in \partial Y$ and there exists a line segment

$$S = \{\lambda a + (1 - \lambda)b : 0 \le \lambda \le 1\}, \qquad b \ne a$$

with $Y \cap S = \varnothing$. Show that a is a regular boundary point of Y.

§23. Countable Topology

In this section we prove the Theorem of Radó which asserts that every Riemann surface has a countable topology. (Clearly this is trivial for compact Riemann surfaces.) Also for later use we construct special exhaustions of non-compact Riemann surfaces.

23.1. Lemma. *Suppose X and Y are topological spaces and $f: X \to Y$ is a continuous, open and surjective mapping. If X has a countable topology, then so does Y.*

PROOF. Let $\mathfrak{U}$ be a countable basis for the topology on X and let

$$\mathfrak{B} = \{f(U) : U \in \mathfrak{U}\}.$$

Then $\mathfrak{B}$ is a countable family of open subsets of Y which we claim is a basis for the topology of Y.

Suppose D is an open subset of Y and $y \in D$. We have to show that there exists $V \in \mathfrak{B}$ with $y \in V \subset D$. Since f is surjective, there exists $x \in X$ with $f(x) = y$. The set $f^{-1}(D)$ is an open neighborhood of x. Hence there exists $U \in \mathfrak{U}$ with $x \in U \subset f^{-1}(D)$. Thus $V := f(U)$ satisfies $y \in V \subset D$. □

23.2. Lemma (Poincaré–Volterra). *Suppose X is a connected manifold, Y is a Hausdorff space with countable topology and $f: X \to Y$ is a continuous, discrete mapping. Then X has a countable topology.*

PROOF. Suppose $\mathfrak{U}$ is a countable basis for the topology on Y. Denote by $\mathfrak{V}$ the collection of all open subsets V of X with the following properties:

(i) V has a countable topology,
(ii) V is the connected component of a set $f^{-1}(U)$ with $U \in \mathfrak{U}$.

(a) We claim that $\mathfrak{V}$ is a basis for the topology on X. Suppose D is an open subset of X with $x \in D$. We have to show that there exists $V \in \mathfrak{V}$ with $x \in V \subset D$. Since f is discrete, there is a relatively compact open neighborhood $W \subset D$ of x so that ∂W does not meet the fiber $f^{-1}(f(x))$. Now $f(\partial W)$ is compact and thus closed and does not contain the point $f(x)$. Hence there exists a $U \in \mathfrak{U}$ with $f(x) \in U$ and $U \cap f(\partial W) = \varnothing$. Let V be the connected component of $f^{-1}(U)$ which contains the point x. Since $V \cap \partial W = \varnothing$, one has $V \subset W$ and hence V has a countable topology, i.e., $V \in \mathfrak{V}$. This verifies claim (a).

(b) Next we claim that for every $V_0 \in \mathfrak{V}$ there exist at most countably many $V \in \mathfrak{V}$ with $V_0 \cap V \neq \varnothing$. For every $U \in \mathfrak{U}$ the connected components of $f^{-1}(U)$ are disjoint. Since V_0 has countable topology it can only meet countably many of these connected components. Since $\mathfrak{U}$ is also countable, the result follows.

(c) Now we show that $\mathfrak{V}$ is countable. Fix $V^* \in \mathfrak{V}$ and define for $n \in \mathbb{N}$ the set $\mathfrak{V}_n \subset \mathfrak{V}$ as follows: $\mathfrak{V}_n$ consists of all $V \in \mathfrak{V}$ such that there are $V_0, V_1, \ldots, V_n \in \mathfrak{V}$ with

$$V_0 = V^*, \quad V_n = V \quad \text{and } V_{k-1} \cap V_k \neq \varnothing \quad \text{for } k = 1, \ldots, n.$$

Since X is connected, $\bigcup_{n \in \mathbb{N}} \mathfrak{V}_n = \mathfrak{V}$. Thus it suffices to show that each $\mathfrak{V}_n$ is countable. We do this by induction. Clearly $\mathfrak{V}_0 = \{V^*\}$ is countable. Suppose we already know that $\mathfrak{V}_n$ is countable. Then it follows directly from (b) that $\mathfrak{V}_{n+1}$ is also countable.

This then proves that X has a countable topology. □

23.3. Theorem (Radó). *Every Riemann surface X has a countable topology.*

PROOF. Suppose U is a coordinate neighborhood on X. Choose two disjoint compact disks $K_0, K_1 \subset U$ and set $Y := X \backslash (K_0 \cup K_1)$. Since the boundary $\partial Y = \partial K_0 \cup \partial K_1$ satisfies the regularity condition of Theorem (22.18), there is a continuous function $u: \bar{Y} \to \mathbb{R}$, which is harmonic on Y, and which satisfies the boundary conditions

$$u \mid \partial K_0 = 0 \quad \text{and } u \mid \partial K_1 = 1.$$

Hence $\omega := d'u$ is a non-trivial holomorphic 1-form on Y. Let f be any holomorphic primitive for $p^*\omega$ on the universal covering $p: \tilde{Y} \to Y$. Since f is not constant, the mapping $f\colon \tilde{Y} \to \mathbb{C}$ satisfies the assumptions of Lemma (23.2). Thus $\tilde{Y}$ has a countable topology. By Lemma (23.1) Y then has a countable topology. Since $X = Y \cup U$, the topology on X is also countable. □

In the proof of Runge's Approximation Theorem we will need to have special exhaustions. These are constructed using a certain hull operator which we now define.

23.4. Definition. Suppose X is a Riemann surface. For any subset $Y \subset X$ let $\mathscr{h}(Y)$ denote the union of Y with all the relatively compact connected components of $X \backslash Y$. An open subset $Y \subset X$ is called *Runge* if $Y = \mathscr{h}(Y)$, i.e., if none of the connected components of $X \backslash Y$ is compact. The following properties can be checked quite easily:

(i) $\mathscr{h}(\mathscr{h}(Y)) = \mathscr{h}(Y)$ for every $Y \subset X$.
(ii) $Y_1 \subset Y_2 \Rightarrow \mathscr{h}(Y_1) \subset \mathscr{h}(Y_2)$.

Remark. If we want to indicate the dependence on X, we will write $\mathscr{h}_X(Y)$ instead of $\mathscr{h}(Y)$. Consider the following example. Let

$$Y := \{z \in \mathbb{C} : 1 < |z| < 2\}.$$

Then Y may be thought of as either a subset of $\mathbb{C}$ or of $\mathbb{C}^*$ and

$$\mathscr{h}_{\mathbb{C}}(Y) = \{z \in \mathbb{C} : |z| < 2\}$$
$$\mathscr{h}_{\mathbb{C}^*}(Y) = Y.$$

23.5. Theorem. *Suppose Y is a subset of a Riemann surface X. Then the following hold*:

(i) *Y closed $\Rightarrow \mathscr{h}(Y)$ closed*
(ii) *Y compact $\Rightarrow \mathscr{h}(Y)$ compact.*

PROOF

(i) Suppose $C_j, j \in J$, are the connected components of $X \backslash Y$. Since $X \backslash Y$ is open and X is a manifold, all the C_j are open. Let J_0 denote the subset of those $j \in J$ such that C_j is compact. Then

$$X \backslash h(Y) = \bigcup \{C_j : j \in J \backslash J_0\}.$$

Clearly this is an open set and thus $\mathscr{h}(Y)$ is closed.

(ii) We may assume $Y \neq \varnothing$. Let U be a relatively compact open neighborhood of Y and suppose the C_j are as above.

Claim (*a*). Every C_j meets $\bar{U}$. Otherwise if some C_j were contained in $X \backslash \bar{U}$, then

$$\bar{C}_j \subset X \backslash U \subset X \backslash Y.$$

Since C_j is a connected component of $X \setminus Y$, this would imply $C_j = \bar{C}_j$. Thus C_j would be both open and closed. But this contradicts the fact that X is connected.

Claim (*b*). Only finitely many C_j meet ∂U. This follows from the fact that ∂U is compact and is covered by the disjoint, open C_j.

The assertion (ii) is now easily proved. Let $C_j, j \in J_0$, be the relatively compact connected components of $X \setminus Y$ and suppose $C_{j_1}, \ldots, C_{j_m}$ are those which meet ∂U. Then by (*a*) all the others are contained in U. Thus

$$\mathscr{h}(Y) \subset U \cup C_{j_1} \cup \cdots \cup C_{j_m}$$

is relatively compact and hence by (i) is in fact compact. □

23.6. Corollary. *Suppose X is a non-compact Riemann surface. Then there is a sequence $K_j, j \in \mathbb{N}$, of compact subsets of X with the following properties:*

(i) $K_j = \mathscr{h}(K_j)$ *for every* j,
(ii) $K_{j-1} \subset \mathring{K}_j$ *for every* $j \geq 1$,
(iii) $\bigcup_{j=0}^{\infty} K_j = X$.

PROOF. Since X has a countable topology, there exists a sequence of compact subsets $K'_0 \subset K'_1 \subset K'_2 \subset \cdots$ of X which cover X. We will construct the sequence K_j by induction. Let $K_0 := \mathscr{h}(K'_0)$. Now suppose $K_1, \ldots, K_m$ with properties (i) and (ii) have already been constructed. There exists a compact set M with $K'_m \cup K_m \subset \mathring{M}$. Set $K_{m+1} := \mathscr{h}(M)$. Then the sequence $K_j, j \in \mathbb{N}$ satisfies (i), (ii) and (iii). □

23.7. Lemma. *Suppose K_1 and K_2 are compact subsets of a Riemann surface X with $K_1 \subset \mathring{K}_2$ and $\mathscr{h}(K_2) = K_2$. Then there exists an open subset Y of X which is Runge and satisfies $K_1 \subset Y \subset K_2$. Moreover one may choose Y so that its boundary is regular in the sense of solving the Dirichlet Problem.*

PROOF. Given $x \in \partial K_2$ there is a coordinate neighborhood U of x which does not meet K_1. In U choose a compact disk D containing x in its interior. Then finitely many such disks, say $D_1, \ldots, D_k$, cover ∂K_2. Set

$$Y := K_2 \setminus (D_1 \cup \cdots \cup D_k).$$

Then Y is open and $K_1 \subset Y \subset K_2$. Let $C_j, j \in J$, be the connected components of $X \setminus K_2$. By assumption they are not relatively compact. Every D_j is connected and meets at least one C_j. Hence no connected component of $X \setminus Y$ is relatively compact, i.e., $Y = \mathscr{h}(Y)$. Finally, by Theorem (22.18) all the boundary points of Y are regular. □

23.8. Theorem. *Suppose Y is a Runge open subset of a Riemann surface X. Then every connected component of Y is also Runge.*

PROOF

(a) Suppose Y_i, $i \in I$, are the connected components of Y. Since Y is open and X is a manifold, all the Y_i are open. Let $A := X \setminus Y$ be the complement of Y. The connected components A_k, $k \in K$, of A are by assumption closed but not compact.

(b) We claim that $\bar{Y}_i \cap A \neq \emptyset$ for every $i \in I$. Otherwise $\bar{Y}_i \subset Y$. Since

$$\bar{Y}_i \cap \bigcup_{j \neq i} Y_j = \emptyset,$$

it would then follow that $\bar{Y}_i = Y_i$. But this contradicts the connectivity of X.

(c) Next we claim that $C \cap A \neq \emptyset$ for every connected component C of $X \setminus Y_i$. Otherwise there would be a $j \neq i$ such that $C \cap Y_j \neq \emptyset$. Since C is closed and Y_j is connected, it would then follow that $\bar{Y}_j \subset C$. By (b) this would imply $C \cap A \neq \emptyset$.

(d) Finally suppose C is a connected component of $X \setminus Y_i$. Then C meets at least one A_k by (c) and thus in fact $C \supset A_k$. Since A_k is not compact, C is also not compact. Hence Y_i is Runge. □

23.9. Theorem. *Suppose X is a non-compact Riemann surface. Then there exists a sequence $Y_0 \Subset Y_1 \Subset Y_2 \Subset \cdots$ of relatively compact Runge domains with $\cup Y_\nu = X$ and so that every Y_ν has a regular boundary with respect to solving the Dirichlet Problem.*

PROOF. The result will follow if we show that for every compact set $K \subset Y$ there exists a Runge domain $Y \Subset X$ which has a regular boundary and contains K.

Given K, we can find a connected compact set $K_1 \supset K$ and a compact set K_2 with $K_1 \subset \mathring{K}_2$. By Lemma (23.7) there is a Runge open set Y_1 with $K_1 \subset Y_1 \subset h(K_2)$ and a regular boundary. Let Y be the connected component of Y_1 which contains K_1. By (23.8) Y is also Runge and by the Remark in (22.14) it has a regular boundary. □

EXERCISES (§23)

23.1. Suppose X is a Riemann surface and $\tilde{X} \to X$ is its universal covering. Show that $\operatorname{Deck}(\tilde{X}/X)$ is countable.

23.2. Let $Y \subset \mathbb{C}$ be open and K a compact connected component of $\mathbb{C} \setminus Y$. Let $(f_\nu)_{\nu \in \mathbb{N}}$ be a sequence of polynomials which converges uniformly on every compact subset of Y. Show that (f_ν) converges uniformly on K.

23.3. Suppose $Y \subset \mathbb{C}$ is an open subset such that every holomorphic function $f \in \mathcal{O}(Y)$ can be approximated uniformly on every compact subset of Y by polynomials. Conclude that $Y = h_{\mathbb{C}}(Y)$.

§24. Weyl's Lemma

In this section we introduce the notion of distributions. These are generalized functions. In the class of distributions differentiation is possible without any restrictions. Hence it is possible to consider solutions of differential equations in the sense of distributions. Now Weyl's Lemma asserts that for Laplace's equation $\Delta u = 0$ both kinds of solutions are the same, i.e., every harmonic distribution is a smooth function in the usual sense which satisfies Laplace's equation.

24.1. Suppose X is an open subset of the complex plane. Recall that $\mathscr{E}(X)$ denotes the vector space of all the infinitely differentiable (with respect to the real coordinates) functions $f: X \to \mathbb{C}$. By the support, denoted $\mathrm{Supp}(f)$, of such a function is meant the closure (in X) of the set $\{x \in X : f(x) \neq 0\}$. Set

$$\mathscr{D}(X) := \{f \in \mathscr{E}(X) : \mathrm{Supp}(f) \text{ is compact in } X\}.$$

Introduce the following notion of convergence in the vector space $\mathscr{D}(X)$. A sequence $(f_\nu)_{\nu \in \mathbb{N}}$ of functions in $\mathscr{D}(X)$ converges to a function $f \in \mathscr{D}(X)$, denoted $f_\nu \underset{\mathscr{D}}{\to} f$, if:

(i) There exists a compact subset $K \subset X$ such that $\mathrm{Supp}(f) \subset K$ and $\mathrm{Supp}(f_\nu) \subset K$ for every $\nu \in \mathbb{N}$.

(ii) For every $\alpha = (\alpha_1, \alpha_2) \in \mathbb{N}^2$ the sequence $D^\alpha f_\nu$ converges uniformly on K to $D^\alpha f$, where D^α denotes the differential operator

$$D^\alpha = \frac{\partial^{\alpha_1 + \alpha_2}}{\partial x^{\alpha_1}\, \partial y^{\alpha_2}}.$$

Thus convergence in $\mathscr{D}(X)$ is a much stronger condition than either pointwise or uniform convergence of sequences of functions.

24.2. Definition. Suppose $X \subset \mathbb{C}$ is open. A *distribution* on X is a continuous linear mapping

$$T: \mathscr{D}(X) \to \mathbb{C}, \qquad f \mapsto T[f].$$

Saying that T is continuous means that if $f_\nu \underset{\mathscr{D}}{\to} f$, then $T[f_\nu] \to T[f]$, where this latter convergence is that of a sequence of complex numbers. Denote by $\mathscr{D}'(X)$ the vector space of all distributions on X.

24.3. Examples

(a) To every continuous function $h \in \mathscr{C}(X)$ is associated a distribution $T_h \in \mathscr{D}'(X)$ as follows. For $f \in \mathscr{D}(X)$ let

$$T_h[f] := \iint_X h(z) f(z)\, dx\, dy, \quad \text{where } z = x + iy.$$

Clearly the map $f \mapsto T_h[f]$ is linear and continuous. If $h_1, h_2 \in \mathscr{C}(X)$ and $T_{h_1}[f] = T_{h_2}[f]$ for every $f \in \mathscr{D}(X)$, then $h_1 = h_2$. Hence the (linear) map

$$\mathscr{C}(X) \to \mathscr{D}'(X), \qquad h \mapsto T_h$$

is injective and one can identify a continuous function on X with its associated distribution.

(b) Suppose $a \in X$. For $f \in \mathscr{D}(X)$ set

$$\delta_a[f] := f(a).$$

This defines a distribution $\delta_a \in \mathscr{D}'(X)$ which is called the *Dirac delta distribution* at the point a. Unlike Example (a), this distribution cannot be represented by a function.

24.4. The Differentiation of Distributions. Suppose $h \in \mathscr{E}(X)$ and $f \in \mathscr{D}(X)$. Then for every $\alpha = (\alpha_1, \alpha_2) \in \mathbb{N}^2$

$$\iint\limits_X h(z) D^\alpha f(z)\, dx\, dy = (-1)^{\alpha_1 + \alpha_2} \iint\limits_X f(z) D^\alpha h(z)\, dx\, dy.$$

This is proved by integrating by parts $(\alpha_1 + \alpha_2)$ times and noting that since f has compact support, all the integrals over the boundary are zero.

Hence, using the notation of Example (24.3.a),

$$T_{D^\alpha h}[f] = (-1)^{|\alpha|} T_h[D^\alpha f], \quad \text{where } |\alpha| := \alpha_1 + \alpha_2.$$

This motivates the following definition. For $T \in \mathscr{D}'(X)$ set

$$(D^\alpha T)[f] := (-1)^{|\alpha|} T[D^\alpha f] \quad \text{for every } f \in \mathscr{D}(X).$$

Since $f_\nu \underset{\mathfrak{D}}{\to} f$ implies $D^\alpha f_\nu \underset{\mathfrak{D}}{\to} D^\alpha f$, the map $D^\alpha T: \mathfrak{D}(X) \to \mathbb{C}$ is continuous, i.e., $D^\alpha T \in \mathfrak{D}'(X)$. This points out that for differentiable functions the derivative in the usual sense and in the sense of distributions is the same.

24.5. Lemma. *Suppose given an open subset $X \subset \mathbb{C}$, a compact subset $K \subset X$ and an open interval $I \subset \mathbb{R}$. Suppose $g: X \times I \to \mathbb{C}$ is an infinitely (real) differentiable function with* $\operatorname{Supp}(g) \subset K \times I$ *and T is a distribution on X. Then the function $t \mapsto T_z[g(z, t)]$ is infinitely differentiable on I and satisfies*

$$\frac{d}{dt} T_z[g(z, t)] = T_z\left[\frac{\partial g(z, t)}{\partial t}\right]. \tag{*}$$

The subscript z indicates that T operates on $g(z, t)$ as a function of z while t is thought of as a parameter. Thus one may interchange the operation of applying a distribution to a function depending on a parameter and differentiation with respect to that parameter.

PROOF. It suffices to prove (*), since repeated application of this result will show the infinite differentiability with respect to t. Since T is linear,

$$\frac{d}{dt} T_z[g(z, t)] = \lim_{h\to 0} \frac{1}{h}(T_z[g(z, t + h)] - T_z[g(z, t)])$$

$$= \lim_{h\to 0} T_z\left[\frac{g(z, t + h) - g(z, t)}{h}\right].$$

For fixed $t \in I$ and sufficiently small $h \in \mathbb{R}^*$ let

$$f_h(z) := \frac{1}{h}(g(z, t + h) - g(z, t)).$$

Then $f_h \in \mathscr{D}(X)$ and

$$f_h \underset{\mathscr{D}}{\rightarrow} \frac{\partial g(\cdot, t)}{\partial t} \quad \text{as } h \to 0.$$

Hence, because T is continuous,

$$\lim_{h\to 0} T[f_h] = T_z\left[\frac{\partial g(z, t)}{\partial t}\right]. \qquad \square$$

The next Lemma asserts that the operation of applying a distribution to a function depending on a parameter may be interchanged with integration with respect to that parameter.

24.6. Lemma. *Suppose X, Y are open subsets of $\mathbb{C}$ and $K \subset X$, $L \subset Y$ are compact subsets. Further suppose $g: X \times Y \to \mathbb{C}$ is an infinitely (real) differentiable function with* Supp$(g) \subset K \times L$. *Then for any distribution T on X*

$$T_z\left[\iint_Y g(z, \zeta)\, d\xi\, d\eta\right] = \iint_Y T_z[g(z, \zeta)]\, d\xi\, d\eta, \qquad \zeta = \xi + i\eta.$$

PROOF. It follows from (24.5) that $T_z[g(z, \zeta)]$ is infinitely differentiable with respect to $\zeta = \xi + i\eta$. Thus the integral on the right hand side is well-defined. Suppose $R \subset \mathbb{C}$ is a rectangle with sides parallel to the axes which contains L. Then the function $g(z, \zeta)$ extends as zero to $K \times R$. For every integer $n > 0$ partition R into n^2 subrectangles $R_{n\nu}$, $\nu = 1, \ldots, n^2$, by subdividing the sides into n equal parts. Choose a point $\zeta_{n\nu}$ in each $R_{n\nu}$. Let F be the area of R. Then the Riemann sums

$$G_n(z) := \frac{F}{n^2} \sum_{\nu=1}^{n^2} g(z, \zeta_{n\nu})$$

converge as $n \to \infty$ to the integral $\iint_Y g(z, \zeta)\, d\xi\, d\eta$. Since Supp$(G_n) \subset K$ for each n,

$$G_n \underset{\mathscr{D}}{\rightarrow} \iint_Y g(\cdot, \zeta)\, d\xi\, d\eta \quad \text{as } n \to \infty.$$

Thus from the continuity of T it follows that

$$T_z\left[\iint_Y g(z, \zeta)\, d\xi\, d\eta\right] = \lim_{n\to\infty} T[G_n] = \iint_Y T_z[g(z, \zeta)]\, d\xi\, d\eta. \qquad \square$$

24.7. The Smoothing of Functions. Choose a function $\rho \in \mathfrak{D}(\mathbb{C})$ with the following properties:

(i) $\mathrm{Supp}(\rho) \subset \{z \in \mathbb{C}\colon |z| < 1\}$.
(ii) ρ is invariant under rotations, i.e., $\rho(z) = \rho(|z|)$ for every $z \in \mathbb{C}$.
(iii) $\iint_{\mathbb{C}} \rho(x + iy)\, dx\, dy = 1$.

For $\varepsilon > 0$ and $z \in \mathbb{C}$ set

$$\rho_\varepsilon(z) := \frac{1}{\varepsilon^2}\, \rho\left(\frac{z}{\varepsilon}\right).$$

Then $\mathrm{Supp}(\rho_\varepsilon) \subset \{z \in \mathbb{C}\colon |z| < \varepsilon\}$ and

$$\iint_{\mathbb{C}} \rho_\varepsilon(x + iy)\, dx\, dy = 1.$$

Denote by $D(z, \varepsilon)$ the open disk with center z and radius ε and by $\bar{D}(z, \varepsilon)$ its closure.

If $U \subset \mathbb{C}$ is an open set, then

$$U^{(\varepsilon)} := \{z \in U\colon \bar{D}(z, \varepsilon) \subset U\}$$

is also open.

Given a continuous function $f\colon U \to \mathbb{C}$, define a new function $\mathrm{sm}_\varepsilon f\colon U^{(\varepsilon)} \to \mathbb{C}$ by

$$(\mathrm{sm}_\varepsilon f)(z) := \iint_U \rho_\varepsilon(z - \zeta) f(\zeta)\, d\xi\, d\eta, \qquad \zeta = \xi + i\eta.$$

Clearly $\mathrm{sm}_\varepsilon f \in \mathscr{E}(U^{(\varepsilon)})$, since one can differentiate under the integral. The function $\mathrm{sm}_\varepsilon f$ is called a *smoothing* of f.

Remark. Naturally the definition depends on the choice of the function ρ.

24.8. Lemma. *Suppose $U \subset \mathbb{C}$ is open, $f \in \mathscr{E}(U)$ and $\varepsilon > 0$.*

(a) *For every $\alpha \in \mathbb{N}^2$*

$$D^\alpha(\mathrm{sm}_\varepsilon f) = \mathrm{sm}_\varepsilon(D^\alpha f).$$

(b) *If $z \in U^{(\varepsilon)}$ and f is harmonic on $D(z, \varepsilon)$, then*

$$(\mathrm{sm}_\varepsilon f)(z) = f(z).$$

PROOF

(a) For $z \in U^{(\varepsilon)}$ translation of the integration variable gives

$$(\mathrm{sm}_\varepsilon f)(z) = \iint\limits_{|\zeta|<\varepsilon} \rho_\varepsilon(\zeta) f(z+\zeta)\, d\xi\, d\eta.$$

Thus

$$D^\alpha(\mathrm{sm}_\varepsilon f)(z) = \iint\limits_{|\zeta|<\varepsilon} \rho_\varepsilon(\zeta) D^\alpha f(z+\zeta)\, d\xi\, d\eta$$

$$= \iint\limits_{U} \rho_\varepsilon(z-\zeta) D^\alpha f(\zeta)\, d\xi\, d\eta = \mathrm{sm}_\varepsilon(D^\alpha f)(z).$$

(b) If f is harmonic on $D(z, \varepsilon)$, then for every $r \in [0, \varepsilon[$ it satisfies the Mean Value Principle (22.4)

$$f(z) = \frac{1}{2\pi}\int_0^{2\pi} f(z + re^{i\theta})\, d\theta.$$

Thus

$$(\mathrm{sm}_\varepsilon f)(z) = \iint\limits_{|\zeta|<\varepsilon} \rho_\varepsilon(\zeta) f(z+\zeta)\, d\xi\, d\eta$$

$$= \iint\limits_{\substack{0 \le r \le \varepsilon \\ 0 \le \theta \le 2\pi}} \rho_\varepsilon(r) f(z + re^{i\theta}) r\, dr\, d\theta$$

$$= \int_0^\varepsilon \rho_\varepsilon(r) r\, dr \cdot 2\pi f(z) = f(z),$$

since

$$1 = \iint\limits_{\mathbb{C}} \rho_\varepsilon(\xi + i\eta)\, d\xi\, d\eta = 2\pi \int_0^\varepsilon \rho_\varepsilon(r) r\, dr.$$ □

24.9. Theorem (Weyl's Lemma). *Suppose U is an open set in $\mathbb{C}$ and T is a distribution on U with $\Delta T = 0$. Then T is a smooth function.*

In other words, if $T\colon \mathscr{D}(U) \to \mathbb{C}$ is a linear functional such that $T[\Delta\varphi] = 0$ for every $\varphi \in \mathscr{D}(U)$, then there exists a function $h \in \mathscr{E}(U)$ with $\Delta h = 0$ and

$$T[f] = \iint\limits_{U} h(z) f(z)\, dx\, dy \quad \text{for every } f \in \mathscr{D}(U).$$

PROOF. Suppose $\varepsilon > 0$ is arbitrary. For $z \in U^{(\varepsilon)}$ the function $\zeta \mapsto \rho_\varepsilon(\zeta - z)$ has compact support in U. Hence

$$h(z) := T_\zeta[\rho_\varepsilon(\zeta - z)]$$

is defined. By (24.5) the function $z \mapsto h(z)$ belongs to $\mathscr{E}(U^{(\varepsilon)})$. Obviously it is enough to prove that for every function $f \in \mathscr{D}(\mathbb{C})$ with $\operatorname{Supp}(f) \subset U^{(\varepsilon)}$ one has

$$T[f] = \iint_{U^{(\varepsilon)}} h(z) f(z)\, dx\, dy. \tag{1}$$

The function $\operatorname{sm}_\varepsilon f$ has compact support in U and by (24.6) one has

$$\begin{aligned} T[\operatorname{sm}_\varepsilon f] &= T_\zeta\left[\iint_U \rho_\varepsilon(\zeta - z) f(z)\, dx\, dy\right] \\ &= \iint_{U^{(\varepsilon)}} h(z) f(z)\, dx\, dy. \end{aligned} \tag{2}$$

By (13.3) there exists a function $\psi \in \mathscr{E}(\mathbb{C})$ with $\Delta\psi = f$. The function ψ is harmonic on $V := \mathbb{C}\backslash\operatorname{Supp}(f)$. Thus by (24.8.b)

$$\psi = \operatorname{sm}_\varepsilon \psi \quad \text{on } V^{(\varepsilon)}.$$

Hence $\varphi := \psi - \operatorname{sm}_\varepsilon \psi$ has compact support in U and by (24.8.a) satisfies

$$\Delta\varphi = \Delta(\psi - \operatorname{sm}_\varepsilon \psi) = \Delta\psi - \operatorname{sm}_\varepsilon \Delta\psi = f - \operatorname{sm}_\varepsilon f.$$

Since $\Delta T = 0$, one has $T[\Delta\varphi] = 0$. Thus

$$T[f] = T[\operatorname{sm}_\varepsilon f + \Delta\varphi] = T[\operatorname{sm}_\varepsilon f].$$

Combining this with (2) then yields (1). □

24.10. Corollary. *Suppose T is a distribution on the open set $U \subset \mathbb{C}$ with $(\partial T/\partial \bar{z}) = 0$. Then T is a holomorphic function on U.*

PROOF. Since $(\partial T/\partial \bar{z}) = 0$,

$$\Delta T = 4 \frac{\partial}{\partial z}\left(\frac{\partial}{\partial \bar{z}} T\right) = 0.$$

Thus $T \in \mathscr{E}(U)$ by (24.9). Because $(\partial T/\partial \bar{z}) = 0$, T is holomorphic. □

Remark. The proof given here for Weyl's Lemma in the plane carries over almost word for word for harmonic functions on $\mathbb{R}^n$. But Weyl's Lemma is only a special case of a general regularity theorem for elliptic differential operators on differentiable manifolds, cf. [35], [43].

EXERCISES (§24)

24.1. Let X be an open subset of $\mathbb{C}$ and $T_\nu \in \mathscr{D}'(X)$, $\nu \in \mathbb{N}$, a sequence of distributions in X. A sequence $(T_\nu)_{\nu \in \mathbb{N}}$ is said to converge to a distribution $T \in \mathscr{D}'(X)$ if

$$T_\nu[\varphi] \to T[\varphi] \quad \text{for every } \varphi \in \mathscr{D}(X).$$

Denote this by $T_\nu \underset{\mathscr{D}'}{\to} T$. Show that if $T_\nu \underset{\mathscr{D}'}{\to} T$, then

$$D^\alpha T_\nu \underset{\mathscr{D}'}{\to} D^\alpha T,$$

for every differential operator

$$D^\alpha = \left(\frac{\partial}{\partial x}\right)^{\alpha_1}\left(\frac{\partial}{\partial y}\right)^{\alpha_2}.$$

24.2. Let $Y \subset \mathbb{C}$ be open. A sequence of continuous functions $f_\nu \colon Y \to \mathbb{C}$ is said to converge weakly to a continuous function $f\colon Y \to \mathbb{C}$ if

$$\iint_Y f_\nu \varphi \, dx\, dy \to \iint_Y f\varphi \, dx\, dy \quad \text{for every } \varphi \in \mathscr{D}(X).$$

Show that if all the f_ν are harmonic (resp. holomorphic) and converge weakly to f, then f is also harmonic (resp. holomorphic).

§25. The Runge Approximation Theorem

The classical Runge Approximation Theorem asserts that on a simply connected domain $Y \subset \mathbb{C}$ every holomorphic function can be approximated, uniformly on compact sets, by functions which are holomorphic on all of $\mathbb{C}$ (and thus by polynomials). This theorem was generalized by Behnke–Stein [51] to arbitrary non-compact Riemann surfaces X. In order to approximate all holomorphic functions on an open subset $Y \subset X$ by functions holomorphic on X, one has to replace the assumption that Y is simply connected by the assumption that no connected component of $X \setminus Y$ is compact. The proof we present is based on a functional analytic proof using Weyl's Lemma which was first given by Malgrange [55].

25.1. Suppose X is a Riemann surface and $Y \subset X$ is an open subset. We would like to introduce the structure of a Fréchet space on the vector space $\mathscr{E}(Y)$ of differentiable functions on Y. To do this, choose a countable family of compact sets $K_j \subset Y$, $j \in J$, with $\bigcup \mathring{K}_j = Y$ and such that each K_j is contained in some coordinate neighborhood (U_j, z_j). For $j \in J$ and $\nu = (\nu_1, \nu_2) \in \mathbb{N}^2$ define a semi-norm $p_{j\nu}\colon \mathscr{E}(Y) \to \mathbb{R}_+$ by

$$p_{j\nu}(f) := \sup_{a \in K_j} |D_j^\nu f(a)|,$$

where

$$D_j^\nu = \left(\frac{\partial}{\partial x_j}\right)^{\nu_1}\left(\frac{\partial}{\partial y_j}\right)^{\nu_2}$$

is the appropriate differential operator relative to the coordinates $z_j = x_j + iy_j$. These countably many semi-norms $p_{j\nu}$ define a topology on $\mathscr{E}(Y)$. A neighborhood basis of zero is given by finite intersections of sets of the form

$$\mathscr{U}(p_{j\nu}, \varepsilon) := \{f \in \mathscr{E}(Y): p_{j\nu}(f) < \varepsilon\}, \qquad \varepsilon > 0.$$

Then convergence $f_n \to f$ with respect to this topology means uniform convergence of the functions and all of their derivatives on every K_j. With this topology $\mathscr{E}(Y)$ is a Fréchet space. One can easily check that this topology is independent of the choice of K_j and (U_j, z_j). On the vector subspace $\mathcal{O}(Y) \subset \mathscr{E}(Y)$ the induced topology coincides with the topology of uniform convergence on compact subsets. For, in the case of holomorphic functions uniform convergence on compact subsets implies the uniform convergence on compact subsets of all the derivatives. Analogously one can introduce the structure of a Fréchet space on the vector space $\mathscr{E}^{0,1}(Y)$ of (0, 1)-forms on Y with differentiable coefficients. An element $\omega \in \mathscr{E}^{0,1}(Y)$ may be written $\omega = f_j\, d\bar{z}_j$ on U_j, where $f_j \in \mathscr{E}(U_j \cap Y)$. Set

$$p_{j\nu}(\omega) := \sup_{a \in K_j} |D_j^\nu f_j(a)|.$$

Then the Fréchet structure is obtained as above from the semi-norms $p_{j\nu}$.

25.2. Lemma. *Suppose Y is an open subset of a Riemann surface X. Then every continuous linear map $T: \mathscr{E}(Y) \to \mathbb{C}$ has compact support, i.e., there exists a compact subset $K \subset Y$ such that*

$$T[f] = 0 \quad \textit{for every } f \in \mathscr{E}(Y) \quad \textit{with } \operatorname{Supp}(f) \subset Y \backslash K.$$

An analogous result is also true for $\mathscr{E}^{0,1}(Y)$.

PROOF. Since T is continuous, there exists a neighborhood $\mathscr{U}$ of zero in $\mathscr{E}(Y)$ such that $|T[f]| < 1$ for every $f \in \mathscr{U}$. By the definition of the topology on $\mathscr{E}(Y)$ there exist elements $j_1, \ldots, j_m \in J$, $\nu_1, \ldots, \nu_m \in \mathbb{N}^2$ and $\varepsilon > 0$, where the notation is the same as in (25.1), such that

$$\mathscr{U}(p_{j_1\nu_1}, \varepsilon) \cap \cdots \cap \mathscr{U}(p_{j_m\nu_m}, \varepsilon) \subset \mathscr{U}.$$

Let $K := K_{j_1} \cup \cdots \cup K_{j_m}$. We now show that if $f \in \mathscr{E}(Y)$ with $\operatorname{Supp}(f) \subset Y \backslash K$, then $T[f] = 0$. Namely for arbitrary $\lambda > 0$,

$$p_{j_1\nu_1}(\lambda f) = \cdots = p_{j_m\nu_m}(\lambda f) = 0.$$

Thus $\lambda f \in \mathscr{U}$ and $|T[\lambda f]| < 1$. But this implies $|T[f]| < 1/\lambda$ for every $\lambda > 0$ and this is possible only if $T[f] = 0$. $\square$

25.3. Lemma. *Suppose Z is an open subset of a Riemann surface X and $S: \mathscr{E}^{0,1}(X) \to \mathbb{C}$ is a continuous linear mapping with $S[d''g] = 0$ for every*

$g \in \mathcal{E}(X)$ with $\operatorname{Supp}(g) \Subset Z$. Then there exists a holomorphic 1-form $\sigma \in \Omega(X)$ such that

$$S[\omega] = \iint_Z \sigma \wedge \omega$$

for every $\omega \in \mathcal{E}^{0,1}(X)$ with $\operatorname{Supp}(\omega) \Subset Z$.

PROOF. Suppose $z: U \to V \subset \mathbb{C}$ is a chart on X which lies in Z. Identify U with V. For $\varphi \in \mathcal{D}(U)$ denote by $\tilde{\varphi}$ any 1-form in $\mathcal{E}^{0,1}(X)$ which equals $\varphi\, d\bar{z}$ on U and zero on $X \backslash U$. Then the mapping

$$S_U: \mathcal{D}(U) \to \mathbb{C}, \varphi \mapsto S[\tilde{\varphi}]$$

is a distribution on U which vanishes on all functions of the form $\varphi = \partial g/\partial \bar{z}$, $g \in \mathcal{D}(U)$, i.e., $\partial S_U/\partial \bar{z} = 0$. Hence by Corollary (24.10) there exists a unique holomorphic function $h \in \mathcal{O}(U)$ with

$$S[\tilde{\varphi}] = \iint_U h(z)\varphi(z)\, dz \wedge d\bar{z} \quad \text{for every } \varphi \in \mathcal{D}(U).$$

Setting $\sigma_U := h\, dz$, we get

$$S[\omega] = \iint_U \sigma_U \wedge \omega$$

for every $\omega \in \mathcal{E}^{0,1}(U)$ with $\operatorname{Supp}(\omega) \Subset U$.

Now if we carry out the same construction with respect to another chart $z': U' \to V'$, then we get a 1-form $\sigma_{U'} \in \Omega(U')$ with the corresponding properties. Hence

$$\iint_U \sigma_U \wedge \omega = \iint_{U'} \sigma_{U'} \wedge \omega$$

for every $\omega \in \mathcal{E}^{0,1}(X)$ with $\operatorname{Supp}(\omega) \Subset U \cap U'$. This implies $\sigma_U = \sigma_{U'}$ on $U \cap U'$. Thus the σ_U piece together to give a 1-form $\sigma \in \Omega(Z)$ such that

$$S[\omega] = \iint_Z \sigma \wedge \omega \tag{*}$$

for every $\omega \in \mathcal{E}^{0,1}(X)$ whose support is compact and lies in a chart inside Z. If $\omega \in \mathcal{E}^{0,1}(X)$ is an arbitrary 1-form with $\operatorname{Supp}(\omega) \Subset Z$, then using a partition of unity one can write $\omega = \omega_1 + \cdots + \omega_n$, where each ω_j satisfies (*). Thus

$$S[\omega] = \sum_{j=1}^{n} S[\omega_j] = \sum_{j=1}^{n} \iint_Z \sigma \wedge \omega_j = \iint_Z \sigma \wedge \omega. \qquad \square$$

25.4. Theorem. *Suppose Y is a relatively compact open Runge subset of a non-compact Riemann surface X. Then for every open subset Y' with $Y \subset Y' \Subset X$ the image of the restriction map $\mathcal{O}(Y') \to \mathcal{O}(Y)$ is dense, where the topology is uniform convergence on compact subsets.*

PROOF. Denote by $\beta: \mathcal{E}(Y') \to \mathcal{E}(Y)$ the restriction map. In order to prove that $\beta(\mathcal{O}(Y'))$ is dense in $\mathcal{O}(Y)$ we can use the Hahn–Banach Theorem (c.f. Appendix B.9). It suffices to show the following. *If $T: \mathcal{E}(Y) \to \mathbb{C}$ is a continuous linear functional with $T \mid \beta(\mathcal{O}(Y')) = 0$, then $T \mid \mathcal{O}(Y) = 0$.*

To prove this, define a linear mapping

$$S: \mathcal{E}^{0,1}(X) \to \mathbb{C}$$

in the following way. By (14.16) given $\omega \in \mathcal{E}^{0,1}(X)$ there exists a function $f \in \mathcal{E}(Y')$ with $d''f = \omega \mid Y'$. Then set

$$S[\omega] := T[f \mid Y].$$

This definition is independent of the choice of the function f. For, if $d''g = \omega \mid Y'$, then $f - g \in \mathcal{O}(Y')$ and thus by assumption $T[(f-g) \mid Y] = 0$. We will now show that S is also continuous. Consider the vector space

$$V := \{(\omega, f) \in \mathcal{E}^{0,1}(X) \times \mathcal{E}(Y'): d''f = \omega \mid Y'\}.$$

Since $d'': \mathcal{E}(Y') \to \mathcal{E}^{0,1}(Y')$ is continuous, V is a closed vector subspace of $\mathcal{E}^{0,1}(X) \times \mathcal{E}(Y')$ and thus is a Fréchet space. Now the projection $\mathrm{pr}_1: V \to \mathcal{E}^{0,1}(X)$ is surjective and thus by the Theorem of Banach is open. Also the mapping $\beta \circ \mathrm{pr}_2: V \to \mathcal{E}(Y)$ is continuous. Since the diagram

$$\begin{array}{ccc} V & \xrightarrow{\beta \circ \mathrm{pr}_2} & \mathcal{E}(Y) \\ {\scriptstyle \mathrm{pr}_1}\downarrow & & \downarrow{\scriptstyle T} \\ \mathcal{E}^{0,1}(X) & \xrightarrow{\quad S \quad} & \mathbb{C} \end{array}$$

is commutative by definition, S is continuous because T is.

By Lemma (25.2) there exists a compact subset $K \subset Y$ with

(1) $T[f] = 0$ for every $f \in \mathcal{E}(Y)$ with $\mathrm{Supp}(f) \subset Y \backslash K$

and a compact subset $L \subset X$ with

(2) $S[\omega] = 0$ for every $\omega \in \mathcal{E}^{0,1}(X)$ with $\mathrm{Supp}(\omega) \subset X \backslash L$.

If $g \in \mathcal{E}(X)$ is a function with $\mathrm{Supp}(g) \Subset X \backslash K$, then $S[d''g] = T[g \mid Y] = 0$. Thus by Lemma (25.3) there exists a holomorphic 1-form $\sigma \in \Omega(X \backslash K)$ such that

$$S[\omega] = \iint_{X \backslash K} \sigma \wedge \omega$$

for every $\omega \in \mathscr{E}^{0,1}(X)$ with $\operatorname{Supp}(\omega) \Subset X \backslash K$. Because of (2) it must be the case that $\sigma | X \backslash (K \cup L) = 0$. Every connected component of $X \backslash \mathscr{h}(K)$ is not relatively compact and hence meets $X \backslash (K \cup L)$. Thus by the Identity Theorem $\sigma | X \backslash \mathscr{h}(K) = 0$, i.e.

(3) $S[\omega] = 0$ for every $\omega \in \mathscr{E}^{0,1}(X)$ with $\operatorname{Supp}(\omega) \Subset X \backslash \mathscr{h}(K)$.

Now suppose $f \in \mathcal{O}(Y)$. We have to show $T[f] = 0$. Since Y is Runge, $\mathscr{h}(K) \subset Y$. Hence there is a function $g \in \mathscr{E}(X)$ with $f = g$ in a neighborhood of $\mathscr{h}(K)$ and $\operatorname{Supp}(g) \Subset Y$. Then $T[f] = T[g | Y]$ by (1) and $T[g | Y] = S[d''g]$ by the definition of S. Since g is holomorphic on a neighborhood of $\mathscr{h}(K)$, one has $\operatorname{Supp}(d''g) \Subset X \backslash \mathscr{h}(K)$ and thus $S[d''g] = 0$ by (3). Collecting these statements together we have $T[f] = 0$ for every $f \in \mathcal{O}(Y)$. □

25.5. The Runge Approximation Theorem. *Suppose X is a non-compact Riemann surface and Y is an open subset whose complement contains no compact connected component. Then every holomorphic function on Y can be approximated uniformly on every compact subset of Y by holomorphic functions on X.*

PROOF. It suffices to consider the case when Y is relatively compact in X. Suppose $f \in \mathcal{O}(Y)$, a compact subset $K \subset Y$ and $\varepsilon > 0$ are given. By (23.9) there exists an exhaustion $Y_1 \Subset Y_2 \Subset \cdots$ of X by Runge domains with $Y_0 := Y \Subset Y_1$. By Theorem (25.4) there is a holomorphic function $f_1 \in \mathcal{O}(Y_1)$ with

$$\|f_1 - f\|_K < 2^{-1}\varepsilon,$$

where $\| \quad \|_K$ denotes the supremum norm on K.

Now using Theorem (25.4) and induction one gets a sequence of functions $f_n \in \mathcal{O}(Y_n)$ with

$$\| f_n - f_{n-1} \|_{\bar{Y}_{n-2}} < 2^{-n}\varepsilon \quad \text{for every } n \geq 2.$$

For every $n \in \mathbb{N}$ the sequence $(f_\nu)_{\nu > n}$ converges uniformly on Y_n. Hence there exists a function $F \in \mathcal{O}(X)$, holomorphic on all of X, which on each Y_n is the limit of the sequence $(f_\nu)_{\nu > n}$. Thus, by construction, $\|F - f\|_K < \varepsilon$. □

25.6. Theorem. *Suppose X is a non-compact Riemann surface. Then given a 1-form $\omega \in \mathscr{E}^{0,1}(X)$ there exists a function $f \in \mathscr{E}(X)$ with $d''f = \omega$.*

PROOF. For every relatively compact open subset $Y \Subset X$ there exists by (14.16) a function $g \in \mathscr{E}(Y)$ with $d''g = \omega | Y$. Now the proof is similar to the proof of Theorem (13.2), namely one uses an exhaustion process.

Suppose $Y_0 \Subset Y_1 \Subset Y_2 \Subset \cdots$ is an exhaustion of X by Runge domains (23.9). By induction on n we will construct functions $f_n \in \mathscr{E}(Y_n)$ such that

(i) $d''f_n = \omega \mid Y_n$,
(ii) $\|f_{n+1} - f_n\|_{Y_{n-1}} \leq 2^{-n}$.

To begin choose any function $f_0 \in \mathscr{E}(Y_0)$ which is a solution of the differential equation $d''f_0 = \omega \mid Y_0$. Now suppose $f_0, \ldots, f_n$ have been constructed. There exists $g_{n+1} \in \mathscr{E}(Y_{n+1})$ with $d''g_{n+1} = \omega \mid Y_{n+1}$. On Y_n one has $d''g_{n+1} = d''f_n$ and thus $g_{n+1} - f_n$ is holomorphic on Y_n. By the Runge Approximation Theorem there exists $h \in \mathcal{O}(Y_{n+1})$ such that

$$\|(g_{n+1} - f_n) - h\|_{Y_{n-1}} \leq 2^{-n}.$$

Set $f_{n+1} := g_{n+1} - h$. Then $d''f_{n+1} = d''g_{n+1} = \omega \mid Y_{n+1}$ and

$$\|f_{n+1} - f_n\|_{Y_{n-1}} \leq 2^{-n}.$$

As in the proof of (13.2) it now follows that the functions f_n converge to a solution $f \in \mathscr{E}(X)$ of the differential equation $d''f = \omega$. □

EXERCISES (§25)

25.1. Let X be a Riemann surface and $S: \mathscr{E}^{(2)}(X) \to \mathbb{C}$ a continuous linear functional such that $S[d'd''g] = 0$ for every $g \in \mathscr{E}(X)$. Prove that there is a harmonic function $h \in \mathscr{E}(X)$ such that

$$S[\omega] = \iint_X h\omega \quad \text{for every } \omega \in \mathscr{E}^{(2)}(X) \text{ with compact support.}$$

25.2. Let $Y \subset \mathbb{C}$ be open. Given any $g \in \mathscr{E}(Y)$ show that there exists an $f \in \mathscr{E}(Y)$ such that

$$\Delta f = g.$$

§26. The Theorems of Mittag-Leffler and Weierstrass

We now consider the problem of constructing meromorphic functions on non-compact Riemann surfaces having prescribed principal parts, resp. having zeros and poles of given orders. These are analogues of the Theorems of Mittag-Leffler and Weierstrass in the complex plane. For compact Riemann surfaces the comparable problems were looked at in sections 18 and 20. While in the compact case there are particular conditions which are necessary in order for a solution to exist (Theorems 18.2 and 20.7), it turns out that in the non-compact case the analogues of the Theorems of Mittag-Leffler and Weierstrass hold without any restriction.

26.1. Theorem. *Suppose X is a non-compact Riemann surface. Then*

$$H^1(X, \mathcal{O}) = 0.$$

PROOF. By the Dolbeault Theorem (15.14) one has $H^1(X, \mathcal{O}) \cong \mathcal{E}^{0,1}(X)/d''\mathcal{E}(X)$. But by Theorem (25.6) $\mathcal{E}^{0,1}(X) = d''\mathcal{E}(X)$ and thus $H^1(X, \mathcal{O}) = 0$. □

Remark. Theorem (26.1) is a special case of Theorem B of Cartan–Serre which is valid on arbitrary n-dimensional Stein manifolds, c.f. [32], [34].

26.2. We now recall the notion of a Mittag–Leffler distribution, c.f. (18.1). Suppose $\mathfrak{U} = (U_i)_{i \in I}$ is an open covering of a Riemann surface X. A family $\mu = (f_i)_{i \in I}$ of meromorphic functions $f_i \in \mathcal{M}(U_i)$ is called a Mittag–Leffler distribution if the differences $f_i - f_j$ are holomorphic on $U_i \cap U_j$, i.e., the functions have the same principal parts. By a solution of μ one means a global meromorphic function $f \in \mathcal{M}(X)$ such that for each $i \in I$ the difference $f - f_i$ is holomorphic on U_i. The family of differences $f_{ij} := f_j - f_i \in \mathcal{O}(U_i \cap U_j)$ defines a cocycle $(f_{ij}) \in Z^1(\mathfrak{U}, \mathcal{O})$. We proved in (18.1) that μ has a solution precisely if this cocycle is a coboundary, i.e., $(f_{ij}) \in B^1(\mathfrak{U}, \mathcal{O})$. Thus by Theorem (26.1) we have the following.

26.3. Theorem. *On a non-compact Riemann surface every Mittag–Leffler distribution has a solution.*

We now turn to the analogue of the Weierstrass Product Theorem. Given a divisor $D\colon X \to \mathbb{Z}$ on a Riemann surface X one would like to find a meromorphic function $f \in \mathcal{M}^*(X)$ which has the same zeros and poles, counting multiplicities, as D, i.e., $(f) = D$, c.f. definitions (16.1) and (16.2). Recall that the notion of a weak solution was defined in (20.1).

26.4. Lemma. *Every divisor D on a non-compact Riemann surface X has a weak solution.*

PROOF

(a) Choose a sequence $K_1, K_2, \ldots$ of compact subsets of X with the following properties:

(i) $K_j = h(K_j)$ for every $j \geq 1$,
(ii) $K_j \subset \mathring{K}_{j+1}$ for every $j \geq 1$,
(iii) $\bigcup_{j \geq 1} K_j = X$.

This is possible by (23.6).

(b) We claim that given $a_0 \in X \backslash K_j$ and a divisor A_0 with $A_0(a_0) = 1$ and $A_0(x) = 0$ for $x \neq a_0$, then there exists a weak solution φ of A_0 with $\varphi | K_j = 1$.

In order to prove the claim, note that since $K_j = \hbar(K_j)$, the point a_0 lies in a connected component U of $X \backslash K_j$ which is not relatively compact. Hence there exists a point $a_1 \in U \backslash K_{j+1}$ and a curve c_0 in U with initial point a_1 and end point a_0. By Lemma (20.5) there is a weak solution φ_0 of the divisor ∂c_0 with $\varphi_0 | K_j = 1$. Repeating the construction gives a sequence of points $a_\nu \in X \backslash K_{j+\nu}$, $\nu \in \mathbb{N}$, curves c_ν in $X \backslash K_{j+\nu}$ from $a_{\nu+1}$ to a_ν and weak solutions φ_ν of the divisors ∂c_ν with $\varphi_\nu | K_{j+\nu} = 1$. Then $\partial c_\nu = A_\nu - A_{\nu+1}$, where A_ν is the divisor which takes the value 1 at a_ν and is zero otherwise. Thus the product $\varphi_0 \varphi_1 \cdot \dots \cdot \varphi_n$ is a weak solution of the divisor $A_0 - A_{n+1}$. The infinite product

$$\varphi := \prod_{\nu=0}^{\infty} \varphi_\nu$$

converges, since on any compact subset of X there are only finitely many factors which are not identically 1. Now φ is the desired weak solution of the divisor A_0.

(c) Now suppose D is an arbitrary divisor on X. For $\nu \in \mathbb{N}$ set

$$D_\nu(x) := \begin{cases} D(x), & \text{if } x \in K_{\nu+1} \backslash K_\nu, \\ 0, & \text{if } x \notin K_{\nu+1} \backslash K_\nu, \end{cases}$$

where $K_0 := \varnothing$. Then

$$D = \sum_{\nu=0}^{\infty} D_\nu.$$

Since D_ν is non-zero only at a finite number of points, by (b) there is a weak solution ψ_ν of the divisor D_ν with $\psi_\nu | K_\nu = 1$. The product

$$\psi := \prod_{\nu=0}^{\infty} \psi_\nu$$

is thus a weak solution of D. □

26.5. Theorem. *On a non-compact Riemann surface X every divisor $D \in \operatorname{Div}(X)$ is the divisor of a meromorphic function $f \in \mathcal{M}^*(X)$.*

PROOF. Since the problem has a solution locally, there exists an open covering $\mathfrak{U} = (U_i)_{i \in I}$ of X and meromorphic functions $f_i \in \mathcal{M}^*(U_i)$ such that the divisor of f_i coincides with D on U_i. We may assume that all the U_i are simply connected. On the intersection $U_i \cap U_j$ the functions f_i and f_j have the same zeros and poles, i.e.,

$$\frac{f_i}{f_j} \in \mathcal{O}^*(U_i \cap U_j) \quad \text{for every } i, j \in I.$$

Now suppose ψ is a weak solution of D. This exists by (26.4). Then $\psi = \psi_i f_i$ on U_i, where the function $\psi_i \in \mathcal{E}(U_i)$ has no zeros. Since U_i is simply connected, there exists a function $\varphi_i \in \mathscr{E}(U_i)$ with $\psi_i = e^{\varphi_i}$, i.e., $\psi = e^{\varphi_i} f_i$ on U_i. Then on $U_i \cap U_j$ one has

$$e^{\varphi_j - \varphi_i} = \frac{f_i}{f_j} \in \mathcal{O}^*(U_i \cap U_j) \qquad (*)$$

and thus $\varphi_{ij} := \varphi_j - \varphi_i \in \mathcal{O}(U_i \cap U_j)$. Since $\varphi_{ij} + \varphi_{jk} = \varphi_{ik}$ on any triple intersection, the family φ_{ij} is a cocycle $(\varphi_{ij}) \in Z^1(\mathfrak{U}, \mathcal{O})$. Because $H^1(X, \mathcal{O}) = 0$, this cocycle splits. Thus there exist holomorphic functions $g_i \in \mathcal{O}(U_i)$ with

$$\varphi_{ij} = \varphi_j - \varphi_i = g_j - g_i \quad \text{on } U_i \cap U_j$$

for every $i, j \in I$. From (*) one gets $e^{g_j - g_i} = f_i/f_j$, i.e.,

$$e^{g_j} f_j = e^{g_i} f_i \quad \text{on } U_i \cap U_j.$$

Hence there exists a global meromorphic function $f \in \mathscr{M}^*(X)$ with $f = e^{g_i} f_i$ on U_i for every $i \in I$. Since f and f_i define the same divisor on U_i, one has $(f) = D$. □

26.6. Corollary. *On every non-compact Riemann surface X there is a holomorphic 1-form $\omega \in \Omega(X)$ which never vanishes.*

PROOF. Suppose g is a non-constant meromorphic function on X and $f \in \mathscr{M}^*(X)$ is a function with divisor $-(dg)$. Then $\omega := f\,dg$ is a holomorphic 1-form on X which has no zeros. □

26.7. Theorem. *Suppose X is a non-compact Riemann surface and $(a_\nu)_{\nu \in \mathbb{N}}$ is a sequence of distinct points on X which has no point of accumulation. Then given arbitrary complex numbers $c_\nu \in \mathbb{C}$ there exists a holomorphic function $f \in \mathcal{O}(X)$ with $f(a_\nu) = c_\nu$ for every $\nu \in \mathbb{N}$.*

PROOF. By Theorem (26.5) there is a function $h \in \mathcal{O}(X)$ which has a zero of order 1 at each a_ν and is otherwise non-zero. For $i \in \mathbb{N}$ let

$$U_i := X \setminus \bigcup_{\nu \neq i} \{a_\nu\}.$$

Then $\mathfrak{U} := (U_i)_{i \in \mathbb{N}}$ is an open covering of X. Define $g_i \in \mathscr{M}(U_i)$ by $g_i := c_i/h$. For $i \neq j$

$$U_i \cap U_j = X \setminus \{a_\nu : \nu \in \mathbb{N}\}.$$

Thus $1/h$ is holomorphic on $U_i \cap U_j$. Hence $(g_i) \in C^0(\mathfrak{U}, \mathscr{M})$ is a Mittag-Leffler distribution on X which by (26.3) has a solution $g \in \mathscr{M}(X)$. Let $f := gh$. On U_i one has

$$f = gh = g_i h + (g - g_i)h = c_i + (g - g_i)h.$$

Since $g - g_i$ is holomorphic on U_i and $h(a_i) = 0$, it follows that $f \in \mathcal{O}(X)$ and $f(a_i) = c_i$ for every $i \in \mathbb{N}$. □

26.8. Corollary. *Every non-compact Riemann surface X is Stein, i.e., the following hold:*

(i) *Given any two points $x, y \in X$, $x \neq y$, then there exists a holomorphic function $f \in \mathcal{O}(X)$ with $f(x) \neq f(y)$.*

(ii) *Given a sequence $(x_n)_{n \in \mathbb{N}}$ in X having no points of accumulation, then there exists a holomorphic function $f \in \mathcal{O}(X)$ with* $\limsup_{n \to \infty} |f(x_n)| = \infty$.

Remark. The Theorems of Mittag–Leffler and Weierstrass for non-compact Riemann surfaces were first proved by H. Florack [54] using the methods developed by Behnke–Stein [51]. The analogues of these problems in several complex variables (the first and second Cousin problems) played an important role in the development of the theory of Stein manifolds (c.f. [53], [59], [61]). Also, the use of cohomology to solve these problems stems from that theory.

Exercises (§26)

26.1. Let X be a non-compact Riemann surface. Prove

$$H^1(X, \Omega) = 0.$$

[*Hint*: Using Corollary (26.6) show that $\Omega \cong \mathcal{O}$.]

26.2. Let X be a non-compact Riemann surface.

(a) Given any $\omega \in \mathcal{E}^{(2)}(X)$ show that there exists $f \in \mathcal{E}(X)$ with

$$d'd''f = \omega.$$

(b) Let $\mathcal{H}$ be the sheaf of harmonic functions on X. Show that

$$H^1(X, \mathcal{H}) = 0.$$

26.3. Show that on a non-compact Riemann surface every meromorphic function is the quotient of two holomorphic functions.

26.4. Let X be a non-compact Riemann surface and suppose $f, g \in \mathcal{O}(X)$ are holomorphic functions which have no common zero.

(a) Show that the following sequence of sheaves is exact

$$0 \to \mathcal{O} \xrightarrow{\alpha} \mathcal{O}^2 \xrightarrow{\beta} \mathcal{O} \to 0,$$

where

$$\alpha(\psi) := (\psi g, -\psi f)$$

$$\beta(\varphi_1, \varphi_2) := \varphi_1 f + \varphi_2 g.$$

(b) Show that there exist holomorphic functions $\Phi, \Psi \in \mathcal{O}(X)$ such that

$$\Phi f + \Psi g = 1.$$

26.5. Let X be a non-compact Riemann surface and let

$$\mathfrak{A} = \sum_{j=1}^{k} \mathcal{O}(X)f_j, \qquad f_j \in \mathcal{O}(X),$$

be a finitely generated ideal in $\mathcal{O}(X)$. Prove that $\mathfrak{A}$ is a principal ideal. [*Hint*: Consider the divisor D on X defined by

$$D(x) = \min_j \operatorname{ord}_x(f_j)$$

and let $f \in \mathcal{O}(X)$ be a holomorphic function with $(f) = D$. Verify that $\mathfrak{A} = \mathcal{O}(X)f$.]

26.6. Let X be a non-compact Riemann surface and D a divisor on X. Prove

(a) $H^1(X, \mathcal{O}_D) = 0$,
(b) $H^1(X, \mathcal{M}) = 0$.

§27. The Riemann Mapping Theorem

The Riemann Mapping Theorem asserts that any simply connected Riemann surface, which is not isomorphic either to $\mathbb{P}^1$ or $\mathbb{C}$, can be mapped biholomorphically onto the unit disk. This means that the universal covering of an arbitrary Riemann surface is always isomorphic to one of three normal forms: the Riemann sphere, the complex plane or the unit disk. The Riemann Mapping Theorem was presented by Riemann in his dissertation in 1851, but not in its most general form and not with a completely acceptable proof. The first complete proofs were given by H. Poincaré and P. Koebe in the year 1907.

27.1. For a Riemann surface X denote by $\mathrm{Rh}^1_{\mathcal{O}}(X) := \Omega(X)/d\mathcal{O}(X)$ the "holomorphic" deRham group, cf. (15.15). If X is simply connected, then every holomorphic 1-form on X has a primitive (10.7) and thus $\mathrm{Rh}^1_{\mathcal{O}}(X) = 0$. We will prove the Riemann Mapping Theorem for Riemann surfaces X satisfying the seemingly more general condition $\mathrm{Rh}^1_{\mathcal{O}}(X) = 0$. However, one consequence of this will be that $\mathrm{Rh}^1_{\mathcal{O}}(X) = 0$ implies X is simply connected.

27.2. Lemma. *Suppose X is a Riemann surface with* $\mathrm{Rh}^1_{\mathcal{O}}(X) = 0$. *Then*

(i) *For every holomorphic function $f: X \to \mathbb{C}^*$ a logarithm and a square root of f exist, i.e., there exist functions $g, h \in \mathcal{O}(X)$ such that $e^g = f$ and $h^2 = f$.*

(ii) *Every harmonic function $u: X \to \mathbb{R}$ is the real part of a holomorphic function $f: X \to \mathbb{C}$.*

PROOF

(i) $f^{-1}\,df$ is a holomorphic 1-form on X. Since $\mathrm{Rh}^1_{\mathcal{O}}(X) = 0$, there exists a function $g \in \mathcal{O}(X)$ with $dg = f^{-1}\,df$. By adding a constant to g if necessary, we may assume that for some point $a \in X$ one has $e^{g(a)} = f(a)$. Now

$$d(fe^{-g}) = (df)e^{-g} - fe^{-g}f^{-1}\,df = 0$$

and thus fe^{-g} is constantly equal to 1. Hence $e^g = f$.

Taking $h := e^{g/2}$ one has $h^2 = f$.

(ii) By Theorem (19.4) there is a holomorphic 1-form $\omega \in \Omega(X)$ with $du = \mathrm{Re}(\omega)$. Since $\Omega(X) = d\mathcal{O}(X)$, one has $du = \mathrm{Re}(dg)$ for some $g \in \mathcal{O}(X)$. Thus $u = \mathrm{Re}(g) + \text{const}$. □

27.3. Theorem. *Suppose X is a non-compact Riemann surface and $Y \Subset X$ is a domain with $\mathrm{Rh}^1_{\mathcal{O}}(Y) = 0$. Suppose also that the boundary of Y is regular with respect to solving the Dirichlet problem. Then there exists a biholomorphic mapping of Y onto the unit disk D.*

PROOF. Choose a point $a \in Y$. By Weierstrass' Theorem (26.5) there exists a holomorphic function g on X which has a zero of first order at a and does not vanish on $X \backslash a$. By Theorem (22.17) there exists a function u: $\bar{Y} \to \mathbb{R}$, continuous on $\bar{Y}$ and harmonic on Y, with

$$u(y) = \log|g(y)| \quad \text{for every } y \in \partial Y. \qquad (*)$$

By Lemma (27.2.ii) u is the real part of a holomorphic function $h \in \mathcal{O}(Y)$. Set

$$f := e^{-h}g \in \mathcal{O}(Y).$$

Now we claim that f maps Y biholomorphically onto the unit disk D. First we will show $f(Y) \subset D$. For $y \in Y\backslash a$ one has

$$|f(y)| = |e^{-h(y)}|\,|g(y)| = e^{\log|g(y)| - u(y)}.$$

Hence the function $|f|$ which is defined on Y can be continued to a continuous function φ: $\bar{Y} \to \mathbb{R}$ which because of (*) is identically equal to 1 on ∂Y. Then the Maximum Principle implies $|f(y)| < 1$ for every $y \in Y$, i.e., $f(Y) \subset D$.

Now we will show that the mapping f: $Y \to D$ is proper. To do this, it suffices to show that for every $r < 1$ the preimage Y_r of the disk $\{z \in \mathbb{C}: |z| \le r\}$ is compact in Y. But

$$Y_r = \{y \in Y : |f(y)| \le r\} = \{y \in \bar{Y} : \varphi(y) \le r\}$$

and thus Y_r is a closed subset of the compact set $\bar{Y}$ and so is compact.

Since f: $Y \to D$ is proper, each value is attained equally often (Theorem 4.24). But the value zero is taken exactly once. Thus f: $Y \to D$ is bijective and hence biholomorphic. □

27.4. The general Riemann Mapping Theorem can be derived from Theorem (27.3) by using an exhaustion process. To do this we require a few additional tools.

Notation. For $r \in]0, \infty]$ let

$$D(r) := \{z \in \mathbb{C}: |z| < r\}.$$

In particular $D(1) = D$ is the unit disk and $D(\infty) = \mathbb{C}$ is the complex plane.

The following observation is a simple consequence of Cauchy's Integral Formula. Suppose $f: D(r) \to D(r')$ is a holomorphic mapping. Then

$$|f'(0)| \leq \frac{r'}{r}.$$

27.5. Lemma. *Suppose $G \subset \mathbb{C}$ is a domain such that $\mathbb{C} \backslash G$ has interior points and suppose $w_0 \in G$. Then the set*

$$\{f \in \mathcal{O}(D): f(D) \subset G \text{ and } f(0) = w_0\}$$

is a compact subset of $\mathcal{O}(D)$ with respect to the topology of uniform convergence on compact subsets.

PROOF. Suppose a is an interior point of $\mathbb{C} \backslash G$. Then the mapping $z \mapsto 1/(z - a)$ takes the domain G biholomorphically onto a subdomain of some disk $D(r)$ with $r < \infty$. Hence the result follows from Montel's Theorem. □

27.6. Theorem. *The set $\mathcal{S}$ of all schlicht ($=$ injective) holomorphic functions $f: D \to \mathbb{C}$ with $f(0) = 0$ and $f'(0) = 1$ is compact in $\mathcal{O}(D)$.*

PROOF

(a) Suppose $(f_n)_{n \in \mathbb{N}}$ is a sequence of functions in $\mathcal{S}$. It suffices to show that there is a subsequence which converges to a function $f \in \mathcal{S}$.

Denote by r_n the maximum radius such that $D(r_n) \subset f_n(D)$. Then $r_n \leq 1$ since the inverse φ_n of f_n maps $D(r_n)$ into D and hence $1 = \varphi_n'(0) \leq 1/r_n$. Choose a point $a_n \in \partial D(r_n)$ with $a_n \notin f_n(D)$ and set $g_n := f_n/a_n$. Then

$$D \subset g_n(D) \quad \text{and } 1 \notin g_n(D).$$

(b) Since $g_n(D)$ is homeomorphic to D and thus is simply connected, there exists a holomorphic function $\psi: g_n(D) \to \mathbb{C}^*$ with $\psi(0) = i$ and $\psi(z)^2 = z - 1$ for every $z \in g_n(D)$. Set $h_n := \psi \circ g_n$. Thus $h_n^2 = g_n - 1$.

We claim that $w \in h_n(D)$ implies $-w \notin h_n(D)$. For, otherwise we would have $w = h_n(z_1)$ and $-w = h_n(z_2)$ for $z_1, z_2 \in D$. But then $w^2 = (-w)^2$ implies $g_n(z_1) = g_n(z_2)$ and since g_n is injective, $z_1 = z_2$. Hence $w = -w$ and thus $w = h_n(z) = 0$. However this implies $g_n(z_1) = 1$, which is a contradiction.

(c) Because $D \subset g_n(D)$, one has $U := \psi(D) \subset h_n(D)$. Thus $(-U) \cap h_n(D) = \varnothing$. By Lemma (27.5) the sequence (h_n) has a convergent subsequence. Since $f_n = a_n(1 + h_n^2)$ and $|a_n| \leq 1$ for every n, the sequence (f_n)

also has a convergent subsequence (f_{n_k}) which converges to some function $f\colon D \to \mathbb{C}$. Clearly, $f(0) = 0$ and $f'(0) = 1$, so f is not constant.

(d) The remaining point is to show that f is schlicht. If this were not the case, then there would exist an $a \in \mathbb{C}$ such that $f - a$ would have at least two zeros in D. Then one could find an $r < 1$ such that $f - a$ has, counting multiplicities, at least $k \geq 2$ zeros in $D(r)$ and does not vanish on $\partial D(r)$. Then

$$k = \frac{1}{2\pi i} \int_{|z|=r} \frac{f'(z)}{f(z) - a}\, dz.$$

Thus every function sufficiently close to f also takes the value a k times, contradicting the fact that every f_{n_k} is schlicht. □

27.7. Lemma. *Suppose $R \in\,]0, \infty]$ and Y is a proper subdomain of $D(R)$ with $0 \in Y$ and $\mathrm{Rh}^1_{\mathcal{O}}(Y) = 0$. Then there exists an $r < R$ and a holomorphic map $f\colon Y \to D(r)$ with $f(0) = 0$ and $f'(0) = 1$.*

Proof. First consider the case $R < \infty$. Without loss of generality we may assume $R = 1$ and thus $Y \subset D$. By assumption there is a point $a \in D \setminus Y$. Define a biholomorphic map $\varphi\colon D \to D$ by

$$\varphi(z) := \frac{z - a}{1 - \bar{a}z}.$$

Then $0 \notin \varphi(Y)$ and thus by Lemma (27.2) there exists $g \in \mathcal{O}(Y)$ with $g^2 = \varphi \mid Y$. Clearly $g(Y) \subset D$. Set

$$\psi(z) := \frac{z - b}{1 - \bar{b}z}, \quad \text{where } b := g(0).$$

The mapping $h := \psi \circ g\colon Y \to D$ satisfies $h(0) = 0$ and

$$\gamma := h'(0) = \psi'(b)g'(0) = \psi'(b)\frac{\varphi'(0)}{2g(0)} = \frac{1}{1 - |b|^2}\frac{1 - |a|^2}{2b}$$

$$= \frac{1 + |b|^2}{2b},$$

since $b^2 = -a$. Thus $|\gamma| > 1$. Now letting $r := 1/|\gamma|$ and $f := h/\gamma$, the map $f\colon Y \to D(r)$ has the desired properties.

The case $R = \infty$ is handled similarly. □

27.8. Lemma. *Suppose X is a non-compact Riemann surface with $\mathrm{Rh}^1_{\mathcal{O}}(X) = 0$ and $Y \subset X$ is a Runge domain. Then $\mathrm{Rh}^1_{\mathcal{O}}(Y) = 0$ as well.*

Proof. Suppose $\omega \in \Omega(Y)$ is an arbitrary holomorphic 1-form on Y. We have to show that ω has a primitive. By Corollary (26.6) choose a holomorphic 1-form ω_0 on X which has no zeros. Then $\omega = f\omega_0$ for some $f \in \mathcal{O}(Y)$.

By the Runge Approximation Theorem there exists a sequence $f_n \in \mathcal{O}(X)$, $n \in \mathbb{N}$, which converges uniformly on compact subsets in Y to f. Hence for every closed curve α in Y the integrals $\int_\alpha f_n\omega_0$ converge to $\int_\alpha \omega$. Since every 1-form $f_n\omega_0$ has a primitive on X, $\int_\alpha f_n\omega_0 = 0$. Thus $\int_\alpha \omega = 0$. Since all the periods of ω vanish, by Theorem (10.15) ω has a primitive. □

27.9. The Riemann Mapping Theorem. *Suppose X is a Riemann surface with $\mathrm{Rh}^1_{\mathcal{O}}(X) = 0$. Then X can be mapped biholomorphically onto either the Riemann sphere $\mathbb{P}^1$, the complex plane $\mathbb{C}$ or else the unit disk D.*

Remark. As was pointed out in (27.1), the assumption $\mathrm{Rh}^1_{\mathcal{O}}(X) = 0$ holds whenever X is simply connected. Since $\mathbb{P}^1$, $\mathbb{C}$ and D are simply connected, the Riemann Mapping Theorem shows that the converse also holds.

PROOF

(a) If X is compact, then every holomorphic function on X is constant and so $d\mathcal{O}(X) = 0$. Hence $\mathrm{Rh}^1_{\mathcal{O}}(X) = 0$ implies $\Omega(X) = 0$, i.e., X has genus 0. By Corollary (16.13) X is biholomorphic to $\mathbb{P}^1$.

(b) Now assume X is non-compact. By Theorem (23.9) there exists an exhaustion $Y_0 \Subset Y_1 \Subset Y_2 \Subset \cdots$ of X by Runge domains Y_n whose boundaries are regular with respect to solving the Dirichlet problem. By Lemma (27.8) $\mathrm{Rh}^1_{\mathcal{O}}(Y_n) = 0$ for every n. Thus by Theorem (27.3) every Y_n is biholomorphic to the unit disk. Choose a point $a \in Y_0$ and a coordinate neighborhood (U, z) of a. Then there exists a real number $r_n > 0$ and a biholomorphic mapping

$$f_n\colon Y_n \to D(r_n)$$

with

$$f_n(a) = 0 \quad \text{and} \quad \frac{df_n}{dz}(a) = 1.$$

(c) Now $r_n \le r_{n+1}$ for every n. To see this note that the mapping

$$h := f_{n+1} \circ f_n^{-1}\colon D(r_n) \to D(r_{n+1})$$

satisfies $h(0) = 0$ and $h'(0) = 1$ and thus by the Remark in (27.4) one has $1 = h'(0) \le r_{n+1}/r_n$. Let

$$R := \lim_{n\to\infty} r_n \in \,]0, \infty].$$

We will now show that X is mapped biholomorphically onto $D(R)$.

(d) We claim that there exists a subsequence $(f_{n_k})_{k\in\mathbb{N}}$ of the sequence $(f_n)_{n\in\mathbb{N}}$ such that for every m the sequence $(f_{n_k} \mid Y_m)_{k\ge m}$ converges uniformly on compact subsets of Y_m. The mapping $z \mapsto f_0^{-1}(r_0 z)$ maps D biholomorphically onto Y_0. Set

$$g_n(z) := \frac{1}{r_0} f_n(f_0^{-1}(r_0 z)), \qquad n \ge 0.$$

Then $g_n: D \to \mathbb{C}$ is a schlicht holomorphic function with $g_n(0) = 0$ and $g_n'(0) = 1$. Hence by Theorem (27.6) there is a subsequence $(f_{n_{0k}})_{k \in \mathbb{N}}$ of the sequence (f_n), which converges uniformly on compact subsets of Y_0. By the same reasoning there is a subsequence $(f_{n_{1k}})$ of this sequence which converges uniformly on compact subsets of Y_1. Repeating this process we get for each m a subsequence $(f_{n_{mk}})$ of the previous subsequence which converges uniformly on compact subsets of Y_m. Set $f_{n_k} := f_{n_{kk}}$. Then the sequence $(f_{n_k})_{k \in \mathbb{N}}$ has the desired property.

Suppose $f \in \mathcal{O}(X)$ is the limit of the sequence (f_{n_k}), i.e. f is that holomorphic function on X which coincides on every Y_m with the limit of the sequence $(f_{n_k} \mid Y_m)_{k \geq m}$. The mapping $f: X \to \mathbb{C}$ is injective and satisfies

$$f(a) = 0 \quad \text{and} \quad \frac{df}{dz}(a) = 1.$$

(e) Finally we claim f maps X biholomorphically onto $D(R)$. Since it is obvious that $f(X) \subset D(R)$, it suffices to show $f: X \to D(R)$ is surjective. Suppose the contrary. Then by Lemma (27.7) there exists an $r < R$ and a holomorphic mapping $g: f(X) \to D(r)$ with $g(0) = 0$ and $g'(0) = 1$. Choose n so large that $r_n > r$. Then the mapping

$$h := g \circ f \circ f_n^{-1}: D(r_n) \to D(r)$$

satisfies $h(0) = 0$ and $h'(0) = 1$. Since $r < r_n$, this is not possible. This contradiction shows that $f: X \to D(R)$ is surjective and thus the proof of the Riemann Mapping Theorem is complete. □

27.10. Suppose X is a Riemann surface and $\tilde{X}$ is its universal covering. Since $\tilde{X}$ is simply connected, we may apply the Riemann Mapping Theorem to it. It is standard to call X *elliptic*, *parabolic* or *hyperbolic* depending on whether its universal covering is isomorphic to $\mathbb{P}^1$, $\mathbb{C}$ or D.

Suppose $G = \operatorname{Deck}(\tilde{X}/X)$ is the group of covering transformations of the universal covering. Every $\sigma \in G$ is an automorphism of $\tilde{X}$, i.e., a biholomorphic mapping of $\tilde{X}$ onto itself. Also the group G acts without fixed points and discretely on $\tilde{X}$, i.e.,

(i) If $\sigma \in G \setminus \{id\}$, then $\sigma x \neq x$ for every $x \in \tilde{X}$.
(ii) For every $x \in \tilde{X}$ the orbit

$$Gx := \{\sigma x : \sigma \in G\}$$

is a discrete subset of $\tilde{X}$.

Property (i) follows since each covering transformation is uniquely determined once one knows the image of any point. Property (ii) holds since the covering $p: \tilde{X} \to X$ is Galois and thus $Gx = p^{-1}(p(x))$.

The Riemann surface X may be thought of as the quotient of $\tilde{X}$ modulo G, i.e., two points of $\tilde{X}$ are identified if one can be transformed into the other by some element of G. Thus every hyperbolic Riemann surface is a quotient of the unit disk D modulo a group of automorphisms of D acting without fixed points and discretely.

27.11. Lemma

(a) *Every automorphism of $\mathbb{P}^1$ has a fixed point.*

(b) *Suppose G is a group of automorphisms of $\mathbb{C}$ which acts discretely and without fixed points. Then G is one of the following:*

(i) $G = \{id\}$.

(ii) *G consists of all translations of the form*

$$z \mapsto z + n\gamma, \qquad n \in \mathbb{Z},$$

where γ is a fixed non-zero complex number.

(iii) *G consists of all translations of the form*

$$z \mapsto z + n\gamma_1 + m\gamma_2, \qquad n, m \in \mathbb{Z}$$

where γ_1 and γ_2 are two fixed complex numbers linearly independent over $\mathbb{R}$.

PROOF

(a) As is well known, the automorphisms of $\mathbb{P}^1$ are linear fractional transformations of the form

$$z \mapsto \frac{az + b}{cz + d}, \qquad ad - bc \neq 0.$$

Every such transformation has at least one fixed point.

(b) The automorphisms of $\mathbb{C}$ are affine linear mappings of the form

$$z \mapsto az + b, \qquad a \in \mathbb{C}^*, \qquad b \in \mathbb{C}.$$

If $a \neq 1$, then this transformation has a fixed point. Thus the group G consists only of translations $z \mapsto z + b$. Let Γ be the orbit of zero under G. Then Γ is a discrete additive subgroup of $\mathbb{C}$ and G consists of all translations $z \mapsto z + b$, where $b \in \Gamma$. Let $V \subset \mathbb{C}$ be the smallest real vector subspace containing Γ. Depending on whether $\dim_{\mathbb{R}} V$ is 0, 1 or 2 one has case (i), (ii) or (iii). This follows from Theorem (21.1). □

27.12. Theorem

(a) *The Riemann sphere $\mathbb{P}^1$ is elliptic.*

(b) *The complex plane $\mathbb{C}$, the punctured plane $\mathbb{C}^*$ and all tori $\mathbb{C}/\Gamma$ are parabolic.*

(c) *Every other Riemann surface is hyperbolic.*

Thus, in particular, a compact Riemann surface is elliptic, parabolic or hyperbolic depending on whether its genus is zero, one or greater than one.

Remark. Compact Riemann surfaces of genus one are also called *elliptic curves.* As this is easily confused with the above terminology, $\mathbb{P}^1$ is seldom called an elliptic Riemann surface.

PROOF. The assertions (a) and (b) are clear. One only has to show that if X is not hyperbolic, then it is isomorphic to one of the surfaces listed in (a) or (b).

Case 1. The universal covering of X is isomorphic to $\mathbb{P}^1$. Then Lemma (27.11.a) implies X itself is isomorphic to $\mathbb{P}^1$.

Case 2. The universal covering of X is isomorphic to $\mathbb{C}$. Then the group G of covering transformations is, by (27.11.b), either (i), (ii) or (iii). In case (i) X is isomorphic to $\mathbb{C}$ and in case (ii) to $\mathbb{C}^*$, for then the covering is isomorphic to

$$\mathbb{C} \to \mathbb{C}^*, \qquad z \mapsto \exp\left(\frac{2\pi i}{\gamma} z\right).$$

Finally in case (iii) X is a torus. □

A simple consequence is the so-called Little Theorem of Picard.

27.13. Theorem. *Suppose $f\colon \mathbb{C} \to \mathbb{C}$ is a non-constant holomorphic function. Then f takes every value $c \in \mathbb{C}$ with at most one exception.*

PROOF. Suppose f did not take two distinct values $a, b \in \mathbb{C}$. By Theorem (27.12) the Riemann surface $X := \mathbb{C}\setminus\{a, b\}$ is hyperbolic. Hence the mapping $f\colon \mathbb{C} \to X$ can be lifted to a mapping $\tilde{f}\colon \mathbb{C} \to \tilde{X}$, where $\tilde{X}$ is the universal covering of X. Since $\tilde{X}$ is isomorphic to the unit disk, it follows from Liouville's Theorem that $\tilde{f}$ and thus also f are constant. Contradiction! □

EXERCISES (§27)

27.1. Let X be a Riemann surface and

$$f_\nu\colon X \to \mathbb{C}\setminus\{0, 1\}, \qquad \nu \in \mathbb{N},$$

be a sequence of holomorphic functions which do not take the values 0 and 1. Suppose there exists a point $x_0 \in X$ such that $(f_\nu(x_0))_{\nu \in \mathbb{N}}$ converges to a point $c \in \mathbb{C}\setminus\{0, 1\}$. Prove that there exists a subsequence $(f_{\nu_k})_{k \in \mathbb{N}}$ which converges uniformly on every compact subset of X to a holomorphic function

$$f\colon X \to \mathbb{C}\setminus\{0, 1\}.$$

[*Hint*: Let $\tilde{X} \to X$ and $D \to \mathbb{C}\setminus\{0, 1\}$ be the universal coverings (D is isomorphic to the unit disk). Consider suitable liftings $\tilde{f}_\nu\colon \tilde{X} \to D$ of the f_ν and use the classical theorem of Montel.]

27.2. Prove the "Big Theorem of Picard:" Let

$$U := \{z \in \mathbb{C} : 0 < |z| < r\}, \qquad \text{where } r > 0,$$

and

$$f: U \to \mathbb{C}$$

a holomorphic function having an essential singularity at the origin. Then f attains every value $c \in \mathbb{C}$ with at most one exception.
[*Hint*: Consider the sequence of functions

$$f_\nu : U \to \mathbb{C}, \qquad f_\nu(z) := f(2^\nu z)$$

and use Ex. 27.1.]

§28. Functions with Prescribed Summands of Automorphy

In §10 we saw that the integration of differential forms on a Riemann surface X gives rise to additively automorphic functions whose summands of automorphy determine a "period homomorphism" $\pi_1(X) \to \mathbb{C}$. Behnke-Stein [51] proved that conversely given any homomorphism $\pi_1(X) \to \mathbb{C}$ on a non-compact Riemann surface X there always exists a holomorphic 1-form having these periods. In this section we prove the Theorem of Behnke–Stein. At the same time we investigate arbitrary functions having non-constant summands of automorphy.

28.1. Cohomology of Groups. Suppose G is a group whose operation is written multiplicatively and A is a *G-module*, i.e., an additive abelian group together with a mapping

$$G \times A \to A, \qquad (\sigma, a) \mapsto \sigma a$$

satisfying the following:

(i) $\sigma(a + b) = \sigma a + \sigma b$,
(ii) $\sigma(\tau a) = (\sigma\tau)a$,
(iii) $\varepsilon a = a$,

for every $\sigma, \tau \in G$ and $a, b \in A$. The identity of G is denoted by ε. A mapping

$$G \to A, \qquad \sigma \mapsto a_\sigma$$

is called a *crossed homomorphism* if

$$a_{\sigma\tau} = a_\sigma + \sigma a_\tau \quad \text{for every } \sigma, \tau \in G.$$

If G operates trivially on A, i.e., $\sigma a = a$ for every $\sigma \in G$, then a crossed homomorphism is nothing but a group homomorphism in the usual sense.

The set of all crossed homomorphisms $G \to A$ has the natural structure of an additive group and will be denoted by $Z^1(G, A)$. Special crossed homomorphisms are obtained in the following way. Suppose $f \in A$ is fixed and

$$a_\sigma := f - \sigma f \quad \text{for every } \sigma \in G.$$

Then

$$\begin{aligned} a_{\sigma\tau} &= f - \sigma\tau f = f - \sigma f + \sigma f - \sigma\tau f = (f - \sigma f) + \sigma(f - \tau f) \\ &= a_\sigma + \sigma a_\tau . \end{aligned}$$

Crossed homomorphisms which arise in this way are called *coboundaries*. They form a subgroup of $Z^1(G, A)$ which will be denoted by $B^1(G, A)$. The quotient

$$H^1(G, A) := Z^1(G, A)/B^1(G, A)$$

is called the 1*st cohomology group* of G with coefficients in the G-module A.

28.2. Summands of Automorphy. Suppose $p\colon Y \to X$ is an unbranched holomorphic covering mapping between Riemann surfaces and $G := \operatorname{Deck}(Y/X)$ is its group of covering transformations. Then $\mathcal{O}(Y)$ is a G-module, if one defines $\sigma f \in \mathcal{O}(Y)$ by $\sigma f := f \circ \sigma^{-1}$ for any $\sigma \in G$ and $f \in \mathcal{O}(Y)$. The differences

$$a_\sigma := f - \sigma f \in \mathcal{O}(Y), \qquad \sigma \in G,$$

are called the *summands of automorphy* of f. By (28.1) the summands of automorphy of f define a crossed homomorphism

$$G \to \mathcal{O}(Y), \qquad \sigma \mapsto a_\sigma .$$

If the covering is Galois (Definition 5.5) and all the summands of automorphy of a function $f \in \mathcal{O}(Y)$ are zero, then the function f lies in the subring $p^*\mathcal{O}(X) \subset \mathcal{O}(Y)$ and thus may be identified with a function on X.

One can do the same thing for the meromorphic functions $\mathcal{M}(Y)$ and the differentiable functions $\mathcal{E}(Y)$.

28.3. Galois Coverings. The notation will be the same as in (28.2). Assume $p\colon Y \to X$ is Galois. Every point $x \in X$ has a connected open neighborhood U such that

$$p^{-1}(U) = \bigcup_{\lambda \in \Lambda} V_\lambda,$$

where the V_λ are disjoint open subsets of Y and the mappings $p \mid V_\lambda \to U$ are homeomorphisms. Now we construct a homeomorphism

$$\varphi\colon p^{-1}(U) \to U \times G,$$

where G has the discrete topology, in the following way. Choose an index $\lambda_0 \in \Lambda$. Then for every $\lambda \in \Lambda$ there is precisely one $\sigma \in G$ such that

$\sigma(V_{\lambda_0}) = V_\lambda$. For $y \in V_\lambda$ set $\varphi(y) := (p(y), \sigma)$. This maps V_λ homeomorphically onto $U \times \{\sigma\}$ and implies that φ is a homeomorphism. The mapping φ is fiber-preserving, i.e., the diagram

$$\begin{array}{ccc} p^{-1}(U) & \xrightarrow{\varphi} & U \times G \\ & {\scriptstyle p}\searrow \quad \swarrow{\scriptstyle \mathrm{pr}_U} & \\ & U & \end{array}$$

is commutative. Moreover φ is compatible with the action of G, i.e., $\varphi(y) = (x, \sigma)$ implies $\varphi(\tau y) = (x, \tau\sigma)$ for every $\tau \in G$. We will call such a fiber-preserving homeomorphism

$$\varphi: p^{-1}(U) \to U \times G$$

which is compatible with the action of G a *G-chart* of the Galois covering $p: Y \to X$. A G-chart has a decomposition $\varphi = (p, \eta)$, where $\eta: p^{-1}(U) \to G$ is a mapping such that

$$\eta(\tau y) = \tau\eta(y) \quad \text{for every } y \in p^{-1}(U) \quad \text{and } \tau \in G.$$

28.4. Theorem. *Suppose X and Y are non-compact Riemann surfaces, $p: Y \to X$ is a holomorphic unbranched Galois covering map and $G = \mathrm{Deck}(Y/X)$ is its group of covering transformations. Then given any crossed homomorphism*

$$G \to \mathcal{O}(Y), \qquad \sigma \mapsto a_\sigma,$$

there exists a holomorphic function $f \in \mathcal{O}(Y)$ having summands of automorphy a_σ.

Remark. Theorem (28.4) asserts that $H^1(G, \mathcal{O}(Y)) = 0$. This is also true for arbitrary Stein manifolds (Stein [62], Serre [59]).

PROOF

(a) Choose an open covering $\mathfrak{U} = (U_i)_{i \in I}$ of X and G-charts

$$\varphi_i = (p, \eta_i): p^{-1}(U_i) \to U_i \times G.$$

Now on $Y_i := p^{-1}(U_i)$ define functions $f_i: Y_i \to \mathbb{C}$ by

$$f_i(y) := a_{\eta_i(y)}(y) \quad \text{for every } y \in Y_i.$$

Clearly f_i is holomorphic on Y_i.

(b) We now claim that $f_i - \sigma f_i = a_\sigma$ on Y_i for every $\sigma \in G$. For $y \in Y_i$ one has by definition

$$(\sigma f_i)(y) = f_i(\sigma^{-1}y) = a_{\eta_i(\sigma^{-1}y)}(\sigma^{-1}y) = a_{\sigma^{-1}\eta_i(y)}(\sigma^{-1}y).$$

The relation $a_{\sigma\tau} = a_\sigma + \sigma a_\tau$ with $\tau := \sigma^{-1}\eta_i(y)$ implies

$$\begin{aligned} a_\sigma(y) &= a_{\sigma\tau}(y) - a_\tau(\sigma^{-1}y) \\ &= a_{\eta_i(y)}(y) - a_{\sigma^{-1}\eta_i(y)}(\sigma^{-1}y) = f_i(y) - (\sigma f_i)(y). \end{aligned}$$

Thus the functions f_i have the desired automorphic behavior on Y_i.

(c) By (b) the differences $g_{ij} := f_i - f_j \in \mathcal{O}(Y_i \cap Y_j)$ are invariant under covering transformations and thus may be considered as elements $g_{ij} \in \mathcal{O}(U_i \cap U_j)$. Obviously $g_{ij} + g_{jk} = g_{ik}$ on any triple intersection and thus the family (g_{ij}) is a cocycle in $Z^1(\mathfrak{U}, \mathcal{O})$. Because $H^1(X, \mathcal{O}) = 0$, this cocycle splits. Hence there exist elements $g_i \in \mathcal{O}(U_i)$ with

$$g_{ij} = g_i - g_j \quad \text{on } U_i \cap U_j.$$

Consider the g_i as functions on Y_i which are invariant under covering transformations. Then the functions

$$\tilde{f}_i := f_i - g_i \in \mathcal{O}(Y_i)$$

also satisfy $\tilde{f}_i - \sigma\tilde{f}_i = a_\sigma$ for every $\sigma \in G$. On any intersection $Y_i \cap Y_j$ one has

$$\tilde{f}_i - \tilde{f}_j = f_i - f_j - (g_i - g_j) = g_{ij} - (g_i - g_j) = 0.$$

Hence the $\tilde{f}_i$ piece together to give a global function $f \in \mathcal{O}(Y)$ with $f - \sigma f = a_\sigma$ for every $\sigma \in G$. □

28.5. Theorem. *Suppose X and Y are Riemann surfaces, $p\colon Y \to X$ is a holomorphic unbranched Galois covering map and $G = \operatorname{Deck}(Y/X)$ is its group of covering transformations. Then given any crossed homomorphism*

$$G \to \mathcal{E}(Y), \qquad \sigma \mapsto a_\sigma,$$

there exists a differentiable function $f \in \mathcal{E}(Y)$ having summands of automorphy a_σ.

Proof. This is proved in the same way as Theorem (28.4), except the sheaf $\mathcal{O}$ is replaced by the sheaf $\mathcal{E}$. This is possible since $H^1(X, \mathcal{E}) = 0$ for every Riemann surface, regardless of whether it is compact or not (Theorem 12.6). □

28.6. Theorem (Behnke–Stein). *Suppose X is a non-compact Riemann surface and*

$$\pi_1(X) \to \mathbb{C}, \qquad \sigma \mapsto a_\sigma,$$

is a group homomorphism. Then there exists a holomorphic 1-form $\omega \in \Omega(X)$ with

$$\int_\sigma \omega = a_\sigma \quad \textit{for every } \sigma \in \pi_1(X).$$

PROOF. For the universal covering $p: \tilde{X} \to X$ one has $\text{Deck}(\tilde{X}/X) = \pi_1(X)$. By Theorem (28.4) there exists a holomorphic function $F \in \mathcal{O}(\tilde{X})$ having the constant summands of automorphy a_σ. Then by Theorem (10.13) the differential dF can be considered as a 1-form on X and it has the periods a_σ. □

28.7. Theorem. *Suppose X is a compact Riemann surface and*

$$\pi_1(X) \to \mathbb{C}, \qquad \sigma \mapsto a_\sigma,$$

is a group homomorphism. Then there is a unique harmonic 1-form $\omega \in \text{Harm}^1(X)$ with

$$\int_\sigma \omega = a_\sigma \quad \textit{for every } \sigma \in \pi_1(X).$$

PROOF. Similar to (28.6) it follows from Theorem (28.5) that there exists a closed 1-form $\tilde{\omega} \in \mathcal{E}^{(1)}(X)$ with

$$\int_\sigma \tilde{\omega} = a_\sigma \quad \text{for every } \sigma \in \pi_1(X).$$

By Theorem (19.12) there exists a harmonic 1-form $\omega \in \text{Harm}^1(X)$ and a function $f \in \mathcal{E}(X)$ with

$$\tilde{\omega} = \omega + df.$$

Clearly $\tilde{\omega}$ and ω have the same periods. The uniqueness follows from (19.8). □

EXERCISES (§28)

28.1. Let X be a Riemann surface. Prove

$$H^1(X, \mathbb{C}) \cong \text{Hom}(\pi_1(X), \mathbb{C}).$$

28.2. Let X be a non-compact Riemann surface and let

$$\text{Rh}^1_{\mathcal{O}}(X) := \Omega(X)/d\mathcal{O}(X)$$

be the "holomorphic" deRham group. Prove

$$H^1(X, \mathbb{C}) \cong \text{Rh}^1_{\mathcal{O}}(X).$$

28.3. Let $g: \mathbb{C} \to \mathbb{C}$ be a holomorphic function. Prove that there exists a holomorphic function $f: \mathbb{C} \to \mathbb{C}$ such that

$$f(z+1) = f(z) + g(z) \quad \text{for every } z \in \mathbb{C}.$$

[*Hint*: Consider the Galois covering $\text{ex}: \mathbb{C} \to \mathbb{C}^*$, $\text{ex}(z) = e^{2\pi i z}$, and construct suitable summands of automorphy

$$\mathbb{Z} \to \mathcal{O}(\mathbb{C}).]$$

§29. Line and Vector Bundles

In the consideration of many problems of analysis on manifolds the following situation arises. To every point x of the manifold X there is associated a vector space E_x which depends in some continuous (or if X is a Riemann surface, holomorphic) way on x. This leads to the notion of a vector bundle on X, which we now make precise.

29.1. Definition. Suppose E and X are topological spaces and $p: E \to X$ is a continuous mapping. Further suppose that every fiber $E_x := p^{-1}(x)$ has the structure of an n-dimensional vector space over $\mathbb{C}$. Then $p: E \to X$, or simply E, is called a *vector bundle* of rank n on X if every point $a \in X$ has an open neighborhood U such that there exists a homeomorphism h of $E_U := p^{-1}(U)$ onto $U \times \mathbb{C}^n$ with the following properties:

(i) h is fiber-preserving, i.e., the following diagram

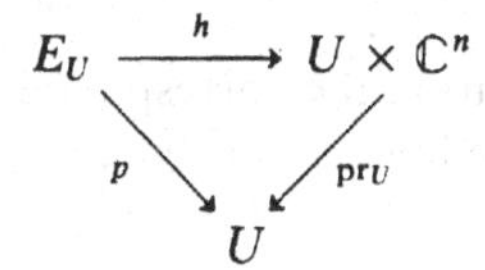

is commutative.

(ii) For every $x \in U$ the mapping $h \mid E_x$ is a vector space isomorphism of E_x onto $\{x\} \times \mathbb{C}^n \cong \mathbb{C}^n$.

The mapping $h: E_U \to U \times \mathbb{C}^n$ is called a *local trivialization* or *linear chart* of E over U. If $\mathfrak{U} = (U_i)_{i \in I}$ is an open covering of X and $h_i: E_{U_i} \to U_i \times \mathbb{C}^n$ are local trivializations, then the family of all the h_i is called an *atlas* of E.

29.2. Definition. A vector bundle of rank n is called *trivial* if there exists a global linear trivialization $h: E \to X \times \mathbb{C}^n$.

Thus a vector bundle is always locally trivial. For local considerations the notion of a vector bundle tells us nothing which is new. It only plays a role when one is dealing with global problems.

29.3. Definition. A *line bundle* is a vector bundle of rank one.

29.4. Theorem. *Suppose $E \to X$ is a vector bundle of rank n on the topological space X and $h_i: E_{U_i} \to U_i \times \mathbb{C}^n$, $i \in I$, is an atlas of E. Then there are unique continuous mappings*

$$g_{ij}: U_i \cap U_j \to GL(n, \mathbb{C}),$$

such that the mappings

$$\varphi_{ij} := h_i \circ h_j^{-1}: (U_i \cap U_j) \times \mathbb{C}^n \to (U_i \cap U_j) \times \mathbb{C}^n$$

satisfy

$$\varphi_{ij}(x, t) = (x, g_{ij}(x)t) \quad \textit{for every } (x, t) \in (U_i \cap U_j) \times \mathbb{C}^n.$$

On $U_i \cap U_j \cap U_k$ *one has the "cocycle relation"*

$$g_{ij} g_{jk} = g_{ik}.$$

Notation. The mappings g_{ij} are called the *transition functions* and the family (g_{ij}) is the cocycle associated to the atlas (h_i).

PROOF. The mapping

$$\varphi_{ij} = h_i \circ h_j^{-1} \colon (U_i \cap U_j) \times \mathbb{C}^n \to (U_i \cap U_j) \times \mathbb{C}^n$$

is a fiber-preserving homeomorphism which, restricted to every fiber, is an isomorphism of vector spaces. Hence for every $x \in U_i \cap U_j$ there exists a matrix $g_{ij}(x) \in \mathrm{GL}(n, \mathbb{C})$ with

$$\varphi_{ij}(x, t) = (x, g_{ij}(x)t).$$

Since φ_{ij} is a homeomorphism, the correspondence $x \mapsto g_{ij}(x)$ is continuous. The relation $g_{ij} g_{jk} = g_{ik}$ follows from the corresponding one for the mappings φ_{ij}. □

29.5. Definition. Suppose X is a Riemann surface, $E \to X$ is a vector bundle of rank n on X and

$$\mathfrak{A} = \{h_i \colon E_{U_i} \to U_i \times \mathbb{C}^n, i \in I\}$$

is an atlas of E. The atlas $\mathfrak{A}$ is said to be *holomorphic* if the associated transition functions

$$g_{ij} \colon U_i \cap U_j \to \mathrm{GL}(n, \mathbb{C})$$

are holomorphic.

Two atlases $\mathfrak{A}$ and $\mathfrak{A}'$ of E are called holomorphically compatible if $\mathfrak{A} \cup \mathfrak{A}'$ is a holomorphic atlas. One can easily check that holomorphic compatibility of atlases is an equivalence relation. An equivalence class of holomorphically compatible atlases is called a *holomorphic linear structure.*

A *holomorphic vector bundle* on a Riemann surface X is a vector bundle $E \to X$ together with a holomorphic linear structure. A holomorphic vector bundle $E \to X$ is called *holomorphically trivial* if its holomorphic linear structure contains an atlas consisting of the single chart $E \to X \times \mathbb{C}^n$.

29.6. Cocycles. Suppose X is a Riemann surface. For U open in X let $\mathrm{GL}(n, \mathcal{O}(U))$ denote the group of all invertible $n \times n$-matrices with coefficients in $\mathcal{O}(U)$. For $V \subset U$ one has the natural restriction mapping $\mathrm{GL}(n, \mathcal{O}(U)) \to \mathrm{GL}(n, \mathcal{O}(V))$. This defines a sheaf $\mathrm{GL}(n, \mathcal{O})$ of groups on X which for $n \geq 2$ is not abelian. If $\mathfrak{U} = (U_i)_{i \in I}$ is an open covering of X, let

$Z^1(\mathfrak{U}, \mathrm{GL}(n, \mathcal{O}))$ denote the set of all 1-cocycles with values in $\mathrm{GL}(n, \mathcal{O})$ with respect to $\mathfrak{U}$, i.e., all families $(g_{ij})_{i,j \in I}$ with

$$g_{ij} \in \mathrm{GL}(n, \mathcal{O}(U_i \cap U_j))$$

and

$$g_{ij} g_{jk} = g_{ik} \quad \text{on } U_i \cap U_j \cap U_k$$

for every $i, j, k \in I$. Note that for $n \geq 2$ the set $Z^1(\mathfrak{U}, \mathrm{GL}(n, \mathcal{O}))$ is not a group with respect to component-wise multiplication.

If $\mathfrak{A}$ is a holomorphic atlas of a vector bundle on X, then the family of transition functions of $\mathfrak{A}$ is a cocycle with values in $\mathrm{GL}(n, \mathcal{O})$. Conversely given such a cocycle one can construct a holomorphic vector bundle. We now prove this.

29.7. Theorem. *Suppose X is a Riemann surface, $\mathfrak{U} = (U_i)_{i \in I}$ is an open covering of X and $(g_{ij}) \in Z^1(\mathfrak{U}, \mathrm{GL}(n, \mathcal{O}))$. Then there exists a holomorphic vector bundle $p\colon E \to X$ of rank n and a holomorphic atlas*

$$\{h_i\colon E_{U_i} \to U_i \times \mathbb{C}^n,\ i \in I\}$$

of E, whose transition functions are the given g_{ij}.

PROOF. Let

$$E' := \bigcup_{i \in I} U_i \times \mathbb{C}^n \times \{i\} \subset X \times \mathbb{C}^n \times I.$$

Give E' the topology induced from $X \times \mathbb{C}^n \times I$, where I has the discrete topology. On E' introduce the following equivalence relation:

$$(x, t, i) \sim (x', t', j) \Leftrightarrow x = x' \quad \text{and } t = g_{ij}(x)t'.$$

Because of the cocycle relation $g_{ij} g_{jk} = g_{ik}$, it is easy to check that this really is an equivalence relation. Let $E := E'/\sim$, together with the quotient topology, and let $\kappa\colon E' \to E$ be the canonical quotient map. Since the equivalence relation is compatible with the projection $E' \to X$, the induced mapping $p\colon E \to X$ is continuous. The fibers $p^{-1}(x)$ have a natural structure of an n-dimensional vector space over $\mathbb{C}$. As well

$$E_{U_i} = p^{-1}(U_i) = \kappa(U_i \times \mathbb{C}^n \times \{i\})$$

and $\kappa \mid U_i \times \mathbb{C}^n \times \{i\} \to E_{U_i}$ is a homeomorphism. Local trivializations $h_i\colon E_{U_i} \to U_i \times \mathbb{C}^n$ can now be defined as the inverses of these homeomorphisms followed by the identifications $U_i \times \mathbb{C}^n \times \{i\} \cong U_i \times \mathbb{C}^n$. By construction the transition functions of the atlas (h_i) are the given g_{ij}. $\square$

29.8. Definition. Suppose $p\colon E \to X$ is a vector bundle on a topological space X and U is a subset of X. A *section* of E over U is a continuous mapping $f\colon U \to E$ with $p \circ f = \mathrm{id}_U$.

The condition $p \circ f = \mathrm{id}_U$ means that f assigns to each $x \in U$ an element $f(x) \in E_x$. If $h_i\colon E_{U_i} \to U_i \times \mathbb{C}^n$ is a local trivialization of E then one can associate to the section f a unique continuous function $f_i\colon U_i \cap U \to \mathbb{C}^n$ such that

$$h_i(f(x)) = (x, f_i(x)) \quad \text{for every } x \in U_i \cap U.$$

The function f_i is called a representation of the section f with respect to the local trivialization h_i.

29.9. Definition. Suppose $p\colon E \to X$ is a holomorphic vector bundle of rank n on the Riemann surface X and $\{h_i\colon E_{U_i} \to U_i \times \mathbb{C}^n,\ i \in I\}$ is an atlas of the holomorphic linear structure on E. A section $f\colon U \to E$ over an open subset $U \subset X$ is said to be *holomorphic* if its representation f_i with respect to every local trivialization h_i is a holomorphic mapping $f_i\colon U_i \cap U \to \mathbb{C}^n$. Of course f_i is to be understood as an n-tuple of holomorphic functions $U_i \cap U \to \mathbb{C}$.

Clearly the definition is independent of the choice of atlas. The set of all holomorphic sections of E over U has the natural structure of a vector space, which we denote by $\mathcal{O}_E(U)$. With the natural restriction mappings one gets the sheaf $\mathcal{O}_E$ of holomorphic sections of E.

Suppose $(g_{ij}) \in Z^1(\mathfrak{U}, \mathrm{GL}(n, \mathcal{O}))$ is the cocycle associated to the atlas $\{h_i\colon E_{U_i} \to U_i \times \mathbb{C}^n,\ i \in I\}$. The representations $f_i\colon U_i \cap U \to \mathbb{C}^n$ of a section $f \in \mathcal{O}_E(U)$ satisfy the relation

$$f_i(x) = g_{ij}(x) f_j(x) \quad \text{for every } x \in U_i \cap U_j \cap U. \qquad (*)$$

Hence $\mathcal{O}_E(U)$ is isomorphic to the vector space of all families $(f_i)_{i \in I}$, with

$$f_i \in \mathcal{O}(U_i \cap U)^n,$$

which satisfy (*). Note $\mathcal{O}_E(U_i)$ is isomorphic to $\mathcal{O}(U_i)^n$. If E is holomorphically trivial, then the sheaf $\mathcal{O}_E$ is isomorphic to $\mathcal{O}^n$.

We now give two important examples of holomorphic line bundles on Riemann surfaces.

29.10. The Holomorphic Cotangent Bundle. Suppose X is a Riemann surface and (U_i, z_i), $i \in I$, is a covering by coordinate neighborhoods. On $U_i \cap U_j$ the function $g_{ij} := dz_j/dz_i$ is holomorphic and does not vanish. Thus the family (g_{ij}) defines a cocycle in $Z^1(\mathfrak{U}, \mathcal{O}^*)$ with respect to the covering $\mathfrak{U} = (U_i)_{i \in I}$. Let $T^*(X)$ be the line bundle associated to the cocycle (g_{ij}). $T^*(X)$ is called the *holomorphic cotangent bundle* or *canonical line bundle* of X. The sheaf of holomorphic sections of $T^*(X)$ is isomorphic to the sheaf Ω of holomorphic 1-forms on X. This isomorphism can be described as follows. Suppose $\omega \in \Omega(U)$. Then on $U_i \cap U$ one may write $\omega = f_i\, dz_i$ with $f_i \in \mathcal{O}(U_i \cap U)$. On $U_i \cap U_j \cap U$ one has $f_i = f_j\, dz_j/dz_i = g_{ij} f_j$. Thus the family (f_i) defines a holomorphic section of $T^*(X)$ over U. Conversely every family (f_i) of holomorphic functions $f_i \in \mathcal{O}(U_i \cap U)$ with $f_i = g_{ij} f_j$ on $U_i \cap U_j \cap U$ gives rise to a 1-form $\omega \in \Omega(U)$ with $\omega = f_i\, dz_i$ on $U_i \cap U$.

29.11. The Line Bundle of a Divisor. Suppose D is a divisor on the Riemann surface X. To D we can associate a holomorphic line bundle E_D such that the sheaf of holomorphic sections of E_D is isomorphic to the sheaf $\mathcal{O}_D$ of meromorphic multiples of $-D$ (cf. 16.4). There exists an open covering $\mathfrak{U} = (U_i)_{i \in I}$ of X and meromorphic functions $\psi_i \in \mathcal{M}(U_i)$ with $(\psi_i) = D$ on U_i. Then

$$g_{ij} := \frac{\psi_i}{\psi_j} \in \mathcal{O}^*(U_i \cap U_j),$$

since ψ_i and ψ_j have the same zeros and poles on $U_i \cap U_j$. The family g_{ij} forms a cocycle $(g_{ij}) \in Z^1(\mathfrak{U}, \mathcal{O}^*)$. Let E_D be the holomorphic line bundle corresponding to this cocycle (cf. Theorem (29.7)).

Suppose U is open in X and $f \in \mathcal{O}_D(U)$, i.e. $(f) \geq -D$ on U. Then there are holomorphic functions $f_i \in \mathcal{O}(U_i \cap U)$ such that $f = f_i/\psi_i$ on $U_i \cap U$. Hence on any intersection $U_i \cap U_j \cap U$ one has

$$\frac{f_i}{\psi_i} = \frac{f_j}{\psi_j} \quad \text{and thus } f_i = g_{ij} f_j.$$

Hence the family (f_i) defines a holomorphic section of E_D over U. Conversely any holomorphic section of E_D over U is given by a family (f_i) of holomorphic functions $f_i \in \mathcal{O}(U_i \cap U)$ with $f_i = g_{ij} f_j$. Then $f_i/\psi_i = f_j/\psi_j$ on $U_i \cap U_j \cap U$. Thus there is a meromorphic function $f \in \mathcal{M}(U)$ with $f = f_i/\psi_i$ on $U_i \cap U$ for every $i \in I$. Therefore $f \in \mathcal{O}_D(U)$.

We will now prove several facts about cohomology with values in the sheaf of holomorphic sections of vector bundles, which are analogous to what we did in §14 for the sheaf $\mathcal{O}$. For the sake of variety we will use different methods this time.

29.12. Lemma. *Suppose X is a Riemann surface, E is a holomorphic vector bundle on X and Y is a relatively compact open subset of X. Then for every open subset $Y_0 \subset Y$ the restriction mapping $H^1(Y, \mathcal{O}_E) \to H^1(Y_0, \mathcal{O}_E)$ is surjective.*

PROOF. There are finitely many open sets $U_i \subset X$, $i = 1, \ldots, r$, which are biholomorphic to open sets in $\mathbb{C}$, with $Y = U_1 \cup \cdots \cup U_r$ and for which there exist holomorphic linear charts $h_i \colon E_{U_i} \to U_i \times \mathbb{C}^n$. For every open subset $V \subset U_i$ one then has

$$H^1(V, \mathcal{O}_E) \cong H^1(V, \mathcal{O})^n = 0,$$

cf. Theorem (26.1). Set

$$Y_k := Y_0 \cup \bigcup_{i=1}^{k} U_i$$

Clearly it suffices to show that the mappings

$$H^1(Y_k, \mathcal{O}_E) \to H^1(Y_{k-1}, \mathcal{O}_E)$$

for $k = 1, \ldots, r$ are surjective. Fix k and let

$$V_i := U_i \cap Y_{k-1} \quad \text{for } i = 1, \ldots, r,$$

$$V_i' := V_i \quad \text{for } i \neq k \quad \text{and } V_k' := U_k.$$

Then $\mathfrak{B} = (V_i)_{1 \leq i \leq r}$ is a Leray covering of Y_{k-1} and $\mathfrak{B}' = (V_i')_{1 \leq i \leq r}$ is a Leray covering of Y_k. Hence $Z^1(\mathfrak{B}, \mathcal{O}_E) = Z^1(\mathfrak{B}', \mathcal{O}_E)$, since $V_i \cap V_j = V_i' \cap V_j'$ for every $i \neq j$. Thus $H^1(\mathfrak{B}', \mathcal{O}_E) \to H^1(\mathfrak{B}, \mathcal{O}_E)$ is surjective. □

29.13. Theorem. *Suppose Y is a relatively compact open subset of a Riemann surface X and E is a holomorphic vector bundle on X. Then $H^1(Y, \mathcal{O}_E)$ is finite dimensional.*

PROOF. There is an open set Y' with $Y \Subset Y' \Subset X$ and open sets $V_i \Subset U_i$, $i = 1, \ldots, r$, in X with the following properties:

(i) $\bigcup_{i=1}^r V_i = Y, \quad \bigcup_{i=1}^r U_i = Y'$.
(ii) Every U_i is biholomorphic to an open subset of $\mathbb{C}$.
(iii) On every U_i there is a holomorphic linear chart $h_i\colon E_{U_i} \to U_i \times \mathbb{C}^n$.

Now $\mathfrak{U} = (U_i)$ and $\mathfrak{B} = (V_i)$ are Leray coverings of Y' resp. Y for the sheaf O_E. By Lemma (29.12) it follows that the restriction mapping $H^1(\mathfrak{U}, \mathcal{O}_E) \to H^1(\mathfrak{B}, \mathcal{O}_E)$ is surjective. This implies that the mapping

$$\varphi\colon C^0(\mathfrak{B}, \mathcal{O}_E) \times Z^1(\mathfrak{U}, \mathcal{O}_E) \to Z^1(\mathfrak{B}, \mathcal{O}_E)$$

$$(\eta, \xi) \mapsto \delta(\eta) + \beta(\xi)$$

is surjective, where $\beta\colon Z^1(\mathfrak{U}, \mathcal{O}_E) \to Z^1(\mathfrak{B}, \mathcal{O}_E)$ is the restriction map. One can make the spaces $Z^1(\mathfrak{U}, \mathcal{O}_E)$, $Z^1(\mathfrak{B}, \mathcal{O}_E)$ and $C^0(\mathfrak{B}, \mathcal{O}_E)$ into Fréchet spaces in the following way. First $\mathcal{O}_E(U_i \cap U_j) \cong \mathcal{O}(U_i \cap U_j)^n$ with the topology of uniform convergence on compact subsets is a Fréchet space. Thus so is $C^1(\mathfrak{U}, \mathcal{O}_E) = \prod_{i,j} \mathcal{O}_E(U_i \cap U_j)$ with the product topology. It is easy to see that $Z^1(\mathfrak{U}, \mathcal{O}_E)$ is a closed subspace of $C^1(\mathfrak{U}, \mathcal{O}_E)$. Thus it is likewise a Fréchet space. The topologies on $Z^1(\mathfrak{B}, \mathcal{O}_E)$ and $C^0(\mathfrak{B}, \mathcal{O}_E)$ are defined similarly. With respect to these topologies the mappings $\delta\colon C^0(\mathfrak{B}, \mathcal{O}_E) \to Z^1(\mathfrak{B}, \mathcal{O}_E)$ and $\beta\colon Z^1(\mathfrak{U}, \mathcal{O}_E) \to Z^1(\mathfrak{B}, \mathcal{O}_E)$ are continuous. Then Montel's Theorem implies that β is even compact. Hence

$$\psi\colon C^0(\mathfrak{B}, \mathcal{O}_E) \times Z^1(\mathfrak{U}, \mathcal{O}_E) \to Z^1(\mathfrak{B}, \mathcal{O}_E)$$

$$(\eta, \xi) \mapsto \beta(\eta)$$

is also compact. By the Theorem of L. Schwartz (cf. Appendix B.11) the mapping

$$\varphi - \psi\colon C^0(\mathfrak{B}, \mathcal{O}_E) \times Z^1(\mathfrak{U}, \mathcal{O}_E) \to Z^1(\mathfrak{B}, \mathcal{O}_E)$$

$$(\eta, \xi) \mapsto \delta\eta$$

as the difference of a surjective and a compact continuous linear operator between Fréchet spaces, has a finite codimensional image. However the image of $\varphi - \psi$ is the vector space $B^1(\mathfrak{B}, \mathcal{O}_E)$ of all coboundaries in $Z^1(\mathfrak{B}, \mathcal{O}_E)$. Thus $H^1(Y, \mathcal{O}_E) \cong H^1(\mathfrak{B}, \mathcal{O}_E)$ is finite dimensional. $\square$

29.14. Corollary. *Suppose E is a holomorphic vector bundle on a compact Riemann surface X. Then $H^1(X, \mathcal{O}_E)$ is finite dimensional.*

29.15. Meromorphic Sections. Suppose E is a holomorphic vector bundle of rank n on the Riemann surface X. Let $U \subset X$ be an open set over which a holomorphic linear chart $h\colon E_U \to U \times \mathbb{C}^n$ exists and let a be a point of U. A section $f \in \mathcal{O}_E(U\backslash\{a\})$ can be represented with respect to this chart by an n-tuple of holomorphic functions $(f_1, \ldots, f_n) \in \mathcal{O}(U\backslash\{a\})^n$. The point a is called a *pole of order m* of f, if all the f_j have either a pole of order $\leq m$ or else a removable singularity at a and at least one f_i does have a pole of order m at a. This definition is independent of the choice of linear chart at a.

By a *meromorphic section* of E over an open subset $Y \subset X$ one means a holomorphic section $f \in \mathcal{O}_E(Y')$ over an open subset $Y' \subset Y$ such that the following hold:

(i) $Y\backslash Y'$ is a discrete subset of Y.
(ii) f has a pole at every $a \in Y\backslash Y'$.

Similar to Theorem (14.12) one can now prove the following.

29.16. Theorem. *Suppose E is a holomorphic vector bundle on a Riemann surface X and Y is a relatively compact open subset of X. Then given any $a \in Y$ there exists a meromorphic section of E over Y which has a pole at a and is holomorphic on $Y\backslash\{a\}$.*

29.17. Corollary. *Every holomorphic vector bundle on a compact Riemann surface has a global meromorphic section which does not vanish identically.*

29.18. Line Bundles and Divisors. Suppose E is a holomorphic line bundle on a Riemann surface X and ψ is a global meromorphic section of E, which does not vanish identically. Then the divisor D of ψ is well-defined. For $a \in X$ let $D(a)$ be the order of ψ at a with respect to a holomorphic linear chart of E on some neighborhood of a. This order is independent of the chart. Now we claim that the sheaf $\mathcal{O}_E$ of holomorphic sections of E is isomorphic to the sheaf $\mathcal{O}_D$ of meromorphic multiples of $-D$. For, if $f \in \mathcal{M}(U)$ with $(f) \geq -D$ on U, then $f\psi$ is a holomorphic section of E over U. Conversely for every section $\varphi \in \mathcal{O}_E(U)$ the quotient $f = \varphi/\psi$ is a well-defined meromorphic function in $\mathcal{M}(U)$ with $(f) \geq -D$ on U.

This may be considered in some sense to be the converse of (29.11).

We are now in the position to interpret the Picard group of a compact Riemann surface as a cohomology group. Recall (cf. 21.6) that the Picard group of a compact Riemann surface X is defined as

$$\mathrm{Pic}(X) = \mathrm{Div}(X)/\mathrm{Div}_P(X).$$

29.19. Theorem. *Let X be a compact Riemann surface. Then there is a natural isomorphism of groups*

$$H^1(X, \mathcal{O}^*) \xrightarrow{\sim} \mathrm{Pic}(X).$$

PROOF

(a) Define a map $\alpha\colon H^1(X, \mathcal{O}^*) \to \mathrm{Pic}(X)$ as follows. Suppose $\xi \in H^1(X, \mathcal{O}^*)$ is represented by a cocycle $(g_{ij}) \in Z^1(\mathfrak{U}, \mathcal{O}^*)$ for some open covering $\mathfrak{U} = (U_i)_{i \in I}$ of X. By Theorem (29.7) there is a line bundle $E \to X$ associated to this cocycle. Then E has a non-trivial meromorphic section f by Corollary (29.17). This section is given by a family of meromorphic functions $f_i \in \mathcal{M}(U_i)$, $i = I$, satisfying

$$f_i = g_{ij} f_j \quad \text{on } U_i \cap U_j.$$

Since f does not vanish identically, we even have $f_i \in \mathcal{M}^*(U_i)$ for every $i \in I$. Let D be the divisor of f, i.e.,

$$D(x) = \mathrm{ord}_x(f_i) \quad \text{if } x \in U_i.$$

Clearly this does not depend on the choice of i with $x \in U_i$. Now define

$$\alpha(\xi) := D \bmod \mathrm{Div}_P(X).$$

In order that α be well-defined, this definition must be independent of the various choices made. For example, let

$$g_{ij} = f_i/f_j = \tilde{f}_i/\tilde{f}_j,$$

where $f_i, \tilde{f}_i \in \mathcal{M}^*(U_i)$. Then $f_i/\tilde{f}_i = f_j/\tilde{f}_j$ on $U_i \cap U_j$. Hence the $f_i/\tilde{f}_i$ piece together to define a global meromorphic function $\varphi \in \mathcal{M}^*(X)$. If D and $\tilde{D}$ are the divisors of $f = (f_i)_{i \in I}$ and $\tilde{f} = (\tilde{f}_i)_{i \in I}$ respectively, we have

$$\tilde{D} = D + (\varphi),$$

and thus D and $\tilde{D}$ differ only by a principal divisor. We leave it to the reader to show that $\alpha(\xi)$ is also independent of the choice of cocycle (g_{ij}) representing ξ. It is easy to see that α is a group homomorphism.

(b) The surjectivity of α follows from (29.11). To prove the injectivity, we have to show that if

$$g_{ij} = f_i/f_j, \qquad f_i \in \mathcal{M}^*(U_i)$$

and the divisor of $(f_i)_{i \in I}$ is principal, then the cocycle (g_{ij}) lies in $B^1(\mathfrak{U}, \mathcal{O}^*)$.

Indeed, let $\varphi \in \mathcal{M}^*(X)$ be a meromorphic function having the same divisor as the family $(f_i)_{i \in I}$. Then

$$g_i := \frac{f_i}{\varphi} \in \mathcal{O}^*(U_i)$$

and $g_{ij} = g_i/g_j$. Hence $(g_{ij}) \in B^1(\mathfrak{U}, \mathcal{O}^*)$. □

EXERCISES (§29)

29.1. Let X be a Riemann surface and $\pi: E \to X$ be a holomorphic vector bundle of rank n on X. A holomorphic subbundle $F \subset E$ of rank k is a subset such that the following holds. For every $x \in X$ there exists a holomorphic local trivialization

$$h: E_U \to U \times \mathbb{C}^n, \qquad E_U = \pi^{-1}(U),$$

of E with $x \in U$ such that

$$h(F_U) = U \times (\mathbb{C}^k \times 0)$$

where $F_U = E_U \cap F$.

(a) Let $f: X \to E$ be a holomorphic section of E which never vanishes. For $x \in X$ define $F_x := \mathbb{C} \cdot f(x) \subset E_x$. Show that

$$F := \bigcup_{x \in X} F_x \subset E$$

is a holomorphic subbundle of E of rank 1.

(b) Let f be a meromorphic section of E over X. Show that there exists a unique subbundle $F \subset E$ of rank 1 such that f is a meromorphic section of F.

29.2. Let $L \to X$ be a line bundle on a compact Riemann surface X. The degree of L is defined as $\deg(L) := \deg(D)$, where D is the divisor of a meromorphic section s of L over X.

(a) Show that deg L is well-defined, i.e., it is independent of the choice of the meromorphic section s.

(b) On $\mathbb{P}^1$ let $\mathfrak{U} = (U_1, U_2)$ be the covering given by

$$U_1 := \{z \in \mathbb{C}: |z| < 1 + \varepsilon\}, \quad U_2 := \{z \in \mathbb{P}^1: |z| > 1 - \varepsilon\}, \quad 0 < \varepsilon < 1.$$

Let L be the holomorphic line bundle on $\mathbb{P}^1$ defined by some given transition function

$$g_{12}: U_1 \cap U_2 \to \mathbb{C}^*.$$

Prove that

$$\deg L := \frac{1}{2\pi i} \int_{|z|=1} \frac{g_{12}'(z)}{g_{12}(z)}\, dz.$$

[*Hint*: First consider the special case $g_{12}(z) = z^k$.]

§30. The Triviality of Vector Bundles

In this section we show that every holomorphic vector bundle on a non-compact Riemann surface is trivial. This will be used in the next section to solve the Riemann–Hilbert problem.

30.1. Theorem. *Suppose E is a holomorphic vector bundle of rank n on a Riemann surface X. Let $\mathfrak{U} = (U_i)_{i \in I}$ be an open covering of X, $h_i: E_{U_i} \to U_i \times \mathbb{C}^n$, $i \in I$, be a holomorphic atlas for E and $(g_{ij}) \in Z^1(\mathfrak{U}, \mathrm{GL}(n, \mathcal{O}))$ be the corresponding cocycle of transition functions. Then the following are equivalent:*

(i) *E is holomorphically trivial.*

(ii) *There exist n global holomorphic sections $F_1, \ldots, F_n$ of E such that for each point $x \in X$ the vectors $F_1(x), \ldots, F_n(x) \in E_x$ are linearly independent.*

(iii) *The cocycle (g_{ij}) splits, i.e., there exists a cochain $(g_i) \in C^0(\mathfrak{U}, \mathrm{GL}(n, \mathcal{O}))$ with*

$$g_{ij} = g_i g_j^{-1} \quad \text{on } U_i \cap U_j \quad \text{for every } i, j \in I.$$

PROOF

(i) ⇒ (ii). Since E is holomorphically trivial, the holomorphic linear structure of E contains a chart $h: E \to X \times \mathbb{C}^n$. Let $e_1, \ldots, e_n$ be the standard unit vectors of $\mathbb{C}^n$. Define sections F_ν, $\nu = 1, \ldots, n$, of E by

$$h(F_\nu(x)) = (x, e_\nu) \quad \text{for every } x \in X.$$

Then the F_ν are holomorphic and linearly independent in every fiber.

(ii) ⇒ (iii). Any section F_ν may be represented relative to any chart h_i as an n-tuple of holomorphic functions $f^i_{\mu\nu} \in \mathcal{O}(U_i)$, $\mu = 1, \ldots, n$. Let g_i be the matrix $(f^i_{\mu\nu})_{1 \le \mu, \nu \le n}$. Then $g_i \in \mathrm{GL}(n, \mathcal{O}(U_i))$, since $F_1, \ldots, F_n$ are linearly independent in each fiber. Moreover on $U_i \cap U_j$ one has

$$g_i = g_{ij} g_j \quad \text{and thus } g_{ij} = g_i g_j^{-1},$$

i.e., the cocycle (g_{ij}) splits.

(iii) ⇒ (i). Using the charts $h_i: E_{U_i} \to U_i \times \mathbb{C}^n$ we will construct a linear chart $h: E \to X \times \mathbb{C}^n$ which is holomorphically compatible with all the h_i.

Suppose $v \in E_{U_i}$ and $h_i(v) =: (x, t)$. Then set $h(v) := (x, g_i^{-1} t)$. This definition is independent of the choice of chart. For, suppose $v \in E_{U_j}$ as well and $h_j(v) =: (x, t')$. Then $t = g_{ij} t' = g_i g_j^{-1} t'$ and thus $g_i^{-1} t = g_j^{-1} t'$. Finally it follows directly from the definition of h that $\{h: E \to X \times \mathbb{C}^n\}$ is holomorphically compatible with the atlas consisting of all the h_i. □

30.2. Lemma. *Suppose X is a non-compact Riemann surface and E is a holomorphic vector bundle on X. If E has a non-trivial global meromorphic section, then E also has a global holomorphic section which has no zeros.*

PROOF. Suppose f is a non-trivial meromorphic section of E over X and $A \subset X$ is the discrete subset consisting of its zeros and poles. Suppose $a \in A$ and $h: E_U \to U \times \mathbb{C}^n$ is a holomorphic linear chart of E on an open neighborhood U of a. Relative to the chart h we may represent f as $(f_1, \ldots, f_n) \in \mathcal{M}(U)^n$. Let $k(a)$ be the minimum of the orders of the functions f_ν at the point a. By Weierstrass' Theorem (26.5) there exists a meromorphic function $\varphi \in \mathcal{M}(X)$ which at each point $a \in A$ has order $-k(a)$ and is holomorphic and non-zero on $X \setminus A$. Then $F := \varphi f$ is a holomorphic section of E which has no zeros. □

30.3. Theorem. *Every holomorphic line bundle E on a non-compact Riemann surface X is holomorphically trivial.*

PROOF. Suppose $\emptyset \neq Y_0 \Subset Y_1 \Subset Y_2 \Subset \cdots$ is a sequence of relatively compact Runge domains in X with $\bigcup Y_\nu = X$. By Theorem (29.16) over every Y_ν there is a meromorphic section. Thus by (30.2) there is also a holomorphic section which does not vanish. Hence E is trivial over each Y_ν by Theorem (30.1). It then follows from the Runge Approximation Theorem that every holomorphic section of E over Y_ν can be approximated uniformly on compact subsets by holomorphic sections of E over $Y_{\nu+1}$. Let $f_0 \in \mathcal{O}_E(Y_0)$ be a section which is not zero at some point $a \in Y_0$. One can now construct a sequence $f_\nu \in \mathcal{O}_E(Y_\nu)$, $\nu \geq 1$, such that $\lim_{\nu\to\infty} f_\nu(a) \neq 0$ and such that for each $\nu \in \mathbb{N}$ the sequence $(f_\mu \mid Y_\nu)_{\mu > \nu}$ converges in $\mathcal{O}_E(Y_\nu)$. Then the limit of the sequence (f_ν) is a section $f \in \mathcal{O}_E(X)$ which does not vanish identically. As above this implies that E is trivial over X. □

30.4. Theorem. *Every holomorphic vector bundle E on a non-compact Riemann surface X is holomorphically trivial.*

PROOF. The theorem will be proved by induction on n, the rank of E. Theorem (30.3) is the case $n = 1$. Now assume the result has been proved for all bundles of rank $n - 1$ and suppose E is a bundle of rank n.

(a) First we assume that there exists a section $F_n \in \mathcal{O}_E(X)$ which does not vanish anywhere. Since E is locally trivial, there exists an open covering $\mathfrak{U} = (U_i)_{i \in I}$ of X with the property that for every $i \in I$ there are sections $F_1^i, \ldots, F_{n-1}^i \in \mathcal{O}_E(U_i)$ such that $F_1^i(x), \ldots, F_{n-1}^i(x), F_n(x)$ are linearly independent for every $x \in U_i$. On any intersection $U_i \cap U_j$ these systems are related to each other in the following way:

$$\begin{pmatrix} F^i \\ F_n \end{pmatrix} = \begin{pmatrix} G^{ij} & a^{ij} \\ 0 & 1 \end{pmatrix} \begin{pmatrix} F^j \\ F_n \end{pmatrix}, \tag{1}$$

where F^i denotes the column vector with entries $F_1^i, \ldots, F_{n-1}^i$, the matrix G^{ij} is an element of $\mathrm{GL}(n-1, \mathcal{O}(U_i \cap U_j))$ and a^{ij} is a column vector with $n - 1$ rows having coefficients in $\mathcal{O}(U_i \cap U_j)$. Then $G^{ij}G^{jk} = G^{ik}$ on

$U_i \cap U_j \cap U_k$. Hence by the induction hypothesis there exist matrices $G^i \in \mathrm{GL}(n-1, \mathcal{O}(U_i))$ with

$$G^{ij} = G^i(G^j)^{-1} \quad \text{on } U_i \cap U_j.$$

Setting $\tilde{F}^i = (G^i)^{-1}F^i$ and using (1) gives

$$\begin{pmatrix} \tilde{F}^i \\ F_n \end{pmatrix} = \begin{pmatrix} 1 & b^{ij} \\ 0 & 1 \end{pmatrix} \begin{pmatrix} \tilde{F}^j \\ F_n \end{pmatrix} \tag{2}$$

for some $b^{ij} \in \mathcal{O}(U_i \cap U_j)^{n-1}$. On $U_i \cap U_j \cap U_k$ one has the relation $b^{ij} + b^{jk} = b^{ik}$. Since $H^1(\mathfrak{U}, \mathcal{O}) = 0$, one can thus find holomorphic column vectors $b^i \in \mathcal{O}(U_i)^{n-1}$ having $(n-1)$ rows with

$$b^{ij} = b^i - b^j \quad \text{on } U_i \cap U_j.$$

Set $\hat{F}^i = \tilde{F}^i - b^iF_n$. Then it follows from (2) that

$$\begin{pmatrix} \hat{F}^i \\ F_n \end{pmatrix} = \begin{pmatrix} \hat{F}^j \\ F_n \end{pmatrix} \quad \text{on } U_i \cap U_j.$$

Hence the $\hat{F}^i$ piece together to form a global $(n-1)$-tuple $(F_1, \ldots, F_{n-1}) \in \mathcal{O}_E(X)^{n-1}$. By construction $F_1(x), \ldots, F_n(x)$ are linearly independent for every $x \in X$. Thus E is holomorphically trivial.

(b) We still have to show that E has a holomorphic section which does not vanish. By Theorem (29.16) and Lemma (30.2) this is the case over any relatively compact domain $Y \subset X$. Thus by (a) one has that E is trivial over Y. As in the proof of (30.3) one can now construct with the help of the Runge Approximation Theorem a non-trivial holomorphic section of E over X. By Lemma (30.2) then E also has a nowhere vanishing holomorphic section. This completes the proof of the theorem. □

30.5. Corollary. *Suppose X is a non-compact Riemann surface. Then*

$$H^1(X, \mathrm{GL}(n, \mathcal{O})) = 0.$$

In particular, $H^1(X, \mathcal{O}^*) = 0$.

PROOF. Now $H^1(X, \mathrm{GL}(n, \mathcal{O})) = 0$ means that for every open covering $\mathfrak{U} = (U_i)$ of X every cocycle $(g_{ij}) \in Z^1(\mathfrak{U}, \mathrm{GL}(n, \mathcal{O}))$ splits. But this is equivalent to the triviality of holomorphic vector bundles on X. □

EXERCISES (§30)

30.1. Show that on any Riemann surface X (compact or not) one has $H^1(X, \mathcal{M}^*) = 0$.
[*Hint*: Use the exact cohomology sequence of Ex. 16.4.]

30.2. Let X be a Riemann surface. For $U \subset X$ open, let $\mathrm{SL}(n, \mathcal{O}(U))$ be the group of all $n \times n$-matrices of determinant 1 with coefficients in $\mathcal{O}(U)$. Together with the natural restriction maps this defines a sheaf $\mathrm{SL}(n, \mathcal{O})$ on X. Prove that on a non-compact Riemann surface X

$$H^1(X, \mathrm{SL}(n, \mathcal{O})) = 0,$$

i.e., for every cocycle $(g_{ij}) \in Z^1(\mathfrak{U}, \mathrm{SL}(n, \mathcal{O}))$ there exists a cochain $(g_i) \in C^0(\mathfrak{U}, \mathrm{SL}(n, \mathcal{O}))$ such that

$$g_{ij} = g_i g_j^{-1} \quad \text{on } U_i \cap U_j.$$

§31. The Riemann–Hilbert Problem

In §11 we saw that the automorphic behavior of a fundamental system of solutions of a linear differential equation on a Riemann surface X gives rise to a homomorphism $T\colon \pi_1(X) \to \mathrm{GL}(n, \mathbb{C})$. This homomorphism associates to each $\sigma \in \pi_1(X)$ the factor of automorphy T_σ by which the fundamental system is multiplied when it is analytically continued along σ. Conversely one may ask if given any homomorphism $T\colon \pi_1(X) \to \mathrm{GL}(n, \mathbb{C})$, there exists a linear differential equation on X such that the automorphic behavior of a fundamental system of solutions is exactly given by the homomorphism T. This is called the Riemann–Hilbert problem. In this section we present the solution of the Riemann–Hilbert problem on non-compact Riemann surfaces using the method of H. Röhrl [57].

31.1. Factors of Automorphy. Suppose $p\colon Y \to X$ is a holomorphic unbranched covering mapping between Riemann surfaces and $G := \mathrm{Deck}(Y/X)$ is its group of covering transformations. A holomorphic mapping $\Phi\colon Y \to \mathrm{GL}(n, \mathbb{C})$ is called multiplicatively automorphic with constant factors of automorphy $T_\sigma \in \mathrm{GL}(n, \mathbb{C})$, $\sigma \in G$, if

$$\sigma\Phi = \Phi T_\sigma \quad \text{for every } \sigma \in G.$$

In this case one can easily show that the correspondence $\sigma \mapsto T_\sigma$ is a group homomorphism $G \to \mathrm{GL}(n, \mathbb{C})$, cf. (11.6). The following theorem is analogous to Theorem (28.4).

31.2. Theorem. *Suppose X and Y are non-compact Riemann surfaces, $p\colon Y \to X$ is a holomorphic unbranched Galois covering map and $G := \mathrm{Deck}(Y/X)$ is its group of covering transformations. Then given any homomorphism*

$$T\colon G \to \mathrm{GL}(n, \mathbb{C}), \qquad \sigma \mapsto T_\sigma,$$

there exists a holomorphic mapping $\Phi\colon Y \to \mathrm{GL}(n, \mathbb{C})$ with the factors of automorphy T_σ, i.e., $\sigma\Phi = \Phi T_\sigma$ for every $\sigma \in G$.

PROOF

(a) There exist an open covering $\mathfrak{U} = (U_i)_{i \in I}$ of X and G-charts

$$\varphi_i = (p, \eta_i): p^{-1}(U_i) \to U_i \times G,$$

cf. (28.3). Now define on $Y_i := p^{-1}(U_i)$ functions Ψ_i: $Y_i \to \mathrm{GL}(n, \mathbb{C})$ by

$$\Psi_i(y) := T_{\eta_i(y)^{-1}} \quad \text{for every } y \in Y_i.$$

Since Ψ_i is locally constant, in particular it is holomorphic.

(b) Suppose $y \in Y_i$ and $\sigma \in G$. Then

$$\sigma\Psi_i(y) = \Psi_i(\sigma^{-1}y) = T_{\eta_i(\sigma^{-1}y)^{-1}} = T_{\eta_i(y)^{-1}\sigma}$$
$$= T_{\eta_i(y)^{-1}} T_\sigma = \Psi_i(y) T_\sigma.$$

Thus the functions Ψ_i exhibit the desired automorphic behavior on Y_i.

(c) By (b) the products $F_{ij} := \Psi_i \Psi_j^{-1} \in \mathrm{GL}(n, \mathcal{O}(Y_i \cap Y_j))$ are invariant under covering transformations. Thus they may be considered as elements $F_{ij} \in \mathrm{GL}(n, \mathcal{O}(U_i \cap U_j))$ and so define a cocycle $(F_{ij}) \in Z^1(\mathfrak{U}, \mathrm{GL}(n, \mathcal{O}))$. Since $H^1(X, \mathrm{GL}(n, \mathcal{O})) = 0$ by (30.5), this cocycle is a coboundary. Thus there exist elements $F_i \in \mathrm{GL}(n, \mathcal{O}(U_i))$ with

$$F_{ij} = F_i F_j^{-1} \quad \text{on } U_i \cap U_j.$$

Now consider the F_i as elements of $\mathrm{GL}(n, \mathcal{O}(Y_i))$ which are invariant under covering transformations and set

$$\Phi_i := F_i^{-1} \Psi_i \in \mathrm{GL}(n, \mathcal{O}(Y_i)).$$

Then $\sigma\Phi_i = F_i^{-1}\sigma\Psi_i = F_i^{-1}\Psi_i T_\sigma = \Phi_i T_\sigma$ for every $\sigma \in G$. On any intersection $Y_i \cap Y_j$,

$$\Phi_i^{-1}\Phi_j = \Psi_i^{-1}F_iF_j^{-1}\Psi_j = \Psi_i^{-1}F_{ij}\Psi_j = \Psi_i^{-1}\Psi_i\Psi_j^{-1}\Psi_j = 1,$$

i.e., $\Phi_i = \Phi_j$. Thus the Φ_i piece together to give a global function $\Phi \in \mathrm{GL}(n, \mathcal{O}(Y))$ with $\sigma\Phi = \Phi T_\sigma$ for every $\sigma \in G$. □

31.3. Corollary. *Suppose X is a non-compact Riemann surface and*

$$T: \pi_1(X) \to \mathrm{GL}(n, \mathbb{C}), \qquad \sigma \mapsto T_\sigma,$$

is a group homomorphism. Then there exists a matrix $A \in M(n \times n, \Omega(X))$ and a fundamental system of solutions of the differential equation $dw = Aw$ on the universal covering of X which has the T_σ as factors of automorphy.

PROOF. By (11.6) one only has to apply Theorem (31.2) to the universal covering $p: \tilde{X} \to X$ of X. □

31.4. Suppose X is a non-compact Riemann surface, $S \subset X$ is a closed discrete subset and $X' := X \backslash S$. Then, in particular, one can apply Corollary (31.3) to X'. But we would like to sharpen the result of the corollary so that the resulting differential equation has at most regular singular points at all the points $a \in S$. In order to be able to carry over the definition in (11.12) to this general case, we first prove the following lemma.

Lemma. *Using the same notation as above, suppose $p\colon Y \to X'$ is the universal covering of X'. Further suppose (U, z) is a coordinate neighborhood of a point $a \in S$ with the following properties:*

(i) *$z(U) \subset \mathbb{C}$ is the unit disk and $z(a) = 0$.*
(ii) *$U \cap S = \{a\}$.*

Suppose Z is any connected component of $p^{-1}(U \backslash a)$. Then $p \mid Z \to U \backslash a$ is the universal covering of $U \backslash a$.

PROOF. By Weierstrass' Theorem (26.5) there exists a holomorphic function $f \in \mathcal{O}(X)$ which has a zero of first order at a but is otherwise non-zero. Then $\omega := df/f$ is a holomorphic 1-form on X'. Let γ be the positively oriented curve in U corresponding to $|z| = \frac{1}{2}$. Then

$$\int_\gamma \omega = \int_\gamma \frac{df}{f} = 2\pi i.$$

Now the mapping $p \mid Z \to U \backslash a$ is a covering map. Thus we may apply Theorem (5.10) to it. If $p \mid Z \to U \backslash a$ were not the universal covering, then this mapping would be isomorphic to the covering

$$D^* \to D^*, \qquad z \mapsto z^k$$

for some positive integer k, where D^* denotes the punctured unit disk. But then there would exist k liftings $\gamma_1, \ldots, \gamma_k$ of γ whose product $c = \gamma_1 \cdots \gamma_k$ is a closed curve. This implies

$$\int_c p^*\omega = \sum_{j=1}^{k} \int_{\gamma_j} p^*\omega = \sum_{j=1}^{k} \int_\gamma \omega = 2k\pi i.$$

But on the other hand $\int_c p^*\omega = 0$, since a primitive for $p^*\omega$ exists on Y. This contradiction proves that $p \mid Z \to U \backslash a$ must be the universal covering. □

Now, using the same notation, suppose $dw = Aw$, where $A \in M(n \times n, \Omega(X'))$, is a linear differential equation on X' and $\Phi \in \mathrm{GL}(n, \mathcal{O}(Y))$ is a fundamental system of solutions. Then the differential equation is said to have a *regular singular point* at $a \in S$ if for every connected component Z of $p^{-1}(U \backslash a)$ the function $\Phi \mid Z$ satisfies the condition given in (11.12).

31.5. Theorem. *Suppose X is a non-compact Riemann surface, S is a closed discrete subset of X and $X' := X \backslash S$. Further suppose a homomorphism*

$$T\colon \pi_1(X') \to \mathrm{GL}(n, \mathbb{C}), \qquad \sigma \mapsto T_\sigma,$$

is given. Then there exists a differential equation $dw = Aw$, *where* $A \in M(n \times n, \Omega(X'))$, *which has a regular singular point at every* $a \in S$, *and a fundamental system of solutions* $\Phi \in \mathrm{GL}(n, \mathcal{O}(Y))$ *of* $dw = Aw$ *on the universal covering* $p\colon Y \to X'$ *of* X' *with the factors of automorphy* T_σ.

PROOF. Suppose $S = \{a_i : i \in I\}$. For every i choose a coordinate neighborhood (U_i, z_i) of a_i satisfying conditions (i) and (ii) of Lemma (31.4). We may assume $0 \notin I$. Let $J := I \cup \{0\}$ and set $U_0 := X'$. Then $\mathfrak{U} := (U_j)_{j \in J}$ is an open covering of X. For $i \neq j$ one has $U_i \cap U_j \subset X'$. Further let $Y_0 := Y$ and $Y_i := p^{-1}(U_i \backslash a_i)$ for every $i \in I$.

By Theorem (31.2) there exists a function $\Psi_0 \in \mathrm{GL}(n, \mathcal{O}(Y_0))$ such that $\sigma\Psi_0 = \Psi_0 T_\sigma$ for every $\sigma \in \pi_1(X')$. For every $i \in I$ there exist, by Theorem (11.10), elements $\Psi_i \in \mathrm{GL}(n, \mathcal{O}(Y_i))$ which have regular singular points and display the same automorphic behavior as $\Psi_0 \mid Y_i$. Hence for $i, j \in I$, $i \neq j$,

$$F_{ij} := \Psi_i \Psi_j^{-1} \in \mathrm{GL}(n, \mathcal{O}(Y_i \cap Y_j))$$

is invariant under covering transformations and thus may be considered as an element $F_{ij} \in \mathrm{GL}(n, \mathcal{O}(U_i \cap U_j))$. For every $j \in J$, let $F_{jj} := 1 \in \mathrm{GL}(n, \mathcal{O}(U_j))$. Then

$$(F_{ij}) \in Z^1(\mathfrak{U}, \mathrm{GL}(n, \mathcal{O}))$$

is a cocycle. Because $H^1(X, \mathrm{GL}(n, \mathcal{O})) = 0$, this cocycle is a coboundary. Thus there exist $F_i \in \mathrm{GL}(n, \mathcal{O}(U_i))$ such that

$$F_{ij} = F_i F_j^{-1} \quad \text{on } U_i \cap U_j.$$

Now, for every $j \in J$, define

$$\Phi_j := F_j^{-1} \Psi_j \in \mathrm{GL}(n, \mathcal{O}(Y_j)).$$

As in (31.2) the Φ_j piece together to form a global function $\Phi \in \mathrm{GL}(n, \mathcal{O}(Y))$ which satisfies $\sigma\Phi = \Phi T_\sigma$ for every $\sigma \in \pi_1(X)$. On $U_i \backslash a_i$ one has $\Phi = F_i^{-1}\Psi_i$. Since Ψ_i has regular singular points and F_i^{-1} is holomorphic on all of U_i, it follows that Φ also has regular singular points. As well Φ is a fundamental system of solutions of the differential equation $dw = Aw$, where $A := d\Phi \cdot \Phi^{-1}$ may be considered as an element $A \in M(n \times n, \Omega(X'))$, since it is invariant under covering transformations. This completes the proof of the theorem. □

EXERCISES (§31)

31.1. Let X and Y be non-compact Riemann surfaces, $p\colon Y \to X$ be an unbranched holomorphic Galois covering and $G := \mathrm{Deck}(Y/X)$. Let

$$a\colon G \to \mathrm{GL}(n, \mathcal{O}(Y)), \qquad \sigma \mapsto a_\sigma$$

be a crossed homomorphism, i.e. a map satisfying

$$a_{\sigma\tau} = a_\sigma(\sigma a_\tau) \quad \text{for every } \sigma, \tau \in G.$$

Prove that there exists a holomorphic matrix $\Phi\colon Y \to \mathrm{GL}(n, \mathbb{C})$ such that

$$a_\sigma = \Phi(\sigma\Phi)^{-1} \quad \text{for every } \sigma \in G.$$

31.2. Let $g\colon \mathbb{C} \to \mathrm{GL}(n, \mathbb{C})$ be a holomorphic invertible matrix. Show that there exists a holomorphic matrix $f\colon \mathbb{C} \to \mathrm{GL}(n, \mathbb{C})$ such that

$$f(z+1) = f(z)g(z) \quad \text{for every } z \in \mathbb{C}.$$

[*Hint*: Consider the Galois covering $\mathrm{ex}\colon \mathbb{C} \to \mathbb{C}^*$, $\mathrm{ex}(z) := e^{2\pi i z}$, and apply Ex. 31.1.]

Appendix

A. Partitions of Unity

Partitions of unity are an important tool in the study of differentiable manifolds and have been used throughout this book. As an aid to the reader we now gather together some of the main facts concerning them. Proofs may be found in the literature, e.g., [40], [43], [45] or [48].

A.1. By the *support* $\operatorname{Supp}(f)$ of a real or complex valued function f on a topological space X is meant the closure of the set $\{x \in X : f(x) \neq 0\}$.

The standard example of a C^∞ function, i.e. an infinitely differentiable function, $g\colon \mathbb{R}^n \to \mathbb{R}$, whose support is the closed ball of radius $\varepsilon > 0$, is given by

$$g(x) := \begin{cases} \exp\left(-\dfrac{1}{\varepsilon^2 - \|x\|^2}\right) & \text{for } \|x\| < \varepsilon \\ 0 & \text{for } \|x\| \geq \varepsilon. \end{cases}$$

Here $\|x\| = (|x_1|^2 + \cdots + |x_n|^2)^{1/2}$ denotes the euclidean norm on $\mathbb{R}^n$. This function can now be used to construct all the other C^∞ functions which we will need.

A.2. An n-dimensional manifold is a Hausdorff topological space X with the property that every point $a \in X$ has an open neighborhood homeomorphic to an open subset of $\mathbb{R}^n$. A homeomorphism $\varphi\colon U \to V$ of an open set $U \subset X$ onto an open set $V \subset \mathbb{R}^n$ is called a chart on X. Two charts $\varphi_i\colon U_i \to V_i$, $i = 1, 2$, are said to be differentiably compatible if the mapping

$$\varphi_2 \circ \varphi_1^{-1}\colon \varphi_1(U_1 \cap U_2) \to \varphi_2(U_1 \cap U_2)$$

and its inverse are both infinitely differentiable. Now one can define *differentiable manifolds* in an analogous way to Riemann surfaces (cf. §1), but replacing biholomorphic compatibility by differentiable compatibility. In particular Riemann surfaces are special 2-dimensional differentiable manifolds.

On a differentiable manifold one has the notion of a differentiable function, i.e., a function which is C^∞ with respect to every chart.

A.3. Definition. Suppose X is a differentiable manifold and $\mathfrak{U} = (U_i)_{i \in I}$ is an open covering of X. Then by a differentiable *partition of unity* subordinate to $\mathfrak{U}$ one means a family $(g_i)_{i \in I}$ of differentiable functions $g_i \colon X \to \mathbb{R}$ with the following properties:

(i) $0 \le g_i \le 1$ for every $i \in I$.
(ii) $\operatorname{Supp}(g_i) \subset U_i$ for every $i \in I$.
(iii) The family of supports $\operatorname{Supp}(g_i)$, $i \in I$, is locally finite, i.e., every point $a \in X$ has a neighborhood V such that

$$V \cap \operatorname{Supp}(g_i) \neq \varnothing \text{ for only finitely many } i \in I.$$

(iv) $\sum_{i \in I} g_i = 1$.

(Because of (iii) the sum in (iv) is well-defined.)

A.4. Theorem. *Suppose X is a differentiable manifold which has a countable topology. Then for every open covering $\mathfrak{U}$ of X there exists a differentiable partition of unity subordinate to $\mathfrak{U}$.*

A.5. Corollary. *Suppose X is a differentiable manifold, K is a compact subset of X and U is an open neighborhood of K. Then there exists a differentiable function $f \colon X \to \mathbb{R}$ such that* $\operatorname{Supp}(f) \Subset U$ *and* $f \mid K = 1$.

PROOF. We may assume that X has a countable topology. Otherwise, just replace X by some relatively compact open neighborhood of K. Now suppose U_1 is a relatively compact open neighborhood of K which is contained in U and let $U_2 := X \backslash K$. There exists a differentiable partition of unity (g_1, g_2) subordinate to the covering $\mathfrak{U} = (U_1, U_2)$. Then $f := g_1$ is the desired function. □

B. Topological Vector Spaces

We now present the notions and facts from functional analysis which we have used. Further details and the proofs may be found, for example, in [44], [47].

B.1. By a vector space we will always means a vector space over the field of complex numbers. A *topological vector space* is a vector space E, together with a topology, such that the operations of addition

$$E \times E \to E, \qquad (x, y) \mapsto x + y,$$

and scalar multiplication

$$\mathbb{C} \times E \to E, \qquad (\lambda, x) \mapsto \lambda x,$$

are continuous maps. In particular, for every $a \in E$, the translation $E \to E$, $x \mapsto a + x$, is a homeomorphism. Thus the topology of E is determined once one knows what a neighborhood basis of zero is. For, if $\mathfrak{B}$ is a neighborhood basis of zero, then the translated sets $a + U$, $U \in \mathfrak{B}$, form a neighborhood basis of a.

B.2. Semi-norms. By a *semi-norm* on a vector space E is meant a mapping $p\colon E \to \mathbb{R}$ with the following properties:

(i) $p(x + y) \leq p(x) + p(y)$ for all $x, y \in E$
(ii) $p(\lambda x) = |\lambda| p(x)$ for all $\lambda \in \mathbb{C}$, $x \in E$.

From (i) and (ii) it follows that $p(x) \geq 0$ for every $x \in E$. If, in fact, $p(x) = 0$ only for $x = 0$, then p is called a *norm.*

A family p_i, $i \in I$, of semi-norms on a vector space E induces a topology on E. For $i_1, \ldots, i_m \in I$, $\varepsilon > 0$, the sets of the form

$$U(p_{i_1}, \ldots, p_{i_m}; \varepsilon) := \{x \in E\colon \max(p_{i_1}(x), \ldots, p_{i_m}(x)) < \varepsilon\}$$

are a neighborhood basis of zero. Note that this topology is Hausdorff precisely if $p_i(x) = 0$ for every $i \in I$ implies $x = 0$.

A topological vector space is said to be *locally convex* if its topology can be induced in the above way by a family of semi-norms.

B.3. Fréchet Spaces. A sequence $(x_n)_{n \in \mathbb{N}}$ of elements in a topological vector space is called a *Cauchy sequence* if for every neighborhood U of zero there exists an $n_0 \in \mathbb{N}$ such that

$$x_n - x_m \in U \quad \text{for every } n, m \geq n_0.$$

A topological vector space E is called a *Fréchet space* if the following hold:

(i) The topology of E is Hausdorff and can be defined by a countable family of semi-norms.
(ii) E is complete, i.e., every Cauchy sequence in E is convergent.

A Fréchet space E is metrizable. For, suppose p_n, $n \in \mathbb{N}$, is a family of semi-norms which defines the topology on E. If for $x, y \in E$ one sets

$$d(x, y) := \sum_{n=0}^{\infty} 2^{-n} \frac{p_n(x - y)}{1 + p_n(x - y)},$$

then $d: E \times E \to \mathbb{R}$ is a metric on E which induces the same topology as the semi-norms p_n, $n \in \mathbb{N}$.

A closed vector subspace $F \subset E$ of a Fréchet space is also a Fréchet space. If E_i, $i \in I$, is a countable family of Fréchet spaces, then $\prod_{i \in I} E_i$ with the product topology is also a Fréchet space.

B.4. A typical example of a Fréchet space is the vector space $\mathcal{O}(X)$ of holomorphic functions on an open set $X \subset \mathbb{C}$ with the topology of uniform convergence on compact subsets. This topology is induced by the semi-norms p_K, where

$$p_K(f) := \sup_{x \in K} |f(x)|,$$

as K runs through the compact subsets of X. This topology is also defined by countably many semi-norms p_{K_n}, where K_n, $n \in \mathbb{N}$, is any sequence of compact subsets of X with $\bigcup_{n \in \mathbb{N}} \mathring{K}_n = X$.

B.5. Banach Spaces, Hilbert Spaces. A complete normed vector space is called a Banach space. Thus a Banach space is a Fréchet space whose topology is defined by a single norm. This is usually denoted $\| \ \|$.

A Hilbert space E is a Banach space whose norm is derived from a scalar product

$$\langle \, , \, \rangle : E \times E \to \mathbb{C},$$

i.e., $\|x\| = \sqrt{\langle x, x \rangle}$.

If A is a vector subspace of a Hilbert space E, then its orthogonal complement

$$A^{\perp} := \{y \in E : \langle y, x \rangle = 0 \quad \text{for every } x \in A\}$$

is a closed vector subspace of E. If A itself is closed, then $E = A \oplus A^{\perp}$.

B.6. Theorem of Banach. *Suppose E and F are Fréchet spaces and $f: E \to F$ is a continuous linear surjective mapping. Then f is open.*

B.7. Corollary. *Suppose E and F are Banach spaces and $f: E \to F$ is a continuous linear surjective mapping. Then there exists a constant $C > 0$ such that for every $y \in F$ there is an $x \in E$ with*

$$f(x) = y \quad \textit{and} \quad \|x\| \le C\|y\|.$$

PROOF. Let $U := \{x \in E : \|x\| < 1\}$. Since by the Theorem of Banach f is open, there exists an $\varepsilon > 0$ such that

$$f(U) \supset V := \{y \in F : \|y\| < \varepsilon\}.$$

Let $C := 2/\varepsilon$. Now suppose $y \in F$ is given. If $y = 0$, choose $x = 0$. Otherwise, $\lambda := \|y\| > 0$. The element $y_1 := (1/\lambda C)y$ lies in V and thus there exists $x_1 \in U$

with $f(x_1) = y_1$. Then for $x := \lambda C x_1$, one has $f(x) = y$ and

$$\|x\| = \lambda C \|x_1\| \leq \lambda C = C\|y\|. \qquad \square$$

B.8. Hahn–Banach Theorem. *Suppose E is a locally convex topological vector space, $E_0 \subset E$ is a vector subspace and $\varphi_0: E_0 \to \mathbb{C}$ is a continuous linear functional. Then there exists a continuous linear functional $\varphi: E \to \mathbb{C}$ such that $\varphi | E_0 = \varphi_0$.*

B.9. Corollary. *Suppose E is a locally convex topological vector space and $A \subset B \subset E$ are vector subspaces. If every continuous linear functional $\varphi: E \to \mathbb{C}$ such that $\varphi | A = 0$ satisfies $\varphi | B = 0$, then A is dense in B.*

PROOF. Suppose A is not dense in B. Then there exists $b_0 \in B$ such that $b_0 \notin \bar{A}$. Let $E_0 := \bar{A} \oplus \mathbb{C} b_0$ and define a linear functional $\varphi_0: E_0 \to \mathbb{C}$ by $\varphi_0(a + \lambda b_0) := \lambda$ for $a \in \bar{A}$, $\lambda \in \mathbb{C}$. It is easy to check that φ_0 is continuous. By the Hahn–Banach Theorem φ_0 extends to a continuous linear functional $\varphi: E \to \mathbb{C}$. Then $\varphi | A = 0$, but $\varphi | B \not\equiv 0$, which is a contradiction. $\square$

B.10. Compact Mappings. A linear mapping $\psi: E \to F$ between two topological vector spaces E and F is called *compact* or *completely continuous*, if there exists a neighborhood U of zero in E such that $\psi(U)$ is relatively compact in F. In particular, a compact linear mapping is continuous.

Example. Suppose X is an open subset of $\mathbb{C}$ and $Y \Subset X$ is a relatively compact open subset of X. Then the restriction mapping

$$\beta: \mathcal{O}(X) \to \mathcal{O}(Y), \qquad f \mapsto f | Y,$$

is compact. One sees this as follows. Since $\bar{Y}$ is compact in X, it follows that

$$U := \left\{ f \in \mathcal{O}(X): \sup_{x \in \bar{Y}} |f(x)| < 1 \right\}$$

is a neighborhood of zero in $\mathcal{O}(X)$. By Montel's Theorem the set

$$M := \left\{ g \in \mathcal{O}(Y): \sup_{y \in Y} |g(y)| \leq 1 \right\}$$

is compact in $\mathcal{O}(Y)$. The claim now follows since $\beta(U) \subset M$.

B.11. Theorem of L. Schwartz. *Suppose E and F are Fréchet spaces and φ, $\psi: E \to F$ are continuous linear mappings such that φ is surjective and ψ is compact. Then the image of the mapping $\varphi - \psi: E \to F$ has finite codimension in F.*

For the proof see [60].

References

(a) Complex Analysis in One Variable

1. Ahlfors, L. V.: *Complex Analysis.* New York: McGraw-Hill, 1966.
2. Behnke, H., and Sommer, F.: *Theorie der analytischen Funktionen einer komplexen Veränderlichen.* 3rd Ed. Berlin–Heidelberg–New York: Springer-Verlag 1965.
3. Cartan, H.: *Elementary Theory of Analytic Functions of One or Several Complex Variables.* Reading, MA: Addison-Wesley, 1963.
4. Hurwitz, A., and Courant, R.: *Funktionentheorie* (with an appendix by H. Röhrl). 4th Ed. Berlin–Heidelberg–New York: Springer-Verlag, 1964.
5. Lang, S.: *Complex Analysis.* Reading, MA: Addison-Wesley, 1977.

The books of Behnke–Sommer and Hurwitz–Courant also contain large sections on Riemann surfaces.

(b) Riemann Surfaces

10. Ahlfors, L. V., and Sario, L.: *Riemann Surfaces.* Princeton, NJ: University Press, 1960.
11. Chevalley, C.: *Introduction to the Theory of Algebraic Functions of One Variable.* Amer. Math. Soc., 1951.
12. Farkas, H. M., and Kra, I.: *Riemann Surfaces.* New York–Heidelberg–Berlin: Springer-Verlag, 1980.
13. Guenot, J., and Narasimhan, R.: *Introduction à la Théorie des Surfaces de Riemann.* Monographies de l'Enseignement Mathématique No. 23, Genève, 1976.
14. Gunning, R. C.: Lectures on Riemann surfaces, *Princeton Math. Notes* **2** (1966).
15. Gunning, R. C.: Lectures on vector bundles over Riemann surfaces, *Princeton Math. Notes* **6** (1967).
16. Gunning, R. C.: Lectures on Riemann surfaces: Jacobi varieties, *Princeton Math. Notes* **12** (1972).

17. Nevanlinna, R.: *Uniformisierung*. Berlin–Heidelberg–New York: Springer-Verlag, 1953.
18. Pfluger, A., *Theorie der Riemannschen Flächen*, Berlin–Heidelberg–New York: Springer-Verlag, 1957.
19. Serre, J-P.: *Groupes Algébriques et Corps de Classes*, Paris: Hermann, 1959.
20. Springer, G.: *Introduction to Riemann Surfaces*. Reading, MA: Addison-Wesley 1957.
21. Weyl, H.: *The Concept of a Riemann Surface*. Reading, MA: Addison-Wesley, 1955.

The book of Hermann Weyl, which first appeared in 1913 (in German), was the first modern presentation of the theory of Riemann surfaces. It is still well worth reading and contains many references to the older literature, particularly that of the nineteenth century. The books of Chevalley and Serre treat the theory of Riemann surfaces from the algebraic standpoint.

(c) Complex Analysis in Several Variables, Complex Manifolds

30. Andreian Cazacu, C.: *Theorie der Funktionen mehrerer komplexer Veränderlichen*. Berlin: Dtsch. Verl. Wiss., 1975.
31. Grauert, H. and Remmert, R.: Theory of Stein spaces. New York–Heidelberg–Berlin: Springer-Verlag, 1979.
32. Gunning, R. C., and Rossi, H.: *Analytic Functions of Several Complex Variables*. Englewood Cliffs, NJ: Prentice-Hall, 1965.
33. Hirzebruch, F.: *Topological Methods in Algebraic Geometry*. Berlin–Heidelberg–New York: Springer-Verlag, 1966.
34. Hörmander, L.: *An Introduction to Complex Analysis in Several Variables*. 2nd Ed. Amsterdam: North-Holland, 1973.
35. Wells, R. O.: *Differential Analysis on Complex Manifolds*. Englewood Cliffs, NJ: Prentice-Hall, 1973. 2nd Ed. New York–Heidelberg–Berlin: Springer-Verlag, 1980.

(d) Topology, Differentiable Manifolds, Functional Analysis

40. Bröcker, T., and Jänich, K.: *Einführung in die Differentialtopologie*. Heidelberger Taschenbücher, Bd. 143. Berlin–Heidelberg–New York: Springer-Verlag, 1973.
41. Godement, R.: *Topologie Algébrique et Théorie des Faisceaux*. Paris: Hermann, 1958.
42. Massey, W. S.: *Algebraic Topology: an Introduction*. New York–Heidelberg–Berlin: Springer-Verlag, 1967.
43. Narasimhan, R.: *Analysis on Real and Complex Manifolds*. Amsterdam: North-Holland, 1968.
44. Schaeffer, H.: *Topological Vector Spaces*. New York: MacMillan, 1966.
45. Schubert, H.: *Topologie*. Stuttgart: Teubner, 1964.
46. Seifert, H., and Threlfall, W.: *Lehrbuch der Topologie*, Leipzig: Teubner 1934. Reprinted Chelsea, 1947 (English trans: Academic Press, 1980).
47. Treves, F.: *Topological Vector Spaces, Distributions and Kernels*. New York–London: Academic Press, 1967.
48. Warner, F.: *Foundations of Differentiable Manifolds and Lie Groups*. Glenview, IL-London: Scott-Foresman, 1971.

(e) Special Topics

50. Ahlfors, L. V.: The complex analytic structure of the space of closed Riemann surfaces, in *Analytic Functions*. Princeton NJ, University Press, 1960, 45–66.
51. Behnke, H., and Stein, K.: Entwicklungen analytischer Funktionen auf Riemannschen Flächen. Math. Ann. **120** (1948) 430–461.
52. Bieberbach, L.: *Theorie der gewöhnlichen Differentialgleichungen auf funktionentheoretischer Grundlage dargestellt*. 2nd Ed. Berlin–Heidelberg–New York: Springer-Verlag, 1965.
53. Cartan, H.: Variétés analytiques complexes et cohomologie. *Colloque sur les fonctions de plusieurs variables*, CBRM: Bruxelles, 1953, 41–55.
54. Florack, Herta: Reguläre und meromorphe Funktionen auf nicht geschlossenen Riemannschen Flächen. *Schriftenreihe Math. Inst. Univ. Münster*, 1 (1948).
55. Malgrange, B.: Existence et approximation des solutions des équations aux dérivées partielles à convolution. *Annales de l'Inst. Fourier* **6** (1955/56) 271–355.
56. Meis, T.: Die minimale Blätterzahl der Konkretisierungen einer kompakten Riemannschen Fläche. *Schriftenreihe Math. Inst. Univ. Münster* **16** (1960).
57. Röhrl, H.: Das Riemann–Hilbertsche Problem der Theorie der linearen Differentialgleichungen. *Math. Ann.* **133** (1957) 1–25.
58. Ross, S. L.: *Differential Equations*. New York: Blaisdell, 1964.
59. Serre, J-P.: Applications de la théorie générale à divers problèmes globaux. *Séminaire H. Cartan, E.N.S.* Paris 1951/52, Exposé 20. Reprinted Benjamin, 1967.
60. —: Deux théorèmes sur les applications complètement continues. *Séminaire H. Cartan, E.N.S.* Paris 1953/54, Exposé 16. Reprinted Benjamin, 1967.
61. —: Quelques problèmes globaux relatifs aux variétés de Stein. Colloque sur les fonctions de plusieurs variables, CBRM: Bruxelles, 1953, 57–68.
62. Stein, K.: Analytische Funktionen mehrerer komplexer Veränderlichen zu vorgegebenen Periodizitätsmodulin und das zweite Cousinsche Problem. *Math. Ann.* **123**, (1951) 201–222.
63. Thimm, W.: Der Weierstra*ß*sche Satz der algebraischen Abhängigkeit von Abelschen Funktionen und seine Verallgemeinerungen. in *Festschrift zur Gedächtnisfeier für Karl Weierstrass 1815–1965* (Hrsg. H. Behnke, K. Kopfermann). Köln and Opladen: Westdeutscher Verlag, 1966.

Symbol Index

Author and Subject Index

Graduate Texts in Mathematics

continued from page ii

97 KOBLITZ. Introduction to Elliptic Curves and Modular Forms. 2nd ed.
98 BRÖCKER/TOM DIECK. Representations of Compact Lie Groups.
99 GROVE/BENSON. Finite Reflection Groups. 2nd ed.
100 BERG/CHRISTENSEN/RESSEL. Harmonic Analysis on Semigroups: Theory of Positive Definite and Related Functions.
101 EDWARDS. Galois Theory.
102 VARDARAJAN. Lie Groups, Lie Algebras and Their Representations.
103 LANG. Complex Analysis. 3rd ed.
104 DUBROVIN/FOMENKO/NOVIKOV. Modern Geometry--Methods and Applications. Part II.
105 LANG. $SL_2(\mathbf{R})$.
106 SILVERMAN. The Arithmetic of Elliptic Curves.
107 OLVER. Applications of Lie Groups to Differential Equations. 2nd ed.
108 RANGE. Holomorphic Functions and Integral Representations in Several Complex Variables.
109 LEHTO. Univalent Functions and Teichmüller Spaces.
110 LANG. Algebraic Number Theory.
111 HUSEMÖLLER. Elliptic Curves.
112 LANG. Elliptic Functions.
113 KARATZAS/SHREVE. Brownian Motion and Stochastic Calculus. 2nd ed.
114 KOBLITZ. A Course in Number Theory and Cryptography.
115 BERGER/GOSTIAUX. Differential Geometry: Manifolds, Curves, and Surfaces.
116 KELLEY/SRINIVASAN. Measure and Integral. Vol. I.
117 SERRE. Algebraic Groups and Class Fields.
118 PEDERSEN. Analysis Now.
119 ROTMAN. An Introduction to Algebraic Topology.
120 ZIEMER. Weakly Differentiable Functions: Sobolev Spaces and Functions of Bounded Variation.
121 LANG. Cyclotomic Fields I and II. Combined 2nd ed.
122 REMMERT. Theory of Complex Functions.
Readings in Mathematics
123 EBBINGHAUS/HERMES et al. Numbers.
Readings in Mathematics
124 DUBROVIN/FOMENKO/NOVIKOV. Modern Geometry--Methods and Applications. Part III.
125 BERENSTEIN/GAY. Complex Variables: An Introduction.
126 BOREL. Linear Algebraic Groups.
127 MASSEY. A Basic Course in Algebraic Topology.
128 RAUCH. Partial Differential Equations.
129 FULTON/HARRIS. Representation Theory: A First Course.
Readings in Mathematics
130 DODSON/POSTON. Tensor Geometry.
131 LAM. A First Course in Noncommutative Rings.
132 BEARDON. Iteration of Rational Functions.
133 HARRIS. Algebraic Geometry: A First Course.
134 ROMAN. Coding and Information Theory.
135 ROMAN. Advanced Linear Algebra.
136 ADKINS/WEINTRAUB. Algebra: An Approach via Module Theory.
137 AXLER/BOURDON/RAMEY. Harmonic Function Theory.
138 COHEN. A Course in Computational Algebraic Number Theory.
139 BREDON. Topology and Geometry.
140 AUBIN. Optima and Equilibria. An Introduction to Nonlinear Analysis.
141 BECKER/WEISPFENNING/KREDEL. Gröbner Bases. A Computational Approach to Commutative Algebra.
142 LANG. Real and Functional Analysis. 3rd ed.
143 DOOB. Measure Theory.
144 DENNIS/FARB. Noncommutative Algebra.
145 VICK. Homology Theory. An Introduction to Algebraic Topology. 2nd ed.